Lecture Notes in Earth Sciences

Edited by Somdev Bhattacharji, Gerald M. Friedman, Horst J. Neugebauer and Adolf Seilacher

14

N. Cristescu H. I. Ene (Eds.)

Rock and Soil Rheology

Proceedings of the Euromech Colloquium 196
September 10–13, 1985
Bucharest, Romania

Springer-Verlag
Berlin Heidelberg New York London Paris Tokyo

Editors

Prof. Dr. Nicolae Cristescu
University of Bucharest, Faculty of Mathematics
Str. Academiei 14, Bucharest 1 Cod 70109, Romania

Dr. Horia I. Ene
INCREST, Department of Mathematics
Bd. Pacii 220, 79622 Bucharest, Romania

ISBN 3-540-18841-X Springer-Verlag Berlin Heidelberg New York
ISBN 0-387-18841-X Springer-Verlag New York Berlin Heidelberg

Printed in Germany

Printing and binding: Druckhaus Beltz, Hemsbach/Bergstr.
2132/3140-543210

CONTENTS

LIST OF CONTRIBUTORS

Y. Arkin
Ministry of Energy and Infrastructure
Geological Survey
30, Malkhe Israel Str.
Jerusalem 95501
Israel.

P. Berest
Laboratoire de Mécanique des Solides
Ecole Polytechnique, F 91128 Palaiseau - Cédex
France.

R. Butterfield
Dept. of Civil Engineering
The University Highfield
Southampton, SO9 5NH
United Kingdom.

B. Cambou
Laboratoire de Mécanique de Solides
Ecole Centrale de Lyon
36 Av. Guy de Collongue
B.P. 163, 69131 Ecully - Cédex
France.

N. Cristescu
Dept. of Mathematics
University of Bucharest
Str. Academiei 14, Bucharest 70109
Romania.

Lucia Drăgușin
Dept. of Mathematics
Polytechnical Institute
Spl. Independentei 313, Bucharest
Romania.

Horia I. Ene
Dept. of Mathematics, INCREST
Bd. Păcii 220, 79622 Bucharest
Romania.

F. J. Gilbert
Laboratoire de Mécanique des Solides
Ecole Polytechnique, F 91128 Palaiseau - Cédex
France.

P. Habib
Laboratoire de Mecanique des Solides
Ecole Polytechnique, F 91128 Palaiseau - Cedex
France.

A. F. L. Hyde
Dept. of Civil Engineering
University of Bradford
Bradford, BD7 1DP
United Kingdom.

M. P. Luong
Laboratoire de Mécanique des Solides
Ecole Polytechnique, F 91128 Palaiseau - Cédex
France.

R. Ribacchi
Inst. di Scienza delle Construzioni
Università di Roma
Via Endossiana 18, Roma,
Italy.

R. Traczyk
Institute of Geotechnique
Technical University of Wrocław
Plac Grunwaldzki 9, 50-370 Wrocław,
Poland.

INTRODUCTION

While the complex mechanical properties of rocks and soils are studied for quite a while, it is only in the last decades that sound established mathematical models were developed based on accurate experimental data. Some rheological properties of geomaterials as for instance creep, were studied for a long time but the experimental data reported were incomplete and, as a consequence, the models developed have missed either the generality necessary for the solving of engineering problems or some of the major specific mechanical properties possessed by these materials as for instance dilatancy and/or compressibility, long term damage etc. Generally, these very particular empirical models were made for a specific test only and therefore are not appropriate for solving problems involving general loading histories. Let us remind that due to the presence of a great number of cracks and/or pores existing in roks and soils, the mechanical behaviour of geomaterials is quite distinct from that of other materials as for instance metals or plastics. That is why rock and soil rheology has some specific aspects. It must also be mentioned that the solving of various problems of rock and soil mechanics posed by modern technology was not possible by using time-independent models, thus the study and development of rehological models become absolutely necessary.

In the last decade or so, very accurate experimantal data became available as a result of the development of experimental techniques and of the growing interest for this field of research in the scientific community. These data, in turn, have made possible the development of genuine models for geomaterials, mainly rheological models, able to describe such properties as creep, dilatancy and/or compressibility during creep, long term damage and failure occurring after various time intervals, slip surface formation etc.

Today it is clear that no accurate constitutive equation for rocks can be formulated unless the dilatancy phenomena and the time effects are not included. Another idea is the need of a better description of the concepts of damage and failure of rocks, again using in someway the concepts of irreversible dilatancy or another related notion.

In soil rheology it is clear that the scale effect may be taken into consideration in order to obtain a corect information from the routine tests. Also in writing the constitutive equations for soils it is neccessary to take into account the microscopic or local phenomena, because there is a great variety of types of saturated or nonsaturated soils, granular or cohesionless soil etc.

The aim of the Euromech Colloquium 196 devoted to Rock and Soil Rheology and therefore that of the present volume too, is to review some of the main results obtained in the last years in this field of research and also to formulate some of the major not yet solved problems which are now under consideration. Exchange of opinions and scientific discussions are quite helpful mainly in those areas where some approaches are controversial and the progress made is quite fast. That is especially true for the rheology of geomaterials, domain of great interest for mining and petroleum engineers, engineering geology, seismology, geophisics, civil engineering, nuclear and industrial waste storage, geothermal energy storage, caverns for sports, culture, telecommunications, storage of goods and foodstuffs (cold, hot and refrigerated storages), underground oil and natural gas reservoirs etc. Some of the last obtained results are mentioned in the present volume.

*
*   *

Irén Némethi dealt with the difficult task of typing a large part of the manuscript using a Rank Xerox 860 word processor; we thank her for the excellent job she did.

Nicolae Cristescu
Horia I. Ene

I. THEORETICAL APPROACH

CHANGE OF SCALE METHODS
APPLIED IN MECHANICS OF SATURATED SOILS

F. GILBERT

RESUME

Deux aspects complémentaires de la description mécanique des sols saturés, utilisant des méthodes de changement d'échelle, sont présentés.

Les méthodes d'homogénéisation pour un milieu polyphasique utilisant un changement d'échelle par convolution spatiale sont tout d'abord étendues sous forme lagrangienne. Les équations de bilan au niveau macroscopique sont ainsi établies à partir du niveau local (échelle des grains). Pour les sols saturés la vitesse de filtration et le tenseur de contrainte effective sont introduits. Nous calculons explicitement le tenseur de viscosité apparente et la force de flottabilité.

Les propriétés de perméabilité en régime harmonique de certains empilements bi-dimensionnels de grains, qui rendent compte schématiquement par leur caractère auto-similaire de la forte hétérogénéité locale des sols, sont ensuite étudiées.

L'intérêt de combiner ces deux types d'approches est souligné.

ABSTRACT

Present work deals with two complementary aspects of mechanical description of saturated soils, using change of scale methods.

Homogenization methods for multiphase media using change of scale by spatial convolution are first extended in Lagrangian form. Balance equations at macroscopic scale are thus established starting from corresponding equations valid at the local level (grains scale). For saturated soils seepage velocity and effective stress tensor are introduced. Apparent viscosity stress tensor and buoyancy force are explicitly calculated.

Permeability properties under harmonic conditions are then analysed for particular two-dimensional grains packings, whose self-similarity accounts schematically for strong local heterogeneity of soils.

Interest of combining these two kinds of approaches is emphasized.

INTRODUCTION

Predicting the macroscopic behaviour of saturated soils involves a lot of difficulties, part of which owing to the multiphase character of such a medium where solid and liquid parts are intimately mixed and interact in an intricate manner under unsteady or cycling loading conditions. They can generally be considered as quasi-homogeneous at macroscopic scale, in the sense that they repeat themselves more or less in the space in a statistical manner, but appear always strongly heterogeneous at small length scale owing to the great complexity of their internal geometry.

General mixture theories (Truesdell and Toupin (1960), Müller (1975), Bowen (1976)) have been used for soils by various authors. But apart from their perhaps too wide generality their usefulness is restricted by the lack of precise geometrical and physical interpretation of the various terms. These theories must be supplemented in any case by numerous phenomenological assumptions as made by Prevost (1980).

Furthermore the essential immiscibility character results in kinematical constraints upon the motions of the species. To account for this phenomenon theories with microstructural content have been developed for porous and granular materials and used in particular by Ahmadi (1980) and Ahmadi and Shahinpoor (1983). Review of theories of immiscible and structured mixtures may be found in Bedford and Drumheller (1983). These theories need additional variables and equations which are hoped to account in a global manner for the geometrical arrangement of the constituents, its influence on the mechanical behaviour and its evolution.

Another macroscopic approach for porous media, due to Biot (1961, 1962 a, 1962 b, 1977), uses a postulated lagrangian formulation following motion of the solid part. A discussion of theoretical and experimental results may be found in Coussy and Bourbie (1984). However the underlying homogenization process involved in such formulation is not very clear except for simple cases.

Direct statistical assumptions (Matheron (1965, 1967), Batchelor (1974)) or the hypothesis of fine periodic structure of the medium (Sanchez-Palencia (1974), Ene and Sanchez-Palencia (1975), Auriault and Sanchez-Palencia (1977), Bensoussan, Lions and Papanicolaou (1978), Sanchez-Palencia (1980),

Auriault (1980), Avallet (1981), Borne (1983)) have been used to study various saturated porous materials.

It is worth noting that part of the results so obtained are in fact much more general. Homogenization processes using change of scale by spatial convolution, as suggested in particular by Marle (1967, 1982), Ene and Melicescu-Receanu (1984) and Gilbert (1984, 1985), are hence well suited for a comprehensive physical description of complex multiphase media as saturated soils.

They allow to define in a natural way macroscopic quantities as semilocal ones linked to local quantities of each phase by accurate equations. This may be viewed as a generalization of previous works by Marle (1965), Slattery (1967, 1969, 1972), Whitaker (1969), Gray and O'Neill (1976), Coudert (1973), Hassanizadeh (1979) and Hadj Hamou (1983) for instance.

This paper is organized as follows. Section 1 is devoted to definition of change of space by spatial convolution on a reference configuration and section 2 deals with basic geometrical and kinematic quantities at macroscopic scale in a multiphase medium. Balance equations at macroscopic scale are established in sections 3 and 4, as well as associated expression of the principle of virtual work.

Application to saturated soils using actual configuration as a reference configuration is made in section 5 and stress tensors of interest are introduced. Apparent viscosity stress tensor is calculated in section 6. Explicit value of the buoyancy force is given and filtration processes are described in section 7.

Unsteady permeability of particular self-similar structures is studied in the last three sections. Hydraulic impedance of a two-dimensional narrow gap between two grains is calculated in section 8. Construction of a sort of compact grains packing is recalled in section 9 and method of solution for permeability properties is explained. Numerical results are presented in section 10 as well as possible application to periodic lattices.

LAGRANGIAN DESCRIPTION OF A MULTIPHASE MEDIUM

1.- SPATIAL CONVOLUTION ON A REFERENCE CONFIGURATION FOR A MULTIPHASE MEDIUM

Let E_Z and E_X be the initial and transformed physical spaces corresponding to local (grains scale) and semi-local descriptions (macroscopic scale). Correspondence between them is made at fixed time t_o through use of a positive even weight function $m(Z)$ whose integral over its bounded support $D(O)$ is equal to 1 (*Fig.1*). It will be supposed in this part that the position at time t of any particle is a continuous function of its position at time t_o : thus sliding is actually excluded.

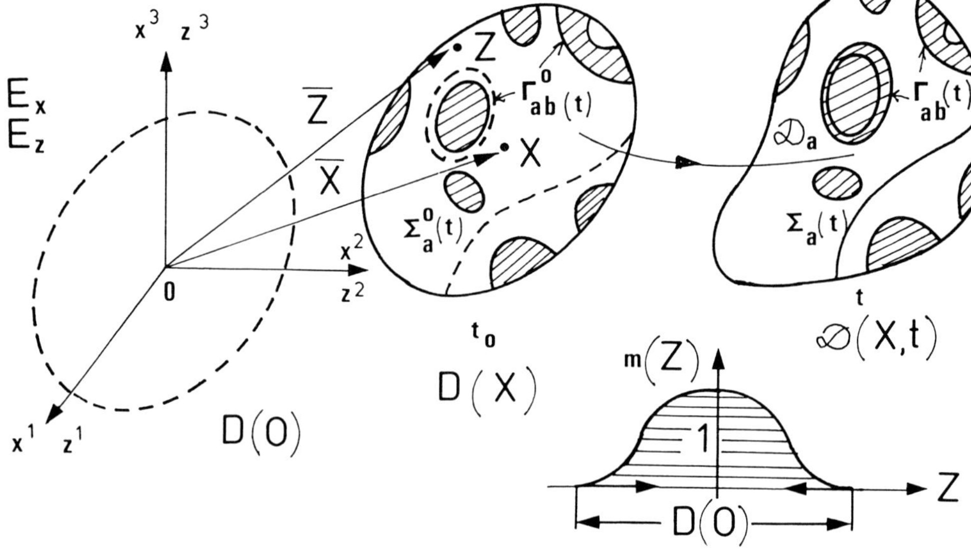

Figure 1 : Change of space by spatial convolution on a reference configuration with a weight function m for a multicomponent medium. Constituent C_a is found at time t in part $\mathcal{D}_a$ of $\mathcal{D}(X,t)$ whose image at time t_o is $D(X)$, the translated of $D(O)$ by vector $\bar{X}$. Contact surface of constituent C_a with the other ones may be expressed as $\Gamma_a = \bigcup \Gamma_{ab}(t)$ $(b \neq a)$ whose images at t_o are $\Gamma^o_{ab}(t)$. (For non reacting media $\Gamma^o_{ab}(t) = \Gamma_{ab}(t_o)$). The same type of notation is used for the internal discontinuity surfaces $\Sigma_a(t)$.

Note that except for particular purposes it is convenient to choose m of class C^N on $\mathbb{R}^3$ (with N being not too small) to ensure a sufficient regularity of macroscopic quantities.

For the separate constituents considered one introduces for every constituent C_a the function $I_a(\bar{Z},t)$ defined as follows : $I_a = 1$ if the particle whose position is $\bar{Z}$ at t_o belong to C_a at time t and $I_a = 0$ otherwise. Hence possible phase changes $C_a \to C_b$ along contact surface $\Gamma_{ab}(t)$, such as freezing of water in the pores (C_a = water, C_b = ice), may be considered. For a non reacting medium I_a is clearly independent of t.

To an additive quantity $\psi_{(a)}(\bar{Z},t)$ relative to C_a is associated an apparent average $<\psi_{(a)}>^a$ at the macroscopic scale by the convolution product (in Lagrange variables)

$$(1.1) \qquad <\Psi_{(a)}>^a = \left(\left(\Psi_{(a)} I_a\right) * m\right)(\bar{X},t) = \int_{D(X)} \Psi_{(a)}(\bar{Z},t) I_a(\bar{Z},t)\, m(\bar{X}-\bar{Z})\, dv_Z .$$

The obtained quantity reflect long trends of the medium regardless of small scale variations. A sort of change of scale is thus achieved (*Fig.2*) for quantities of interest.

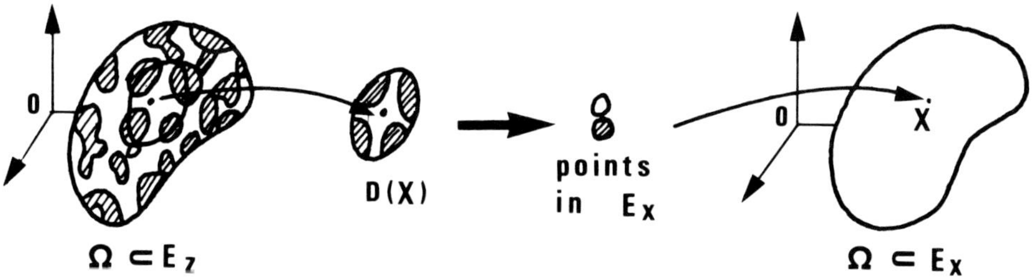

Figure 2 : Macroscopic volume Ω viewed in E_Z and E_X

For instance if $\psi_{(a)} \equiv 1$ one gets quantity

$$(1.2) \qquad \phi_o^a(\bar{X},t) = <1>^a = (I_a * m)(\bar{X},t)$$

which is to be interpreted as the volume fraction at t_o around $\bar{X}$ of the particles initially in D(X) which at time t belong to C_a.

Derivatives of the semi-local quantities $< \psi_{(a)} >^a$ are given by

$$(1.3)\quad \nabla_X < \psi_{(a)} >^a = < \nabla_Z \psi_{(a)} >^a + \left([\![\psi_{(a)}]\!] \otimes \bar{\nu}_o \delta_{\Sigma_a^o(t)} \right) * m - \left(\psi_{(a)} \otimes \bar{n}_{o(a)} \delta_{\Gamma_a^o(t)} \right) * m$$

$$(1.4)\quad \overbrace{< \rho_o(\bar{Z}) \varphi_{(a)}(\bar{Z},t) >^a}^{\prime} = < \rho_o(\bar{Z}) \varphi'_{(a)}(\bar{Z},t) >^a + \left(q^o_{m(a)} [\![\varphi_{(a)}]\!] \delta_{\Sigma_a^o(t)} \right) * m + \left(\tilde{c}_{o(a)} \varphi_{(a)} \delta_{\Gamma_a^o(t)} \right) * m$$

where ∇ denotes the gradient in the reference configuration and the prime $'$ the material derivative in E_X or E_Z ; $\bar{n}_{o(a)}$ is the unit vector normal to $\Gamma_a^o(t)$ (corresponding to the unit vector normal to $\Gamma_a(t)$ and directed outwards C_a at t) and $\bar{\nu}_o$ the unit vector normal to $\Sigma_a^o(t)$. The brackets $[\![\;\;]\!]$ denote the corresponding discontinuity along $\Sigma_a(t)$; $q^o_{m(a)}$ is the mass flux of C_a crossing $\Sigma_a(t)$ per unit time and per unit area of the reference configuration and $\tilde{c}_{o(a)}$ the mass of C_a created by phase changes on $\Gamma_a(t)$ per unit time and per unit area of the reference configuration. Local mass per unit volume in the reference configuration is denoted by $\rho_o(\bar{Z})$.

Formulas (1.3) and (1.4) use the surface distribution δ_S on $\mathbb{R}^4$ (S(t) a surface varying with time t) defined by the following equality valid for any test function $f(\bar{Z}, t)$ with compact support in space-time

$$(1.5)\quad (\delta_S , f) = \int_{-\infty}^{+\infty} \int_{S(t)} f(\bar{Z},t)\, dA_t(\bar{Z})\, dt$$

where $dA_t(\bar{Z})$ is the area element of S(t). For further explanations the reader is advised to refer to the abovementioned references or to Estrada and Kanwal (1980);these references give similar formulas in Euler variables, instead of Lagrange variables used here.

2.- MASSES, POSITIONS AND VELOCITIES IN E_X

The average mass of C_a at time t per unit volume at t_o around $\bar{X}$ is in E_X

$$(2.1)\quad m_o^a(\bar{X},t) = < \rho_o(\bar{Z}) >^a = \left(\rho_o(\bar{Z}) I_a(\bar{Z},t) \right) * m \quad .$$

Total mass at time t per unit volume at time t_o around $\bar{X}$ is clearly equal to $\sum_a m_o^a(\bar{X},t)$. It appears as a quantity much more regular than local mass per unit volume which is a discontinuous function in E_z at the grains scale.

It is convenient to define average velocity $\bar{u}_a(\bar{X},t)$ for constituent C_a by reference to the apparent average (1.1) of the momentum density

$$m_o^a(\bar{X},t)\,\bar{u}_a(\bar{X},t) = \langle \rho_o(\bar{Z})\,\bar{u}_{(a)}(\bar{Z},t) \rangle^a \tag{2.2}$$

where $\bar{u}_{(a)}(\bar{Z},t)$ is the local velocity of C_a. A mean position $\bar{x}_a^+(\bar{X},t)$ for C_a is obtained inserting in formula (2.2) positions instead of velocities. Motion of C_a in E_x is described by

$$\dot{\bar{x}}_a(\bar{X},t) = \bar{u}_a(\bar{X},t) \tag{2.3a}$$

$$\bar{x}_a(\bar{X},t_o) = \bar{X}_a(\bar{X}) \tag{2.3b}$$

where $\bar{X}_a(\bar{X})$ is the value of $\bar{x}_a^+$ at time t_o. For non reacting media the two functions $\bar{x}_a(\bar{X},t)$ and $\bar{x}_a^+(\bar{X},t)$ are equal as suggested by intuition. This property is known to be false (except at particular time t_o) for reacting media.

Average acceleration $\bar{\gamma}_a(\bar{X},t)$ for C_a in E_x is taken as material derivative of $\bar{u}_a(\bar{X},t)$. Average displacement of C_a is defined by

$$\bar{\xi}_a(\bar{X},t) = \bar{x}_a(\bar{X},t) - \bar{X}_a(\bar{X}) \tag{2.4}$$

and hence $\bar{X}$ is to be interpreted as a reference position. Gradient of deformation for C_a in E_x is

$$\underset{\sim}{\gamma}_a(\bar{X},t) = \underset{\sim}{\nabla}\bar{x}_a(\bar{X},t) \quad . \tag{2.5}$$

Note that at time t_o the gradient of deformation for C_a is $\underset{\sim}{\nabla}\bar{X}_a$, which differs slightly from the unit tensor $\underset{\sim}{I}$. More precisely if local mass per unit volume at t_o takes an uniform value for C_a one gets for each constituent

$$\underset{\sim}{\nabla}\bar{X}_a = \underset{\sim}{I} - \frac{1}{\phi_o^a}\underset{\sim}{Y}^a \tag{2.6}$$

where the geometrical tensors $\underset{\sim}{Y}^a(\bar{X})$ related to contact surfaces $\Gamma_a(t_o)$ at time t_o by

(2.7) $$\underset{\sim}{Y}^a(\bar{X}) = \left(\left(\bar{Z}\otimes\bar{n}_{o(a)}\right)\delta_{\Gamma_a(t_o)}\right)*m + \bar{X}_a\otimes\nabla\phi_o^a(\bar{X},t_o)$$

introduce a kind of departure of the medium from macro-homogeneity at scale of D(O).

3.- BALANCE EQUATIONS FOLLOWING MOTIONS OF THE SPECIES

Balance of mass for C_a is obtained through application of formula (1.4) to definition (2.1) as

(3.1) $$\dot{m}_o^a(\bar{X},t) = \overset{v}{c}{}_o^a(\bar{X},t) = \left(\tilde{c}_{o(a)}\,\delta_{\Gamma_a^o(t)}\right)*m$$

where the mass production rate $\overset{v}{c}{}_o^a$ of C_a per unit volume of the reference configuration converts surface reactions in E_z into volume reactions in E_x. As a direct consequence of its definition one gets

(3.2) $$\sum_a \overset{v}{c}{}_o^a(\bar{X},t) = 0$$

which expresses conservation of mass for the whole medium following the motions of the various species.

Balance equation of momentum for C_a in E_x is obtained using the apparent average (1.1) of the local balance equation. This procedure ensures automatically the compatibility between the two considered descriptions. One gets after some calculation material derivative of the momentum as

(3.3) $$m_o^a\bar{\gamma}_a + \overset{v}{c}{}_o^a\bar{u}_a = m_o^a\bar{F}_a + \bar{R}_o^a + \overline{\mathrm{div}}_o\,\underset{\sim}{B}^a$$

where the body force per unit mass $\bar{F}_a$, the interaction force per unit volume of the reference configuration $\bar{R}_o^a$ and the apparent Boussinesq stress tensor $\underset{\approx}{B}^a$ for C_a are defined in terms of corresponding local quantities by

(3.4) $$m_o^a\bar{F}_a(\bar{X},t) = \langle\rho_o\bar{F}_{(a)}\rangle^a$$

(3.5) $$\bar{R}_o^a(\bar{X},t) = \left(\left(\underset{\sim}{B}_{(a)}\cdot\bar{n}_{o(a)} + \tilde{c}_{o(a)}\bar{u}_{(a)}\right)\delta_{\Gamma_a^o(t)}\right)*m$$

(3.6) $$\underset{\sim}{B}^a(\bar{X},t) = \langle\underset{\sim}{B}_{(a)}\rangle^a \quad .$$

Observe that the interaction terms $\bar{R}_o^a$ appear quite naturally and that (neglecting surface tension effects and adding formulas (3.5))

$$\sum_a \bar{R}_o^a(\bar{X},t) = 0 \qquad (3.7)$$

The particularly simple form of equation (3.6) is to be noted and compared with corresponding expressions obtained for Cauchy stress tensors at macroscopic scale used in fully Eulerian descriptions (Marle, 1982, Ene and Melicescu-Receanu 1984, Gilbert 1984). Vanishing of the velocity fluctuation terms obtained here is clearly related to the better definition used for the system to be considered. To formula (3.3) is associated a simple form of the principle of virtual work.

Evaluation in E_Z and E_X at time t_o and for a given macroscopic volume Ω of, say, the momentum of the solid part, yields slightly different results. Comparison has to be made between the two quantities $\bar{P}^{(s)}$ and $\bar{P}^s$

$$\bar{P}^{(s)} = \int_\Omega I_s(\bar{Z},t_o)\,\rho_o(\bar{Z})\,\bar{u}_{(s)}(\bar{Z},t_o)\,dv_Z \qquad (3.8)$$

$$\bar{P}^{s} = \int_\Omega m_o^s(\bar{X},t_o)\,\bar{u}_s(\bar{X},t_o)\,dv_X \qquad (3.9)$$

One can show (Gilbert 1984) that the difference $\bar{P}^s - \bar{P}^{(s)}$ involves integrals over the two small volumes C_+ and C_- (*Fig.3*) obtained respectively by applying to Ω the Serra transforms (Matheron, 1967, Serra, 1982) through dilation and erosion by the symmetrical volume D(0). Hence relative difference is negligible if D(0) is small enough with respect to Ω.

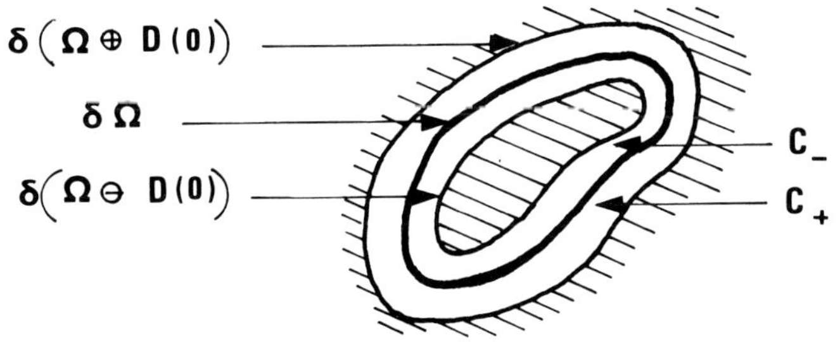

Figure 3 : Volumes contributing to the difference $\bar{P}^s - \bar{P}^{(s)}$. Note that for a symmetrical volume D(0) the two Serra transforms are given by the mentioned Minkowski pseudo-addition and pseudo-subtraction.

Note that rigorous equations in E_X as (3.3) involve momentum defined by equation (3.9) and not by equation (3.8) (which is the right definition in E_Z only).

4.- BALANCE EQUATIONS FOLLOWING MOTION OF THE SOLID PART

One is generally essentially interested by motion relative to solid part. We shall suppose the solid matrix to be chemically inert and for sake of simplicity cartesian orthonormal co-ordinates will be used in the sequel.

Geometrical situation is depicted in figure 4 in E_X.

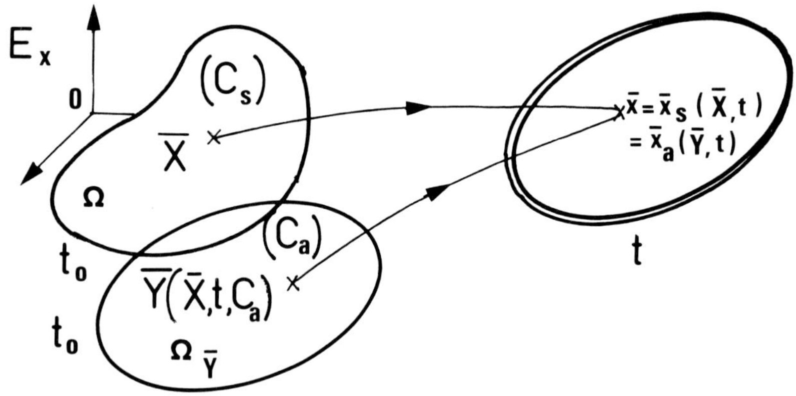

Figure 4 : Comparison in E_X between the motion $\bar{x}_s(\bar{X},t)$ of C_s and the motion of another constituent C_a. Points of C_a belonging at time t to a certain macroscopic volume in E_X, whose motion is given by $\bar{u}_s(\bar{X},t)$, belong at fixed time t_o to the variable volume $\Omega_{\bar{Y}}(\Omega, t, C_a)$ given by $\Omega_{\bar{Y}} = \{\bar{Y} / \bar{x}_a(\bar{Y},t) = \bar{x}_s(\bar{X},t), \bar{X} \in \Omega\}$

Let Ω be a macroscopic volume in E_X given at time t_o and moving by assumption with velocity $\bar{u}_s(\bar{X},t)$ of the solid part C_s. Corresponding mass and momentum of C_a at time t following motion $\bar{x} = \bar{x}_s(\bar{X},t)$ of the solid part are (*Fig.4*)

$$(4.1) \qquad \mathfrak{M}^a(\Omega,t) = \int_\Omega m^a(\bar{X},t)\, dv_X$$

$$(4.2) \qquad \bar{P}^a(\Omega,t) = \int_\Omega m^a(\bar{X},t)\, \bar{u}_a(\bar{x},t)\, dv_X \quad .$$

With the dot $\bullet$ denoting derivative following in E_X motion of the solid part one gets for balance of mass after transformation of formulas (3.1)

(4.3) $$\dot{\mathfrak{M}}^a(\Omega,t) = \int_\Omega \dot{m}^a(\bar{X},t)\,dv_X$$

(4.4a) $$\dot{m}^a(\bar{X},t) = \dot{m}^a_{ch}(\bar{X},t) + \dot{M}^a(\bar{X},t) \quad , \text{ for } a \neq s$$

(4.4b) $$m^s(\bar{X},t) = cte = m^s_o(\bar{X})$$

where m^a_{ch} and M^a are the masses of C_a acquired by phase changes and convection between t_o and t, per unit volume of the reference configuration Ω. Calculation yields

(4.5) $$\dot{m}^a_{ch}(\bar{X},t) = \check{c}^a_o(\bar{Y},t)\cdot\frac{\text{dét}\,\gamma_s(\bar{X},t)}{\text{dét}\,\gamma_a(\bar{Y},t)}$$

(4.6) $$M^a(\bar{X},t) = -\,\text{div}_o\,\overline{M}^a(\bar{X},t)$$

with

(4.7) $$\dot{\overline{M}}^a(\bar{X},t) = m^a(\bar{X},t)\,\gamma_s^{-1}(\bar{X},t)\cdot(\bar{u}_a-\bar{u}_s)(\bar{x},t)$$

and thus classical Biot's structure is recovered. Quantity $M^a(\bar{X},t)$ for a fluid C_a is called fluid accumulation.

Balance of momentum of C_a is expressed by

(4.8) $$\dot{\bar{P}}^a(\Omega,t) = \int_\Omega\left(m^a(\bar{X},t)\bar{\gamma}_a(\bar{x},t)+\dot{m}^a_{ch}(\bar{X},t)\bar{u}_a(\bar{x},t)\right)dv_X - \int_{\partial\Omega}\bar{u}_a(\bar{x},t)\left(\dot{\overline{M}}^a\cdot\bar{n}_o\right)dA_o$$

where the surface integral corresponds to the open character of volume Ω with respect to constituent C_a. Volume integral in formula (4.8) takes another form after transformation of formula (3.3) into (note that simplification for a = s is obvious)

(4.9) $$m^a\bar{\gamma}_a + \dot{m}^a_{ch}\,\bar{u}_a = m^a\,\bar{F}_a + \text{dét}\,\gamma_s\,.\,\bar{R}^a + \overline{\text{div}_o}\,\underset{\sim}{T}^a$$

where

(4.10) $$\bar{R}^a(\bar{X},t) = \frac{\bar{R}^a_o(\bar{Y},t)}{\text{dét}\,\gamma_a(\bar{Y},t)}$$

is the interaction force for C_a at time t per unit volume of the actual configuration.

The apparent stress tensors $\underset{\sim}{T}^a$ for the various C_a are calculated as (note that for $a = s$, $\underset{\sim}{T}^s = \underset{\sim}{B}^s$)

$$(4.11) \quad \underset{\sim}{T}^a = \frac{\det \gamma_s(\bar{X},t)}{\det \gamma_a(\bar{Y},t)} \cdot \underset{\sim}{B}^a(\bar{Y},t) . \underset{\sim}{\gamma}_a^T(\bar{Y},t) . \underset{\sim}{\gamma}_s^{T-1}(\bar{X},t) \quad .$$

The principle of virtual work applied to Ω, considering different virtual displacement fields $\delta\bar{x}_a$ for the various constituents, states then

$$(4.12) \quad \sum_a \left\{ \delta W^a_{surface} + \sum_{b \neq a} \delta W^a_b + \delta W^a_{body} + \delta W^a_{internal} + \delta J^a \right\} = 0$$

with for any constituent C_a surface forces applied to the transform of boundary $\partial\Omega$ only due to C_a itself

$$(4.13a) \quad \delta W^a_{surface} = \int_{\partial\Omega} \delta\bar{x}_a . \underset{\sim}{T}^a . \bar{n}_o dA_o \quad .$$

Contact forces due to influence of the other constituents C_b must no be counted twice and hence are taken in account by volume integrals only

$$(4.13b) \quad \delta W^a_b = \int_\Omega \delta\bar{x}_a . \det\gamma_s(\bar{X},t)\bar{R}^{ab}(\bar{x},t) dv_X \quad , \bar{R}^a = \sum_{b \neq a} \bar{R}^{ab} \quad .$$

Virtual works of body forces, internal forces and inertia forces are expressed by

$$(4.13c) \quad \delta W^a_{body} = \int_\Omega \delta\bar{x}_a . m^a(\bar{X},t) \bar{F}_a(\bar{x},t) dv_X$$

$$(4.13d) \quad \delta W^a_{internal} = - \delta W^a_{def} = - \int_\Omega \underset{\sim}{T}^a : \underset{\sim}{\nabla}^T(\delta\bar{x}_a) dv_X$$

$$(4.13e) \quad \delta J^a = - \int_\Omega \delta\bar{x}_a . \left(m^a(\bar{X},t)\bar{\gamma}_a(\bar{x},t) + \dot{m}^a_{ch}(\bar{X},t)\bar{u}_a(\bar{x},t) \right) dv_X \quad .$$

These equations provide a useful basis for comparison with Biot's theories.

One can see in figure 4 a little difficulty unavoidable with this formulation : at time $t = t_o$ correspondence is made by $\bar{x}$ between the phases as $\bar{X}_a = \bar{X}_s$ and thus $\bar{Y} \neq \bar{X}$. It introduces a small distorsion in the evaluation of mass and momentum of C_a.

APPLICATION TO SATURATED SOILS USING REACTUALIZED REFERENCE CONFIGURATION

5.- STRESS TENSORS

Let us use the particular choice $t_o = t$: fixed reference configuration coincides with actual configuration (note that the description is slightly different from a fully Eulerian one).

It is then a simple matter to show that sliding between grains is now allowed (see also Gilbert 1984). Two constituents are to be distinguished for a saturated soil : the solid part C_s and the pore fluid C_f. Actual porosity n is defined as $\phi_o^f(\bar{X}, t_o)$ and masses m_o^s and m_o^f may be written as $(1-n)\rho_s$ and $n\rho_f$ respectively, with obvious notations for densities at macroscopic scale.

Balances of momentum (3.3) read (no mass production term)

$$(5.1a) \qquad (1-n)\,\rho_s\,\bar{\gamma}_s = (1-n)\,\rho_s\,\bar{g} + \bar{R} + \overline{\mathrm{div}}\,\underset{\sim}{\sigma}^s$$

$$(5.1b) \qquad n\,\rho_f\,\bar{\gamma}_f = n\,\rho_f\,\bar{g} - \bar{R} + \overline{\mathrm{div}}\,\underset{\sim}{\sigma}^f$$

where $\bar{g}$ is the acceleration of gravity and $\bar{R}$ the interaction force (per unit volume) exerted on the solid part by the fluid part ; $\underset{\sim}{\sigma}^s$ and $\underset{\sim}{\sigma}^f$ are the symmetrical apparent averages of corresponding local Cauchy stress tensors. Formulas (5.1) are similar to formulas postulated in general mixture theories. Note that physical meaning of each term is explicitly known here.

It is however convenient to use in soil mechanics other stress tensors. One can define in an obvious way for the fluid part an average fluid pressure p and an apparent viscosity stress tensor $\underset{\sim}{\tau}^f$ by

$$(5.2) \qquad p = \langle p_{(f)} \rangle^f / n$$

$$(5.3) \qquad \underset{\sim}{\tau}^f = \langle \underset{\sim}{\tau}_{(f)} \rangle^f$$

where $p_{(f)}$ and $\underset{\sim}{\tau}_{(f)}$ are the local fluid pressure and the local viscosity stress tensor. Of interest for the solid part is the modified stress tensor

$\underset{\sim}{\sigma}^{sd}$ insensitive to any uniform translation of local stresses along the pressure axis

$$(5.4) \qquad \underset{\sim}{\sigma}^{sd} = \underset{\sim}{\sigma}^{s} + (1-n)\, p\, \underset{\sim}{I} \quad .$$

Hence effective stress tensor $\underset{\sim}{\sigma}^{\prime}$ is given by

$$(5.5) \qquad \underset{\sim}{\sigma}^{\prime} = \underset{\sim}{\sigma} + p\, \underset{\sim}{I} = \underset{\sim}{\sigma}^{sd} + \underset{\sim}{\tau}^{f}$$

where the stress tensor $\underset{\sim}{\sigma}$ for the whole medium is the sum of $\underset{\sim}{\sigma}^{s}$ and $\underset{\sim}{\sigma}^{f}$. Dynamic equations (5.1) are now written in a more useful form

$$(5.6a) \qquad (1-n)\, \rho_s \bar{\gamma}_s = (1-n)\rho_s \bar{g} + \bar{b} - (1-n)\, \overline{\mathrm{grad}}\, p + \overline{\mathrm{div}}\, \underset{\sim}{\sigma}^{sd}$$

$$(5.6b) \qquad n\, \rho_f\, \bar{\gamma}_f = n\, \rho_f\, \bar{g} - \bar{b} - n\, \overline{\mathrm{grad}}\, p + \overline{\mathrm{div}}\, \underset{\sim}{\tau}^{f}$$

where vector $\bar{b}$, which will appear useful, is given by

$$(5.7) \qquad \bar{b} = \bar{R} + p\, \overline{\mathrm{grad}}\, n \quad .$$

6.- EXPLICIT CALCULATION OF THE APPARENT VISCOSITY STRESS TENSOR

The (relative) seepage velocity $\bar{U}(\bar{x},t)$ is defined as

$$(6.1) \qquad \bar{U} = n(\bar{u}_f - \bar{u}_s)$$

at a given point in E_x. For incompressible constituents of uniform densities ρ_s and ρ_f formulas (1.3) yield for consolidation problems

$$(6.2) \qquad \mathrm{div}\, \bar{U} = -\, \mathrm{div}\, \bar{u}_s$$

which expresses conservation of total volume of the medium.

Let us now consider an incompressible newtonian fluid of uniform dynamic viscosity η flowing with seepage velocity $\bar{U}(\bar{x},t)$ through a packing of rigid grains having different velocities and spin vectors, so that solid part moves with average velocity $\bar{u}_s(\bar{x},t)$. Using formulas (1.3) with $t_o = t$ one gets for the apparent viscosity stress tensor

$$(6.3) \qquad \underset{\sim}{\tau}^{f} = 2\eta \left[\underset{\sim}{D}(\bar{U}) + \underset{\sim}{D}(\bar{u}_s) \right]$$

where $\underset{\sim}{D}$ denotes the symmetrical part of the gradient. Note that formula (6.3) is valid for any geometry of the porous medium and for any fields

$\bar{U}$ and $\bar{u}_s$. When solid part is at rest or moves with uniform velocity it reduces to the result given by Gilbert (1984).

The corresponding term in formula (5.6b) is (Δ = Laplacian)

$$\overline{\text{div}\ \underset{\sim}{\tau}^f} = \eta\ \Delta\left(\bar{U} + \bar{u}_s\right) \tag{6.4}$$

and is thus found to be negligible (with respect to $\bar{b}$) for practical applications. Although fluid movement is generally governed essentially by viscosity, corresponding macroscopic terms disappear : $\underset{\sim}{\sigma}^{\nabla}$ and $\underset{\sim}{\sigma}^{sd}$ are almost equal and fluid stress is correctly represented in E_x by a simple pressure $p(\underset{\sim}{\sigma}^f \simeq - np\underset{\sim}{I})$.

7.- FILTRATION PROCESSES

The interaction term $\bar{R}$ is to be splitted in three parts : a static one owing to the possible macroscopic inhomogeneity of the soil (called "buoyancy" force), a kinematic dissipative one (drag force) and a dynamic one corresponding to inertial coupling between fluid and solid parts (virtual mass effect)

$$\bar{R} = \bar{R}_{static} + \bar{R}_{kin.} + \bar{R}_{dyn.} \quad . \tag{7.1}$$

In this section dynamic term $\bar{R}_{dyn.}$ will be disregarded. Neglecting variations of fluid density at scale of D(O) one gets (Gilbert 1984) for the static part

$$\bar{R}_{static} = - p\ \overline{\text{grad}}\ n + \rho_f \bar{g}.\ \underset{\sim}{Y}^f \tag{7.2}$$

where influence of the geometrical tensor $\underset{\sim}{Y}^f$ given by (2.7) is very small, at least in mean value (see formula (2.6)). Equation (7.2) then reads

$$\bar{R}_{static} \simeq - p\ \overline{\text{grad}}\ n \quad . \tag{7.3}$$

Hence in that case $\bar{b}$ (formula (5.7)) and not $\bar{R}$ (as postulated on intuitive grounds in certain mixture theories) equals zero. Note the particularly simple expression of the buoyancy force (7.3) and its obvious geometrical interpretation. Estimates in (5.6b) of vector $\bar{b}$, or $\bar{R} - \bar{R}_{static}$, yield Darcy's law under various forms. Note that an estimate of $\bar{b}$ is naturally not available for any porous medium under any flow condition.

Slow stationary filtration of an incompressible newtonian fluid through a fixed stationary random porous matrix yields (Marle 1967), as D(0) grows, a symmetrical intrinsic permeability tensor $\underset{\sim}{k}$ given by

(7.4) $$\underset{\sim}{k}^{-1} = \frac{1}{n^2} < \frac{\partial M_{ih}}{\partial z^j} \cdot \frac{\partial M_{il}}{\partial z^j} >^f \bar{e}_h \otimes \bar{e}_l$$

as a function of the stationary random tensor $\underset{\sim}{M}(z)$ which maps $\bar{u}_{(f)}(z)$ as a function of $\bar{u}_f$.

One can also treat by this method the corresponding case of spatially periodic slow stationary flow through a periodic fixed matrix. It is convenient here to choose for m the discontinuous function equal to $1/|D|$ in the basic period ($|D|$ being the volume of the basic period of the lattice) and to 0 elsewhere. Equation (5.6b) then reads

(7.5) $$\bar{b} = n(\rho_f\bar{g} - \overline{\text{grad}}\ p) \quad .$$

However $\bar{b}$ and $\overline{\text{grad}}$ p are not constants whatever $\bar{x}$. It appears necessary to use a double averaging process, which eliminates the preceding fluctuations, by introducing

(7.6) $$\bar{B} = \bar{b} * m$$

(7.7) $$P = p * m = ((p_{(f)}I_f) * m) * m \quad .$$

Classical variational structure is then recovered yielding Darcy's law with a symmetrical intrinsic permeability tensor $\underset{\sim}{k}$

(7.8) $$\bar{U} = -\frac{1}{\eta}\underset{\sim}{k}\ (\overline{\text{grad}}\ P - \rho_f\bar{g}) \quad .$$

Denoting by $\lambda(\bar{z})$ a D-periodic function and by $\bar{\xi}$ a constant vector one has for the various pressures

(7.9a) $$p_{(f)}(\bar{z}) = \bar{\xi}.\bar{z} + \lambda(\bar{z}) + cte$$

(7.9b) $$p(\bar{x}) = \bar{\xi}.\overline{X_f(\bar{x})} + cte$$

(7.9c) $$P(\bar{x}) = \bar{\xi}.\bar{x} + cte$$

Observe (Gilbert 1984) that fluctuations of p around P are thus related to geometry only, through the periodic abovementioned tensor $\underset{\sim}{Y}^f$ whose value for a periodic medium is shown to be

(7.10) $$\underset{\sim}{Y}^f = (n-n_1)\bar{e}_1\otimes\bar{e}_1 + (n-n_2)\bar{e}_2\otimes\bar{e}_2 + (n-n_3)\bar{e}_3\otimes\bar{e}_3$$

where n is the (constant) volume porosity and n_i the variable surface porosity of planes $z^i = x^i \pm \frac{1}{2}|\bar{\ell}_i|$ (*Fig.* 5). The difference between p and P is small when the elementary period contains many grains since surface porosities become progressively equal to volume porosity as geometrical disorder in the period grows.

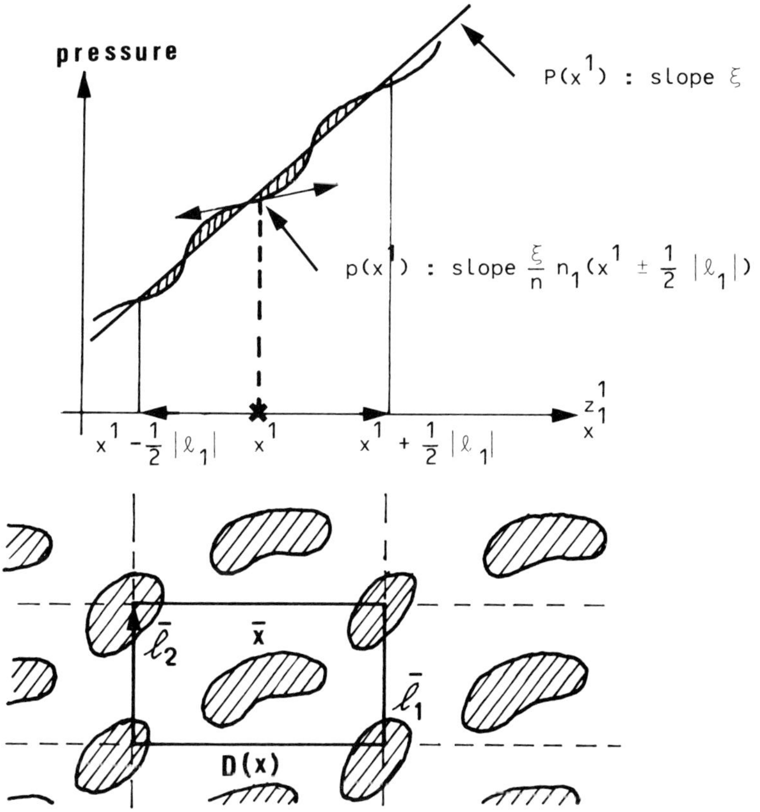

Figure 5 : Average pressures p and P in a particular periodic medium for $\bar{\xi} = \xi\bar{e}_1$. The vectors $\bar{\ell}_i = |\bar{\ell}_i|\bar{e}_i$ are basic vectors of the periodic lattice.

Note that the pressure P may be identified with the first term p_o of the asymptotic development of the pressure in successive powers of the small parameter ε, which is postulated in the theory of homogenization of fine periodic structures.

HARMONIC FLOW THROUGH A PARTICULAR FIXED STRUCTURE

8.- HYDRAULIC IMPEDANCE OF A TWO-DIMENSIONAL NARROW GAP

To obtain more complete information about behaviour of a saturated porous medium one must use postulates or analyse by numerical methods particular structures of interest, which allow to go further. Such an example is presented here concerning harmonic flow of an incompressible newtonian fluid through narrow gaps between the various parts of a fixed solid matrix. For sake of simplicity problem is studied in two dimensions ; grains are roughly schematized as parallel cylinders.

Let us consider a narrow gap of minimum width $2h_{ij}$ between two locally regular motionless cylinders of parallel axes with radii of curvature a_i and a_j respectively in the vicinity of the narrowing (*Fig.6*).

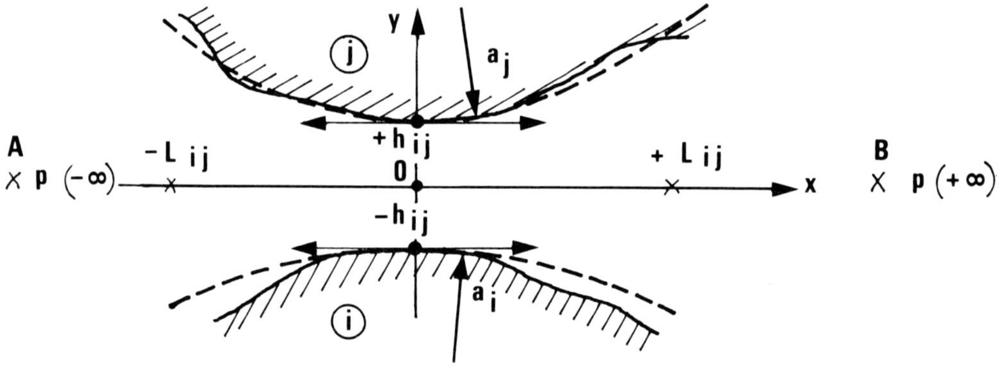

Figure 6 : Two-dimensional gap between locally regular cylinders

An equivalent diameter is introduced for the gap by

(8.1) $$d_{ij} = 4(a_i^{-1} + a_j^{-1})^{-1}$$

Quantity h_{ij} gives a natural width scale ; an appropriate length scale is given by

(8.2) $$L_{ij} = (h_{ij}.d_{ij})^{1/2}$$

and we investigate asymptotic hydrodynamic behaviour of the gap when the grains are very close one to the other, that is when parameter α is very small with respect to one

(8.3) $$\alpha = \frac{h_{ij}}{d_{ij}} \downarrow 0 .$$

Significative scaling is obviously obtained by using dimensionless co-ordinates $\tilde{x}$ and $\tilde{y}$ (inner variables)

(8.4) $$x = L_{ij}\tilde{x} \quad , \quad y = h_{ij}\tilde{y} .$$

Inner asymptotic expansion of Navier-Stokes equations for $\alpha\downarrow 0$ is calculated, matching it on outer conditions expressed at the infinity in inner variables

(8.5) $$p \to p(\pm\infty) = p_o \mp \frac{P}{2} \cos \omega t \quad , \quad \text{as} \quad \tilde{x} \to \pm\infty$$

where p_o is a constant pressure and P the amplitude of periodic pressure drop $P \cos\omega t$ between left and right sides of the gap. Scalings are thus introduced for pressure, time and the two velocity components and classical principle of minimum degeneracy is used.

For sufficiently small pressure drops stationary inertia forces are to be neglected and two-dimensional flow rate $q(t)$ (in $m^2.s^{-1}$) past the gap, which is a periodic function of time, is proportional to P.

Denote by $\hat{q}$ and $\hat{p}$ the complex quantities associated to periodic flow rate and pressure drop. An easy but somewhat tedious calculation yields

(8.6) $$\frac{\hat{p}}{\hat{q}} = Z_{ij} = \frac{9\pi}{16} \cdot \eta \cdot \frac{d_{ij}^{1/2}}{h_{ij}^{5/2}} \cdot F\left(\frac{2h_{ij}}{\delta}\right)$$

where function F is a complex valued integral of the ratio β of minimum width $2h_{ij}$ of the gap to "skin depth" δ at circular frequency ω, given by

(8.7) $$\delta = \left(\frac{2\eta}{\rho_f \omega}\right)^{1/2}$$

We call Z_{ij} hydraulic impedance of the gap and F reduced impedance. It is represented in figure 7 in log-log plot as a function of the dimensionless parameter β. Purely viscous solution of the problem is recovered for $\beta \to 0$ (all inertia terms dropped) as well as solution for an inviscid fluid for $\beta \to +\infty$ (asymptotic expansions are easily calculated for the cases $\beta = 0$ and $\beta = +\infty$).

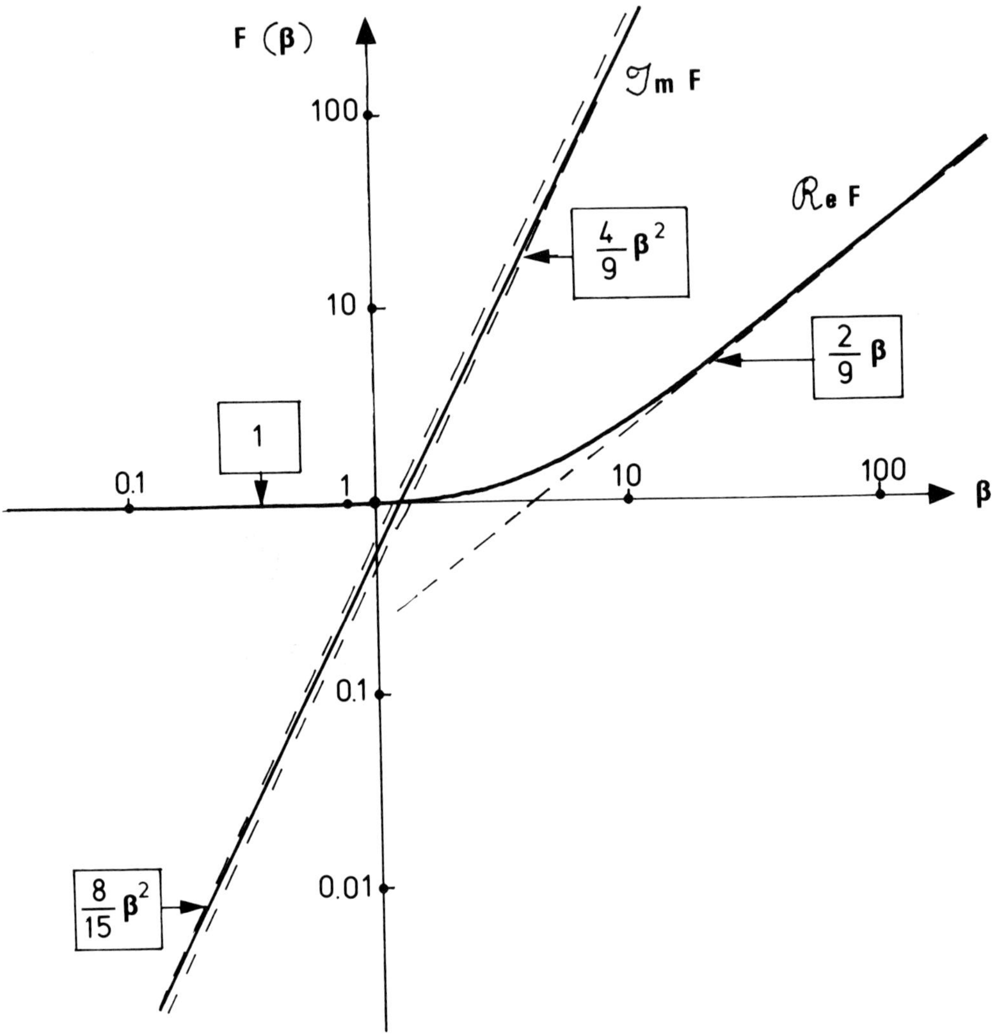

Figure 7 : Reduced impedance F as a function of the parameter β. Dashed lines represent approximations valid for $\beta \to 0$ or $\beta \to +\infty$ obtained by expansion of F. Note their accuracy : for $\beta \leqslant 10^{-2}$ or $\beta \geqslant 10^2$ relative error is less than 1%. Imaginary part of F is such that, whatever β, $0.4444\ \beta^2 < \mathrm{Jm}F < 0.5333\ \beta^2$.

Behaviour of the pressure in the limit $\alpha \downarrow 0$ was carefully investigated : it appears to be independent of $\tilde{y}$ and tends very rapidly, as $|\tilde{x}|$ grows, towards upstream and downstream values given by formula (8.5). Thus an excellent overlap is obtained between inner expansion and outer conditions.

It is worth noting link between such calculation and Biot's simple approximation of viscodynamic operator (Biot 1961) which predicts

(8.8) $$\frac{\hat{p}}{\hat{q}} = c_1 + ic_2\omega$$

where c_1 and c_2 are real positive constants related to dissipation rate and relative kinetic energy.

Extension to flow under unsteady conditions other than harmonic ones is straightforward using Carson-Laplace transforms of flow rate and pressure drop. Flow law with memory effects over pressure drop is obtained as expected.

9.- APPLICATION TO PERMEABILITY OF A LEIBNIZ PACKING OF CYLINDERS UNDER HARMONIC REGIME

Preceding results are used to compute permeability of particular self-similar compact packings of two-dimensional grains, called Leibniz packings and obtained through a slight modification of Appollonius construction (Mandelbrot, 1982). Such packings, which were studied in particular by Adler (1985) and Gilbert and Adler (1985a,b), are build up as follows.

Consider three cylinders with parallel axes, which are almost tangent and have arbitrary radii a_1, a_2, a_3. The gap $2h_{ij}$ between any two of them is assumed to be a fixed small fraction ε' of the radius of the smallest of the two adjacent cylinders

(9.1) $$2h_{ij} = \varepsilon' . \min(a_i, a_j)$$

and hence parameter $\alpha(i,j)$ lies between $\varepsilon'/8$ and $\varepsilon'/4$. These three cylinders constitute the generation number $n = 0$.

It is possible to insert between them a fourth cylinder of radius a_4 almost tangent to them, so that relation (9.1) still holds for gaps around it. This cylinder constitutes the generation number $n = 1$.

By the same process, in a second step $n = 2$, three new cylinders numbered from 5 to 7 are inserted in the three interstices created at step $n = 1$, relation (9.1) still holding. This construction is continued in an obvious way (*Fig.8*) with smaller and smaller cylinders and gaps ; total number of cylinders N_n after step n rapidly increases as n increases.

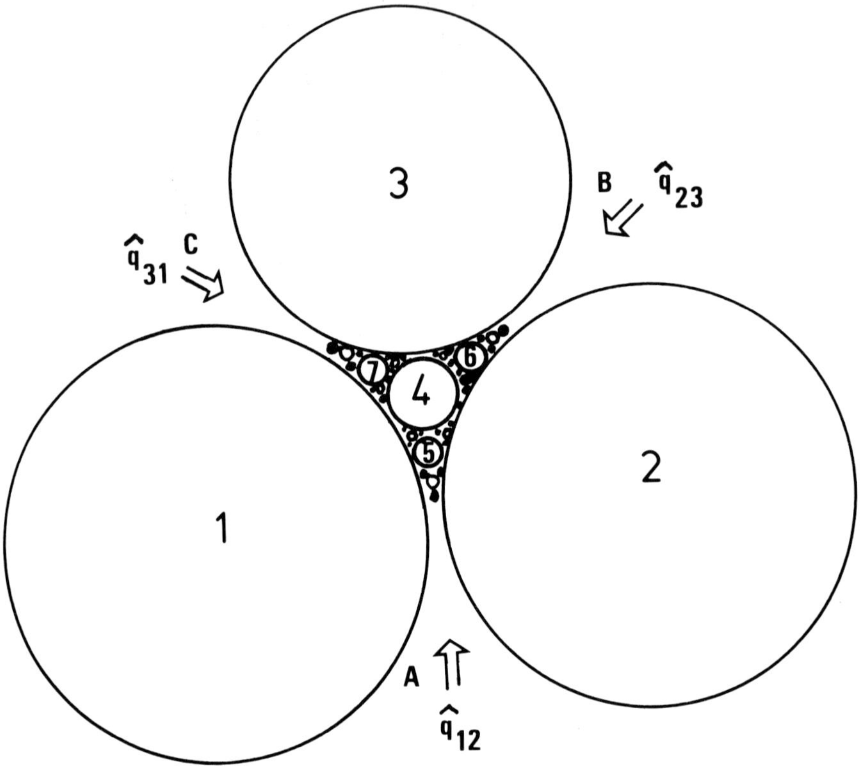

Figure 8 : Leibniz packing containing cylinders of generations 0 to 4. Harmonic flow rates $\hat{q}_{12}$, $\hat{q}_{23}$ *and* $\hat{q}_{31}$ *(with* $\hat{q}_{12}+\hat{q}_{23}+\hat{q}_{31}=0$*) are assigned at the boundary of the packing. Note the obvious self-similarity of the packing.*

Suppose that harmonic flow rates $\hat{q}_{12}$, $\hat{q}_{23}$ and $\hat{q}_{31}$, whose sum is equa to zero, are assigned at the boundary of the packing. Hence pressure drc are observed between points A, B, C and complex pressure drops $\hat{p}_{12}$ and $\hat{p}$ are defined by

(9.2a) $$\mathcal{R}e\ \hat{p}_{12} = (p_A - p_C)(t)$$

(9.2b) $$\mathcal{R}e\ \hat{p}_{23} = (p_B - p_C)(t)$$

Clearly, for $n = 0$ (*Fig.9*) relationship between flow rates $\hat{q}_{12}$ and $\hat{q}_{23}$ and pressure drops $\hat{p}_{12}$ and $\hat{p}_{23}$ is that of a star with hydraulic impe dances $Z_{12}^{(o)}$, $Z_{23}^{(o)}$ and $Z_{31}^{(o)}$ of values Z_{12}, Z_{23} and Z_{31} given by formula (8.6).

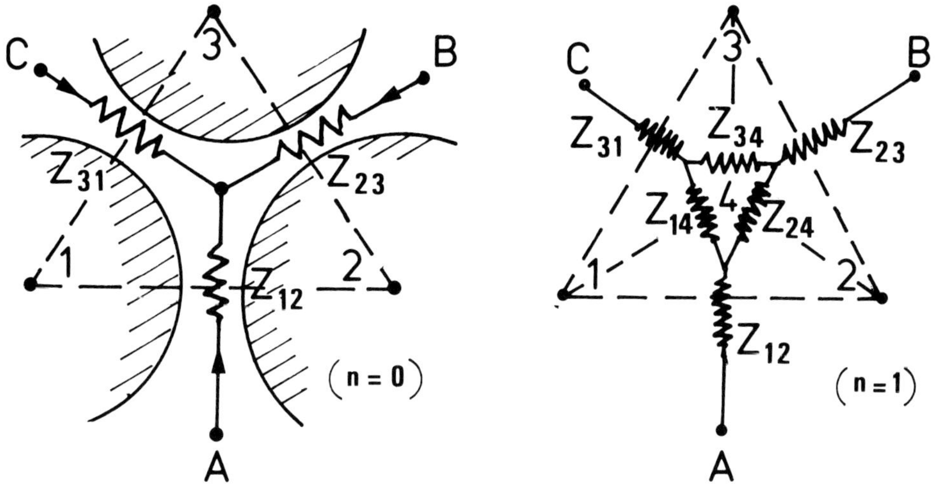

Figure 9 : Equivalent electrical description of the packing for n = 0 and n = 1. Primal graph associated to the packing is shown by dashed lines.

For greater values of n the packing always appears as a network of hydraulic impedances. Its topology is that of dual graph associated to the packing, the primal graph being obtained as follows : vertices correspond to centres of the cylinders and an edge relates two vertices when corresponding cylinders are close one to the other. Edges of the dual graph hence correspond to the gaps of the packing (*Fig.9*). It is a simple matter to state that, whatever n, there is a star equivalent to the whole packing, so that for fixed n

$$(9.3)\qquad \begin{pmatrix} \hat{p}_{12} \\ \hat{p}_{23} \end{pmatrix} = \begin{pmatrix} Z_{12}^{(n)} + Z_{31}^{(n)} & Z_{31}^{(n)} \\ Z_{31}^{(n)} & Z_{23}^{(n)} + Z_{31}^{(n)} \end{pmatrix} \begin{pmatrix} \hat{q}_{12} \\ \hat{q}_{23} \end{pmatrix}$$

where $Z_{12}^{(n)}$, $Z_{23}^{(n)}$ and $Z_{31}^{(n)}$ are the hydraulic impedances of the equivalent star at step n.

A program was written by us at the Laboratoire de Mécanique des Solides to compute physical properties of such packings. Problems concerning many grains (for n = 10, N_n = 29527 grains) are reduced in a recursive manner to much simpler ones which are easily solved. For permeability the matrix procedure uses only the classical triangle-star transformation for networks already used by Adler (1985) for steady conditions. A more intricate problem was studied by Gilbert and Adler (1985a, b) using the abovementioned program.

10.- NUMERICAL RESULTS

A few numerical results are presented in this section. Calculations were performed up to generation $n = 10$ for various configurations ; real and imaginary parts of the hydraulic impedances of the equivalent star at step n obey (for n not too small) power laws as functions of the total number of cylinders N_n, as expected (*Fig. 10*). One gets with a good accuracy

(10.1) $$\mathcal{R}e\, Z_{ij}^{(n)} \simeq c_{ij}\left(\frac{a_2}{a_1}, \frac{a_3}{a_1}\right) \cdot \eta \cdot a_4^{-2} \cdot \varepsilon'^{-5/2} \cdot N_n^{\xi}$$

(10.2) $$\mathcal{I}m\, Z_{ij}^{(n)} \simeq c \cdot \rho_f \cdot \omega \cdot \varepsilon'^{-1/2} \cdot N_n^{\zeta}$$

where c_{12}, c_{23} and c_{31} are dimensionless functions of the ratios a_2/a_1 and a_3/a_1 and c is a dimensionless constant approximately equal to 3.60. Numerical evaluation of the two exponents ξ and ζ yields very different values

(10.3) $$\xi \simeq 2.01 \qquad , \qquad \zeta \simeq 0.469 \quad .$$

Observe that, for n not too small, $\mathcal{R}e Z_{ij}^{(n)}$ is almost independent of ρ_f and ω and $\mathcal{I}m Z_{ij}^{(n)}$ of viscosity η. This is closely related to formula (8.6) and properties of reduced impedance F. Dependence upon parameter ε' is explained in the same manner and is valid for sufficiently low values of ε'.

Closer examination of the recursive procedure reveals that behaviours of $\mathcal{R}e Z_{ij}^{(n)}$ and $\mathcal{I}m Z_{ij}^{(n)}$ as functions of N_n must be, in some sense, independent one from the other as n grows. Explanation of this phenomenon is the relative smallness of the parameters β corresponding to gaps created at step n. Note that initial configuration is always reminded for $\mathcal{R}e Z_{ij}^{(n)}$ only ; $\mathcal{I}m Z_{ij}^{(n)}$ appears to be isotropic and size invariant.

Scaling arguments may be used to obtain estimates of the exponents ξ and ζ, as they are independent from initial configuration. The packing of generation n whose three first cylinders have radii (a_1, a_2, a_3) may be viewed as the sum of the three packings (a_1, a_2, a_4), (a_4, a_2, a_3) and (a_1, a_4, a_3) of generation $n - 1$. Putting for instance $a_1 = a_2 = a_3$ and neglecting differences between the various c_{ij} coefficients one gets,

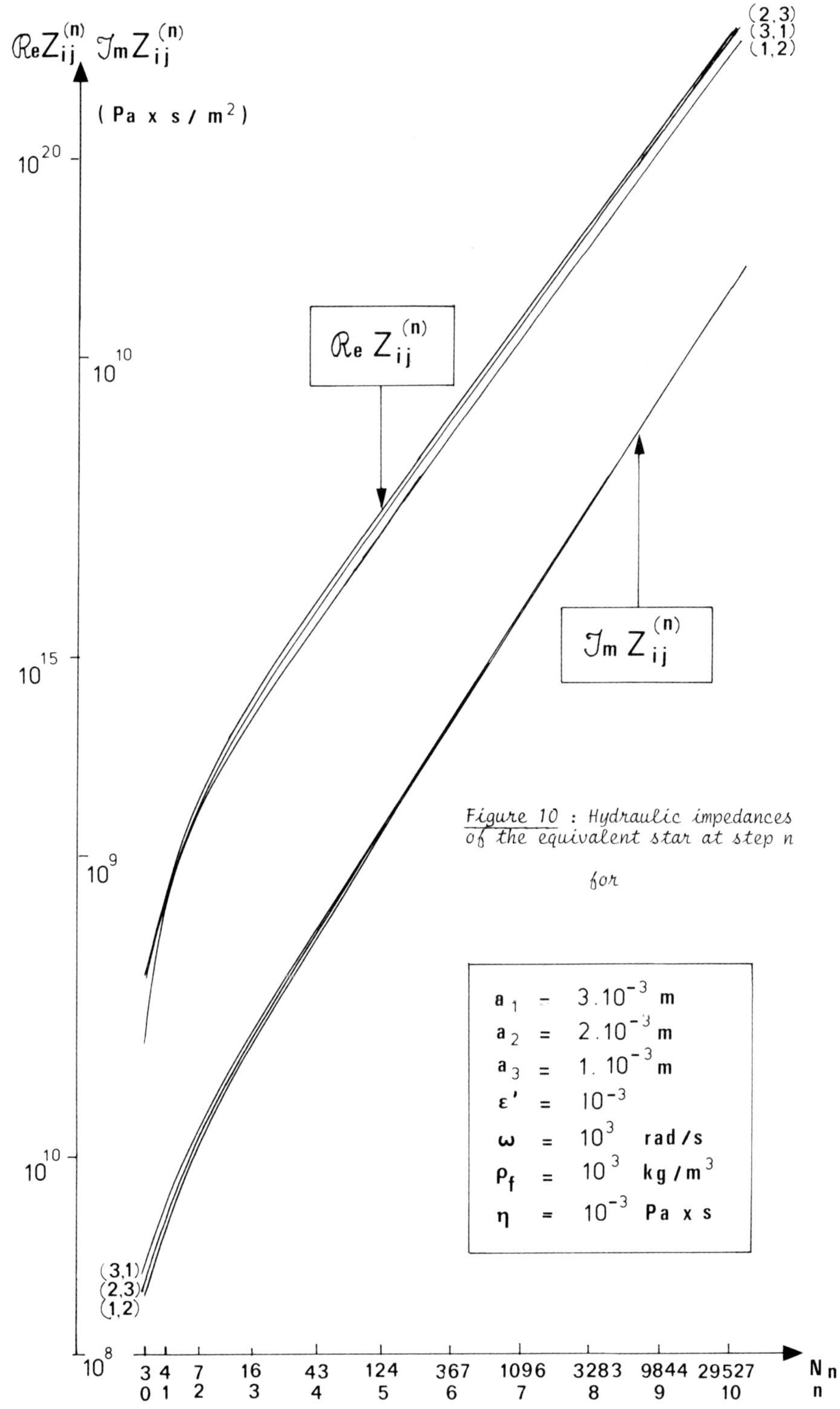

Figure 10 : Hydraulic impedances of the equivalent star at step n

for

$a_1 = 3 \cdot 10^{-3}$ m
$a_2 = 2 \cdot 10^{-3}$ m
$a_3 = 1 \cdot 10^{-3}$ m
$\varepsilon' = 10^{-3}$
$\omega = 10^3$ rad/s
$\rho_f = 10^3$ kg/m^3
$\eta = 10^{-3}$ Pa $\times$ s

using formulas (10.1) and (10.2) for the four packings involved, the following estimates of the exponents

(10.4) $\xi \simeq 2.11$, $\zeta \simeq 0.465$

which are in excellent agreement with numerical estimates given by (10.3), taking into account the simplifying assumption of isotropy for $\mathcal{R}e Z_{ij}^{(n)}$.

Preceding estimate of ζ is also obtained theoretically for a network of pure inductances having the same value for all gaps regardless of the radii of the cylinders involved. Note that it is more or less the case here : it explains the observed isotropic and size invariant character of $\mathcal{I}m Z_{ij}^{(n)}$.

Neglecting conductance of gaps which do not involve one of the three first cylinders yields for ξ the value 2.16.

Obtained results are interesting to compare with Biot's type approximation.

For particular values of the ratios a_2/a_1 and a_3/a_1 one can draw periodic lattices of such packings, whose permeability properties are easily studied starting from preceding results.

CONCLUSION

Empirical approaches use the notion of representative elementary volume (REV) and postulate average balance equations on intuitive grounds. Convolution methods give a rigorous form to these estimates and allow to introduce in a consistent and natural way macroscopic quantities of interest and balance equations for complex heterogeneous media. Explicit physical meaning of the various quantities used is known and thus they provide a useful framework for the discussion of constitutive relations.

Obtained equations, as they are supplemented by constitutive relations, reflect the fact that soils generally exhibit statistical quasi-homogeneity at macroscopic scale.

Relationships between this work and other theoretical approaches, as general mixture theories, Biot's theory or homogenization of fine periodic structures are also to be mentioned.

The shortcoming of the formulation used is naturally connected with the difficulties encountered to express precise constitutive relations. These difficulties arise from the simplification of the geometrical description of the medium involved by this change of scale process, which obviously implies lack of information.

To allow to go further, obtained results must be supplemented by direct macroscopic postulates, geometrical assumptions as local periodicity depending upon a small parameter or use of simple cell models representative of the medium.

In this framework analysis of simple self-similar structures, which account for the strongly heterogeneous character of soils at small length scale, is of particular interest. Despite their apparent great complexity these structures often lead to tractable calculations. Another sort of change of scale is thus achieved.

It seems to us that combining these two methods of change of scale accounting for the two abovementioned essential geometrical aspects of soils will appear fruitful for the study of soil rheology.

REFERENCES

ADLER, P.M., (1985), *Transport processes in fractals I* , Int. J. Multiphase Flow, (under press).

AHMADI, G., (1980), *On mechanics of saturated granular materials* , Int. J. Non-Linear Mechanics, vol.15, pp. 251-262.

AHMADI, G. and SHAHINPOOR, M., (1983), *A continuum theory for fully saturated porous elastic materials*, Int. J. Non-Linear Mechanics, vol.18, n°1, pp. 223-234.

AURIAULT, J.L. and SANCHEZ-PALENCIA, E., (1977), *Etude du comportement macroscopique d'un milieu poreux saturé déformable*, Journal de Mécanique, vol.16, n°4, pp. 576-603.

AURIAULT, J.L., (1980), *Dynamic behaviour of a porous medium saturated by a newtonian fluid*, Int. J. Engng. Sci., vol.18, pp. 775-785.

AVALLET, C., (1981), *Comportement dynamique de milieux poreux saturés déformables*, Thèse 3ème cycle Mécanique, Grenoble, sept. 1981.

BATCHELOR, G.K., (1974), *Transport properties of two-phase materials with random structure*, Annual Review of Fluid Mechanics, vol.6, pp. 227-255.

BEDFORD, A. and DRUMHELLER, D.S., (1983), *Theories of immiscible and structured mixtures*, Int. J. Engng. Sci., vol.21, n°8, pp. 863-960.

BENSOUSSAN, A., LIONS, J.L., PAPANICOLAOU, G., (1978), *Asymptotic analysis for periodic structures*, North-Holland, Amsterdam.

BIOT, M.A., (1961), *Théorie généralisée de la propagation acoustique dans un solide poreux dissipatif*, Colloque Int. CNRS n°111, "La Propagation des Ebranlements dans les Milieux Hétérogènes", Marseille, sept. 1961, pp. 57-65.

BIOT, M.A., (1962a), *Mechanics of deformation and acoustic propagation in porous media*, J. Appl. Phys., vol.33, n°4, pp. 1482-1498.

BIOT, M.A., (1962b), *Generalized theory of acoustic propagation in porous dissipative media*, J. Acoust. Soc. Am., vol.34, n°9, pp. 1254-1264.

BIOT, M.A., (1977), *Variational lagrangian thermodynamics of nonisothermal finite strain mechanics of porous solids and thermomolecular diffusion*, Int. J. Solids Structures, vol.13, pp. 579-597.

BORNE, L., (1983), *Contribution à l'étude du comportement dynamique des milieux poreux saturés déformables : étude de la loi de filtration dynamique*, Thèse Docteur-Ingénieur, Grenoble, sept. 1983.

BOWEN, R.M., (1976), *Theory of mixtures*, Continuum Physics, vol.3, A.C. Eringen Ed., New-York, Academic Press.

COUDERT, J.F., (1973), *Théorie macroscopique des écoulements multiphasiques en milieu poreux*, Revue Inst. Fr. du Pétrole, vol.28, n°2, mars-avril 1973, pp. 171-183 and vol.28, n°3, mai-juin 1973, pp. 373-398.

COUSSY, O. and BOURBIE, T., (1984), *Propagation des ondes acoustiques dans les milieux poreux saturés*, Revue Inst. Fr. du Pétrole, vol.39, n°1, janv-fév.1984, pp. 47-66.

ENE, H.I. and SANCHEZ-PALENCIA, E., (1975), *Equations et phénomènes de surface pour l'écoulement dans un modèle de milieu poreux*, Journal de Mécanique, vol.14, n°1, pp. 73-108.

ENE, H.I. and MELICESCU-RECEANU, M., (1984), *On the viscoelastic behaviour of a porous saturated medium*, Int. J. Engng. Sci., vol.22, n°3, pp. 243-246.

ESTRADA, R. and KANWAL, R.P., (1980), *Applications of distributional derivatives to wave propagation*, J. Inst. Maths Applics, vol.26, pp. 39-63.

GILBERT, F., (1984), *Description des sols saturés par une méthode d'homogénéisation*, Ecole d'Hiver CNRS-IMG "Rhéologie des Géomatériaux", Aussois, France, 33 p.

GILBERT, F., (1985), *A mechanical description of saturated soils*, International Symposium Physical Basis and Modelling of Finite Deformation of Aggregates, Paris, sept-oct 1985, 16 p.

GILBERT, F. and ADLER, P.M., (1985a), *Spin viscosity of heterogeneous suspensions*, 5th International Conference on Surface and Colloid Science, Clarkson, Potsdam, N.Y., june 24-28, 1985.

GILBERT, F. and ADLER, P.M., (1985b), *Rotation tensor of fractal suspensions*, to appear in J. Coll. Int. Sc.

GRAY, W.G. and O'NEILL, K., (1976), *On the general equations for flow in porous media and their reduction to Darcy's law*, Water Resources Research, vol.12, n°2, april 1976, pp. 148-154.

HADJ-HAMOU, A., (1983), *Contribution à l'étude du comportement des sols pulvérulents sous chargements cyclique et dynamique*, Thèse Docteur-Ingénieur, Paris, ENPC, déc. 1983.

HASSANIZADEH, M., (1979), *Macroscopic description of multi-phase systems - A thermodynamic theory of flow in porous media*, Ph. D. Thesis, Princeton, sept. 1979.

MANDELBROT, B.B., (1982), *The fractal geometry of nature*, Freeman, San-Francisco.

MARLE, C.M., (1965), *Application des méthodes de la thermodynamique des processus irréversibles à l'écoulement d'un fluide à travers un milieu poreux*, Bulletin Rilem, nouvelle série, vol.20, pp. 107-117.

MARLE, C.M., (1967), *Ecoulements monophasiques en milieux poreux*, Revue Inst. Fr. du Pétrole, vol.22, n°10, oct. 1967, pp. 1471-1509.

MARLE, C.M., (1982), *On macroscopic equations governing multiphase flow with diffusion and chemical reactions in porous media*, Int. J. Engng. Sci., vol.20, n°5, pp. 643-662.

MATHERON, G., (1965), *Les variables régionalisées et leur estimation*, Thèse Docteur-ès-Sciences Appliquées, Paris.

MATHERON, G., (1967), *Eléments pour une théorie des milieux poreux*, Masson et Cie, Paris.

MULLER, I., (1975), *Thermodynamics of mixtures of fluids*, Journal de Mécanique, vol.14, n°2, pp. 267-303.

PREVOST, J.H., (1980), *Mechanics of continuous porous media*, Int. J. Engng. Sci., vol.18, pp. 787-800.

SANCHEZ-PALENCIA, E., (1974), *Comportement local et macroscopique d'un type de milieux physiques hétérogènes*, Int. J. Engng. Sci., vol.12, pp. 331-351.

SANCHEZ-PALENCIA, E., (1980), *Non-homogeneous media and vibration theory*, Lectures notes in Physics, 127, Springer.

SERRA, J., (1982), *Image analysis and mathematical morphology*, Academic Press.

SLATTERY, J.C., (1967), *Flow of viscoelastic fluids through porous media*, A.I.Ch.E. Journal, vol.13, n°6, nov. 1967, pp. 1066-1071.

SLATTERY, J.C., (1969), *Single-phase flow through porous media*, A.I.Ch.E. Journal, vol.15, n°6, nov. 1969, pp. 866-872.

SLATTERY, J.C., (1972), *Momentum, energy and mass transfer in continua*, Mc Graw Hill.

TRUESDELL, C. and TOUPIN, R.A., (1960), *The classical field theory*, Handbuch der Physik, III, 1, S. Flügge ed., Springer.

WHITAKER, S., (1969), *Advances in theory of fluid motion in porous media*, Industrial and Engineering Chemistry, vol.61, n°12, dec. 1969, pp. 14-28.

THE USE OF THE HOMOGENIZATION METHOD TO DESCRIBE THE VISCOELASTIC BEHAVIOUR OF A POROUS SATURATED MEDIUM

Horia I. ENE

INCREST, Department of Mathematics
Bd. Pacii 220, 79622 Bucharest, Romania

Abstract. Using the homogenization method, we obtain the constitutive equation for a mixture formed by a viscoelastic skeleton and a viscous incompressible fluid. The macroscopic constitutive equation give us the effective stress tensor as the difference between the mean value of the stress tensor in the skeleton and the pore pressure multiplied by the porosity. The motion of the fluid is described by a Darcy's law with memory depending on the pressure gradient and the inertia force. It is also deduced the form of the conservation of mass and momentum.

1. INTRODUCTION

1.1. Generalities

In the general framework of the homogenization method [1, 2] we consider the problem of the motion of a mixture formed by a viscoelastic skeleton and a viscous incompressible fluid. The geometric distribution of the solid and fluid parts is periodic, with small periods. The dimensions of the periods are then associated with the small parameter ε.

It is well known that a great variety of problems can arise if the orders of magnitude of the coefficients are very different, or if the topological properties of the mixture are different [2]. In the case of the vibration of a mixture of an elastic body and a viscous barotropic fluid, it appears that the macroscopic stress tensor is given also by a viscoelastic law with memory [2, 3], but depending only on the strain tensor.

In our case we consider that the solid part is connected, as well as the fluid one, and that the viscosity of the fluid is small (a slightly viscous fluid). In fact it is well known [2, 4] that Darcy's law holds only in the case of large viscosity and small velocity, orsmall viscosity and possible large velocity. More precisely, if the smaller magnitude is of order ε^2 then the larger is of order ε^0. As a consequence of the fact that the displacement vector in the solid part is of order ε^0, we take the velocity in the fluid part of the same order and the viscosity of the fluid of order ε^2.

1.2. Mixture of a viscoelastic solid with a viscous fluid

In the solid part of the mixture, the equations are:

$$(1.1)\quad \rho_s \frac{\partial^2 u_i}{\partial t^2} - \frac{\partial \sigma^s_{ij}}{\partial x_j} = f_i$$

$$(1.2)\quad \sigma^s_{ij} = a^s_{ijkh}\, e_{kh}(\mathbf{u}) + b^s_{ijkh}\, e_{kh}\frac{\partial \mathbf{u}}{\partial t}$$

$$e_{kh}(\mathbf{u}) = \frac{1}{2}\left(\frac{\partial u_k}{\partial x_h} + \frac{\partial u_h}{\partial x_k}\right)$$

where $\mathbf{f}$ is the exterior body force, and the coefficients a^s_{ijkh}, b^s_{ijkh}, satisfy the usual properties of symmetry and positivity;

$$(1.3)\quad a^s_{ijkh} = a^s_{jikh} = a^s_{jihk} = a^s_{khij}$$

$$(1.4)\quad a^s_{ijkh}\, e_{ij}\, e_{kh} \geq \alpha e_{ij}\, e_{ij}; \qquad \alpha > 0$$

and similar relations for b^s_{ijkh}.

In the fluid part, the equations are:

$$(1.5)\quad \rho_f \frac{\partial^2 u_i}{\partial t^2} - \frac{\partial \sigma^f_{ij}}{\partial x_j} = f_i$$

$$(1.6)\quad \sigma^f_{ij} = -p\,\delta_{ij} + \varepsilon^2\,\mu\,\delta_{ik}\,\delta_{jh}\,e_{kh}(\mathbf{v})$$

$$(1.7)\quad \operatorname{div} \mathbf{v} = 0, \qquad \mathbf{v} = \frac{\partial \mathbf{u}}{\partial t}.$$

Moreover, at the interface between the solid and the fluid, we must have the continuity of displacement and stress:

$$(1.8)\quad [\mathbf{u}] = 0, \qquad [\sigma_{ij} n_j] = 0.$$

We must adjoin initial and boundary conditions:

$$(1.9)\quad \mathbf{u} = 0 \quad \text{on } \partial\Omega$$

$$(1.10)\quad \mathbf{u} = \frac{\partial \mathbf{u}}{\partial t} = 0 \quad \text{for } t = 0$$

where Ω is the domain occupied by the mixture, and is formed by Ω_s and Ω_f.

The variational formulation of the problem (1.1)+(1.5)+(1.8)+ +(1.9)+(1.10) is: find $\mathbf{u}$, function of t with values in $H^1_0(\Omega)$, such that:

$$(1.11)\quad \int_\Omega \rho \frac{\partial^2 u_i}{\partial t^2} w_i \, dx + a(\mathbf{u},\mathbf{w}) + b(\frac{\partial \mathbf{u}}{\partial t}, \mathbf{w}) - \int_{\Omega_f} p \operatorname{div} \mathbf{w} dx = \int_\Omega \mathbf{f} \, \mathbf{w} \, dx$$

$$\mathbf{w} \quad H^1_o(\Omega)$$

$$(1.12)\quad \operatorname{div} \mathbf{v} = 0 \quad \text{in } \Omega_f$$

$$(1.13)\quad a(\mathbf{u}, \mathbf{w}) = \int_{\Omega_s} a^s_{ijkh} \, e_{kh}(\mathbf{u}) \, e_{ij}(\mathbf{w}) \, dx$$

$$(1.14)\quad b(\mathbf{v}, \mathbf{w}) = \int_\Omega b_{ijkh} \, e_{kh}(\mathbf{v}) \, e_{ij}(\mathbf{w}) \, dx$$

$$(1.15)\quad b_{ijkh} = \begin{cases} b^s_{ijkh} & \text{in } \Omega_s \\ \varepsilon^2 \mu \, \delta_{ik} \, \delta_{jh} & \text{in } \Omega_f \, . \end{cases}$$

1.3. Two-scale asymptotic process

We consider a parallelipipedic period Y of the space of variables y_i $(i = 1, 2, 3)$ formed by a fluid part Y_f and a solid one Y_s, separated by a smooth boundary Γ. We look for Y-periodic coefficients in the variable $y = x/\varepsilon$: $\rho^\varepsilon(x) \equiv \rho(x/\varepsilon)$, $a^\varepsilon_{ijkh}(x) \equiv a_{ijkh}(x/\varepsilon)$ and $b^\varepsilon_{ijkh}(x) \equiv b_{ijkh}(x/\varepsilon)$.

In order to study the asymptotic process $\varepsilon \to 0$, we assume that the appropiate asymptotic expasion in the solid part is analogous to that of viscoelastic mixture, and in the fluid part analogous to that of flow through porous media. But the first term of the displacement vector expansion in the solid does not depend on y in the solid region. On the contrary, it does in the fluid region. It is then natural to introduce the relative displacement of the fluid with respect to the solid: $\mathbf{u}^r(x, y, t) = \mathbf{u}^o(x, y, t) - \mathbf{u}^o(x, t)$, and consequently we search for a two-scale asymptotic expansion suitable in $\Omega_{\varepsilon f}$ as well as in $\Omega_{\varepsilon s}$:

$$(1.16)\quad \mathbf{u}^\varepsilon(x, t) = \mathbf{u}^o(x,t) + \mathbf{u}^r(x,y,t) + \varepsilon \, \mathbf{u}^1(x,y,t) + \ldots$$

where $y = x/\varepsilon$ and all functions are Y-periodic in y. The vector $\mathbf{u}^r$ takes values in $H^1(Y)$ and is zero on Y_s and on Γ.

For the pressure we have also:

$$(1.17)\quad p^\varepsilon(x,t) = p^o(x,t) + \varepsilon p^1(x,y,t) + \ldots$$

Now, with standard notation in homogenization theory, for fixed ε, the problem (1.11), (1.12) may be considered for $\mathbf{u}^\varepsilon$ and p^ε.

2. MACROSCOPIC EQUATIONS

2.1. Balance of mass

If we replace (1.16) into (1.12) we have

(2.1) $\mathrm{div}_y\, \mathbf{v}^r = 0$

(2.2) $\mathrm{div}_x(\mathbf{v}^o + \mathbf{v}^r) + \mathrm{div}_y \mathbf{v}^1 = 0 \quad \text{in } \Omega_{\varepsilon f}.$

Note that (2.2) only holds in the fluid part. But using the fact that

$$\int_Y \mathrm{div}_y\, \mathbf{v}^1\, dy = \int_{\partial Y} \mathbf{v}^1\, \mathbf{n}\, ds = 0$$

we take the mean value of (2.2) over Y and we have:

(2.3) $$n\, \mathrm{div}_x\, \mathbf{v}^o + \mathrm{div}_x\, \tilde{\mathbf{v}}^r = (1/|Y|) \int_{Y_s} \mathrm{div}_y\, \mathbf{v}^1\, dy; \qquad n = \frac{|Y_f|}{|Y|}$$

which is the balance of mass.

After that, taking test functions depending on ε in the form

(2.4) $$\mathbf{w}(x) = \mathbf{w}^o(x) + \mathbf{w}^r(x,y) + \varepsilon \mathbf{w}^1(x,y); \qquad \mathrm{div}_y\, \mathbf{w}^r = 0$$

at order ε^o, from the equation analogous to (1.11), using (1.16) and (1.17) we obtain:

(2.5) $$\int_\Omega \rho^\varepsilon \frac{\partial^2(u_i^o + u_i^r)}{\partial t^2}(w_i^o + w_i^r)dx + \int_{\Omega_{\varepsilon s}} a^s_{ijkh}\left(\frac{\partial u_k^o}{\partial x_h} + \frac{\partial u_k^1}{\partial y_h}\right)\left(\frac{\partial w_i^o}{\partial x_j} + \right.$$

$$\left. + \frac{\partial w_i^1}{\partial y_j}\right) dx - \int_{\Omega_{\varepsilon f}} p^o(\mathrm{div}_x \mathbf{w}^o + \mathrm{div}_x \mathbf{w}^r + \mathrm{div}_y \mathbf{w}^1)\, dx +$$

$$+ \int_{\Omega_{\varepsilon s}} b^s_{ijkh} \frac{\partial}{\partial t}\left(\frac{\partial u_k^o}{\partial x_h} + \frac{\partial u_k^1}{\partial y_h}\right)\left(\frac{\partial w_i^o}{\partial x_j} + \frac{\partial w_i^1}{\partial y_j}\right) dy +$$

$$+ \int_{\Omega_{\varepsilon f}} \mu \frac{\partial}{\partial t} \frac{\partial u_i^r}{\partial y_j} \frac{\partial w_i^r}{\partial y_j} dx = \int_\Omega f_i(w_i^o + w_i^r)dx.$$

2.2. Relative velocity

The relative motion of the fluid may be obtained if we take in (2.5) $\mathbf{w}^o = \mathbf{w}^1 = 0$, $\mathbf{w}^r = \theta(x)\boldsymbol{\omega}(x/\varepsilon)$, $\theta \in \mathcal{D}(\Omega)$, $\mathrm{div}_y\, \boldsymbol{\omega} = 0$, $\boldsymbol{\omega}$ Y-periodic and zero on Y_s. To this end, it is also useful to modify the corresponding pressure term in (2.5) into

$$\int_\Omega (\mathrm{grad}_x p^o + \mathrm{grad}_y p^1)\mathbf{w}^r dx.$$

Then (2.5) gives:

$$\int_{\Omega_{\varepsilon f}} \rho_f \frac{\partial^2 (u_i^o + u_i^r)}{\partial t^2} \omega_i \, \theta \, dx + \int_{\Omega_{\varepsilon r}} (\mathrm{grad}_x \, p^o + \mathrm{grad}_y \, p^1) \, \boldsymbol{\omega} \, \theta \, dx +$$

$$+ \int_{\Omega_{\varepsilon f}} \mu \frac{\partial}{\partial t} \frac{\partial u_i^r}{\partial y_j} \frac{\partial \omega_i}{\partial y_j} \theta \, dx = \int_{\Omega} f_i \, \omega_i \, \theta \, dx$$

and for $\varepsilon \to 0$ we have the local problem for the relative velocity:

$$\text{(2.6)} \quad \int_{Y_f} \rho_f \left(\frac{\partial v_i^o}{\partial t} + \frac{\partial v_i^r}{\partial t}\right) \omega_i dy + \int_{Y_f} \frac{\partial p^o}{\partial x_i} \omega_i \, dy + \mu \int_{Y_f} \frac{\partial v_i^r}{\partial y_j} \frac{\partial \omega_i}{\partial y_j} dy = \int_{Y_f} f_i \, \omega_i dy.$$

If we define the space $V_y = \{\mathbf{u};\ \mathbf{u} \in H^1(Y_f),\ \mathbf{u}|_\Gamma = 0,\ \mathrm{div}_y \, \mathbf{u} = 0,$ Y-periodic $\}$ and H_y, the completion of V_y for the norm associated with the scalar product

$$(\mathbf{u}, \mathbf{w})_{H_y} = \int_{Y_f} u_i \, w_i \, dy$$

we obtain the evolution problem: find $\mathbf{v}^r$, function of t, with values in V_y, such that:

$$\text{(2.7)} \quad \begin{aligned} &\rho \left(\frac{\partial \mathbf{v}^r}{\partial t}, \omega\right)_{H_y} + \mu (\mathbf{v}^r, \omega)_{V_y} = \left(f_i - \frac{\partial p^o}{\partial x_i} - \rho_f \frac{\partial v_i^o}{\partial t}\right) \int_{Y_f} \omega_i \, dy \qquad \forall \boldsymbol{\omega} \in V_y. \\ &\mathbf{v}^r(0) = 0 \end{aligned}$$

If we introduce the vectors $\boldsymbol{\Phi}^i$ (i = 1, 2, 3), elements of H_y defined by

$$\text{(2.8)} \quad \int_{Y_f} \omega_i dy = (\boldsymbol{\Phi}^i, \boldsymbol{\omega})_{H_y} \qquad \forall \boldsymbol{\omega} \in H_y$$

and the selfadjoint operator A_1 of H_y associated by the representation theorem with the form $(\mathbf{v}, \mathbf{w})_{V_y}$, (2.7) becomes:

$$\text{(2.9)} \quad \begin{aligned} &\rho \frac{\partial \mathbf{v}^r}{\partial t} + \mu \, A_1 \, \mathbf{v}^r = \left(f_i - \frac{\partial p^o}{\partial x_i} - \rho \frac{\partial v_i^o}{\partial t}\right) \boldsymbol{\Phi}^i \\ &\mathbf{v}^r(0) = 0. \end{aligned}$$

The solution of (2.9), by standard semigroup theory, is:

$$\text{(2.10)} \quad \mathbf{v}^r(t) = \rho^{-1} \int_0^t e^{-\rho^{-1} \mu A_1 (t-s)} \boldsymbol{\Phi}^i \left(f_i - \frac{\partial p^o}{\partial x_i} - \rho \frac{\partial v_i^o}{\partial t}\right)(s) \, ds.$$

Taking the mean value of (2.10) we have the macroscopic relative velocity:

$$(2.11)\quad \tilde{v}_k^r(t) = \int_0^t g_{ki}(t-s)\left(f_i - \frac{\partial p^o}{\partial x_i} - \rho\frac{\partial v_i^o}{\partial t}\right)(s)\,ds$$

$$(2.12)\quad g_{ki}(\xi) = \rho^{-1}\left(e^{-\rho^{-1}\mu A_1 \xi}\,\boldsymbol{\Phi}^i, \boldsymbol{\Phi}^k\right)_{H_y}.$$

REMARK 2.1. (2.10) gives $\mathbf{v}^r(t)$ as a functional of exterior body forces, gradient presure and inertia term. The mean value (2.11) contains a well-defined function of ξ, $g_{ki}(\xi)$ which decreases exponentially as $\xi \to \infty$, and $g_{ki} = g_{ik}$. The proof is similar as in the case of acoustics in porous media [2].

2.3. Stress tensor

In order to study the local state in the solid, we take in (2.5) $\mathbf{w}^o = \mathbf{w}^r = 0$, $\mathbf{w}^1 = \theta(x)\boldsymbol{\omega}(x,y)$, $\theta \in D(\Omega)$, $\boldsymbol{\omega}$ Y-periodic. In the same way, the asymptotic process $\varepsilon \to 0$, gives us:

$$(2.13)\quad \begin{aligned} &\int_{Y_s}\left(a^s_{ijkh} + b^s_{ijkh}\frac{\partial}{\partial t}\right)\frac{\partial u^1_k}{\partial y_h}\frac{\partial \omega_i}{\partial y_j}\,dy + \int_{Y_s} a^s_{ijkh}\frac{\partial u^o_k}{\partial x_k}\frac{\partial \omega_i}{\partial y_j}\,dy + \\ &+ \int_{Y_s} b^s_{ijkh}\frac{\partial}{\partial t}\frac{\partial u^o_k}{\partial x_h}\frac{\partial \omega_i}{\partial y_j}\,dy + p^o\int_{Y_s}\delta_{ij}\frac{\partial \omega_i}{\partial y_j}\,dy = 0. \end{aligned}$$

Note that $p^o(x,t)$ is defined in Ω (does not depend on y). In fact we continue $p\varepsilon$ in the solid part, with the periodicity condition, and we use that $\int_Y \operatorname{div}_y \boldsymbol{\omega}\,dy = 0$. If we introduce the space $\tilde{V}_y$ of functions from $H^1(Y_s)$ with zero mean value and the scalar product

$$(2.14)\quad (\mathbf{u},\mathbf{v})_{\tilde{V}_y} = \int_{Y_s} b^s_{ijkh}\frac{\partial u_k}{\partial y_h}\frac{\partial v_i}{\partial y_j}\,dy$$

and $A_2 \in L(\tilde{V}_y,\tilde{V}_y)$, $\mathbf{m}^{kh} \in \tilde{V}_y$, $\mathbf{n}^{kh} \in \tilde{V}_y$, $\boldsymbol{\Psi} \in \tilde{V}_y$ by:

$$(2.15)\quad (A_2\mathbf{u}^1,\boldsymbol{\omega}) = \int_{Y_s} a^s_{ijkh}\frac{\partial u^1_k}{\partial y_h}\frac{\partial \omega_i}{\partial y_j}\,dy$$

$$(2.16)\quad (\mathbf{m}^{kh},\boldsymbol{\omega})_{\tilde{V}_y} = \int_{Y_s} a^s_{ijkh}\frac{\partial \omega_i}{\partial y_j}\,dy$$

$$(2.17)\quad (\mathbf{n}^{kh},\boldsymbol{\omega})_{\tilde{V}_y} = \int_{Y_s} b^s_{ijkh}\frac{\partial \omega_i}{\partial y_j}\,dy$$

$$(2.18)\quad (\boldsymbol{\Psi},\boldsymbol{\omega})_{\tilde{V}_y} = \int_{Y_s}\delta_{ij}\frac{\partial \omega_i}{\partial y_j}\,dy$$

the relation (2.13) is equivalent to:

$$(2.19)\quad \left(\frac{\partial \mathbf{u}^1}{\partial t} + A_2\mathbf{u}^1 + \mathbf{m}^{kh}\frac{\partial u_k^o}{\partial x_h} + \mathbf{n}^{kh}\frac{\partial}{\partial t}\frac{\partial u_k^o}{\partial x_h} + p^o\mathbf{\Psi},\ \omega\right)_{V_y} = 0 \qquad \forall\,\omega \in \tilde{V}_y.$$

Thus the first factor in (2.19) must be zero.

$$(2.20)\quad \frac{\partial \mathbf{u}^1}{\partial t} + A_2\mathbf{u}^1 = -\mathbf{m}^{kh}\frac{\partial u_k^o}{\partial x_h} - \mathbf{n}^{kh}\frac{\partial}{\partial t}\frac{\partial u_k^o}{\partial x_h} - p^o\mathbf{\Psi}$$

$$\mathbf{u}^1(0) = 0.$$

The solution of (2.20) is:

$$(2.21)\quad \mathbf{u}^1 = -\mathbf{n}^{kh}\frac{\partial u_k^o}{\partial x_h} + \int_o^t e^{-A_2(t-s)}\left(\mathbf{r}^{kh}\frac{\partial u_k^o}{\partial x_h} - p^o\mathbf{\Psi}\right)(s)\,ds$$

$$(2.22)\quad \mathbf{r}^{kh} = A_2\mathbf{n}^{kh} - \mathbf{m}^{kh}$$

$$(2.23)\quad \frac{\partial \mathbf{u}^1}{\partial t} = -\mathbf{n}^{kh}\frac{\partial}{\partial t}\frac{\partial u_k^o}{\partial x_h} + \mathbf{r}^{kh}\frac{\partial u_k^o}{\partial x_h} - p^o\,\mathbf{\Psi} -$$

$$- \int_o^t A_2 e^{-A_2(t-s)}\left(\mathbf{r}^{kh}\frac{\partial u_k^o}{\partial x_h} - p^o\mathbf{\Psi}\right)(s)\,ds.$$

REMARK 2.2. The right hand side of the equation (2.23) is well defined as function of $\mathbf{u}^o$ and p^o.

The macroscopic stress tensor is defined as the mean value of

$$(2.24)\quad \sigma_{ij}^o = \left(a_{ijkh}^s + b_{bijkh}^s\frac{\partial}{\partial t}\right)\left(\frac{\partial u_k^o}{\partial x_h} + \frac{\partial u_k^1}{\partial y_h}\right).$$

If we introduce the coefficients and the functions:

$$(2.25)\quad \alpha_{ijkh}^o = \left[a_{ijkh} - a_{ijep}\frac{\partial}{\partial y_p}(\mathbf{n}^{kh})_e + b_{ijep}\frac{\partial}{\partial y_p}(\mathbf{r}^{kh})_e\right]^\sim$$

$$(2.26)\quad \alpha_{ijkh}^1 = \left[b_{ijkh} - b_{ijep}\frac{\partial}{\partial y_p}(\mathbf{n}^{kh})_e\right]^\sim$$

$$(2.27)\quad \alpha_{ij}^2 = \left[b_{ijkh}\frac{\partial \Psi_k}{\partial y_h}\right]^\sim$$

$$(2.28)\quad g_{ijkh}(\xi) = \left[a_{ijep}\frac{\partial}{\partial y_p}\left(e^{-A_2\xi}\,\mathbf{r}^{kh}\right)_e - b_{ijep}\frac{\partial}{\partial y_p}\left(A_2\,e^{-A_2\xi}\,\mathbf{r}^{kh}\right)_e\right]^\sim$$

$$(2.29)\quad g_{ij}^*(\xi) = \left[b_{ijep}\frac{\partial}{\partial y_p}\left(A_2\,e^{-A_2\xi}\,\Psi\right)_e - a_{ikep}\frac{\partial}{\partial y_p}\left(e^{-A_2\xi}\,\Psi\right)_e\right]^\sim$$

we have

$$(2.30)\quad \tilde{\sigma}_{ij}^o = \alpha_{ijkh}^o\,e_{kh}(\mathbf{u}^o) + \alpha_{ijkh}^1\,e_{kh}\left(\frac{\partial \mathbf{u}^o}{\partial t}\right) - \alpha_{ij}^2\,p^o +$$

$$+ \int_o^t g_{ijkh}(t - s)\, e_{kh}(\mathbf{u}^o)(s)\, ds + \int_o^t g^*_{ij}(t - s)\, p^o(s)\, ds.$$

REMARK 2.3. The constitutive equation (2.30) contains an elastic term α^o, a viscoelastic term with instantaneous memory α^1, a pressure term α^2, and two terms with long memory g, g^*, functions of strain and pressure. Because A_2 is a positive defined operator, $g(\xi)$ decays exponentially for $\xi \to \infty$ (also for $g^*(\xi)$). The strain stress law (2.30) is very different than (1.2).

2.4. Balance of momentum

Now it is easy to obtain the balance of momentum. For this we take in (2.5) $\mathbf{w} = \mathbf{w}^1 = 0$. Then, for $\varepsilon \to 0$ we have :

$$(2.31)\quad \int_\Omega \tilde{\rho}\, \frac{\partial^2(u_i^o + u_i^r)}{\partial t^2} w_i^o\, dx + \int_\Omega \tilde{\sigma}_{ij}^o \frac{\partial w_i^o}{\partial x_j} dx - n\int_\Omega p^o \operatorname{div} \mathbf{w}^o\, dx =$$

$$= \int_\Omega f_i\, w_i^o\, dx \qquad \forall\, \mathbf{w}^o \in H_o^1(\Omega)$$

$$\int_\Omega \tilde{\sigma}_{ij}^o \frac{\partial w_i^o}{\partial x_j} dx - \int_\Omega n\, p^o \operatorname{div} \mathbf{w}^o\, dx =$$

$$= -\int_\Omega \frac{\partial}{\partial x_j}(\tilde{\sigma}_{ij}^o - n\, p^o\, \delta_{ij}) w_i^o\, dx = -\int_\Omega \frac{\partial \sigma_{ij}^T}{\partial x_j} w_i^o\, dx$$

$$(2.32)\quad \sigma_{ij}^T = \tilde{\sigma}_{ij}^o - n\, p^0\, \delta_{ij}$$

$$(2.33)\quad \tilde{\rho}\frac{\partial^2 u_i^o}{\partial t^2} + \rho_f \frac{\partial v_i^r}{\partial t} - \frac{\partial \sigma_{ij}^T}{\partial x_j} = f_i .$$

REMARK 2.4. σ_{ij}^T is the effective (or total) stress tensor [5,6]. In the same time (2.32) proves that in the effective stress tensor appears the pore pressure multiplied by the porosity.

3. CONCLUSION

The macroscopic (or homogenized) motion of the mixture may be described by the displacement vector in the solid $\mathbf{u}^o(x,t)$, the pore pressure in the fluid $p^o(x,t)$ and the mean value of the relative velocity $\tilde{\mathbf{v}}^r$. These quantites satisfy the equations (2.3) conservation of mass, (2.11), Darcy's law, and (2.33), conservation of momentum, the effective stress tensor being defined by (2.32) and (2.30).

The coefficients α_{ijkh}^o and a_{ijkh}^1 and the function g_{ijkh} are the same as in the case of homogenization in viscoelasticity [2]. In the particular case of an elastic skeleton our results reduce to those obtained in [2], but the conversation of mass is different. In fact it was proved that a Darcy's law of the from (2.11) is not only a consequence of the compressibility of the fluid, but appears as valid for incompressible fluids.

REFERENCES

1. **A. Bensoussan, J. L. Lions** and **G. Papanicolaou,** *Asymptotic Analysis for Periodic Structure,* North-Holland, Amsterdam (1978).
2. **E. Sanchez - Palencia,** *Topics in Non-Homogeneous Media and Vibration Theory,* Lecture Notes in Physics **127,** Springer-Verlag Berlin (1980).
3. **Th. Levy,** *Intern. J. Enging. Sci.* **17,** p.1005-1014 (1979).
4. **H. I. Ene et E. Sanchez - Palancia,** *Jour. Mecan.* **14,** p.73-108 (1975).
5. **M. A. Biot,** *Indiana Univ. Math. J.* **21**(7), p.597 (1972).
6. **M. A. Biot,** *J.Geoph. Res.* **78**(23), p.4924 (1973).

A STATICAL MICROMECHANICAL DESCRIPTION OF YIELDING IN COHESIONLESS SOIL

B. CAMBOU *

1. Introduction

At present, the description of the behavior of cohesionless soil is not sufficiently realistic and accurate in complex loadings, especially with reorientation of principal stress axes. One reason of this limited success seems to be the lack of micromechanical analyses.

The purpose of this work is to propose a micro statical analysis of a simple granular material to provide a better understanding of the microphenomena and of the fundamental variables which lead to irrecoverable strains. Particular interest is taken in loadings with reorientation of principal axes.

This study is limited to a bidimensional granular material made of cylindrical particles in an irregular array.

2. Microstructural analysis

In a granular medium, the fundamental mechanism of irrecoverable strains is the sliding between particles. The following condition can be written for each contact point (Fig.1):

(1) $$|T^k| \leq \mu N^k$$

μ is the friction coefficient of the material of particles.

Relation (1) between the components of contact forces leads us to the conclusion that a statical analysis seems to be the more appropriate to describe irrecoverable strains.

The macro statical variable is the stress tensor ($\sigma_{\alpha\beta}$) and the microstatical variables are the components of contact forces (F_α^k).

These variables are described by the distributions defined for each possible orientation ($\vec{n}$) (Fig.2).

$P(\vec{n})$ is the probability of occurence of a contact orientation in the range $\beta \pm d\beta$ (Fig.1).

$F_\alpha^k(\vec{n})$ are the components of forces applied on contact points defined by their orientations $\vec{n}$.

In a volume V the number of these contacts is : $\mathcal{N} V P(\vec{n}) \, d\beta$
it is assumed to be sufficiently large.

$\mathcal{N}$ is the number of contact points per unit area.

* Professor Ecole Centrale de Lyon F-69410 Ecully FRANCE

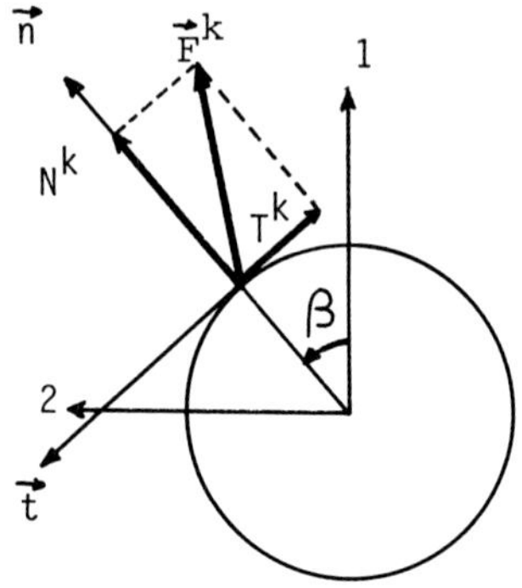

Fig. 1 : Definition of forces $\vec{F}^k$ at contact point k

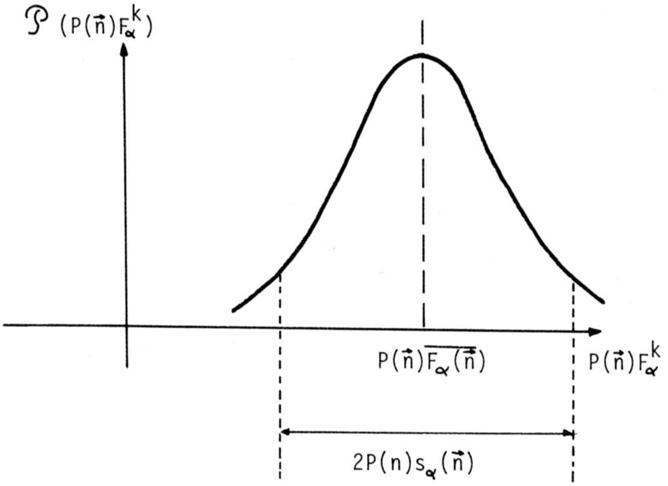

Fig. 2 : Definition of the distribution of $P(\vec{n})F_\alpha^k$ for a given orientation $\vec{n}$

These distributions are defined by their statistic moments.
The accuracy of the description increases with the number of moments taken into account.

A simplified description based on the first moment (mean values) is first proposed.

3. Statistical first-order description.

(Yielding mechanism n°1)

The microstatical variables are described by the mean values of the distributions (Fig.2) $P(\vec{n})$ $\overline{F_\alpha(n)}$.

3.1. Relations between micro and macro statical variables.

Weber {16} , Cristoffersen and al {7} have demonstrated that :

(2) $$\sigma_{\alpha\beta} = \frac{1}{V} F_\alpha^k \ell_\beta^k$$

Summation or subscript k is extended to all contacts in V volume (Fig.1).

For cylindrical particles with an average diameter $\bar{D}$, assuming that β and D are independent variables, relation (2) can be written as :

(3) $$\sigma_{\alpha\beta} = \mathcal{N}\bar{D} \int_0^\pi P(\vec{n}) \overline{F_\alpha(\vec{n})}\, n_\beta \, d\beta$$

This relation points out the product $P(\vec{n})$ $\overline{F_\alpha(\vec{n})}$ which has been chosen as the microstatical variable.

In previous works {1} {2} {3} {4} {5}, it has been shown that this variable can be expressed in principal axes by :

(4) $$P(\vec{n}) F_\Delta(\vec{n}) = \mathcal{N}\bar{D}\, \sigma_\Delta\, f_\Delta(\vec{n})$$

Capital greek subscript Δ means that this relation is only valid in the principal axes.(and without summation on subscript Δ)
In expression (4) σ_Δ defines the actual loading and $\mathcal{N}\bar{D} f_\Delta$ the internal state of the material.

From relation (3) and (4) it is easy to show that f_Δ must satisfy the following conditions :

(5) $$\int_0^\pi f_\Delta\, n_\Gamma \, d\beta = \delta_{\Delta\Gamma}$$

It can be noted that $f_\alpha(\vec{n})$ expressed in any axis system can be defined easily from $f_\Delta(\vec{n})$.

The value of $f_\Delta(\vec{n})$ for an isotropic material subjected to an isotropic loading is noted $f_\Delta^{(0)}(\vec{n})$.

It can be shown {1} that :

$$f_\Delta^{(0)}(\vec{n}) = (2/\pi)\, n_\Delta$$

The internal state defined by $f_\Delta^{(0)}$ is named "*the virgin state*".

3.2. Discretization of the internal state.

$f_\alpha(\vec{n})$ defined by (4) characterizes the internal state of the granular material.

In agreement with Leckie and Onat {8} , we have proposed {3} to express this function by the following expansion :

$$f_\alpha(\vec{n}) = \Psi_{\alpha\beta}\, n_\beta + \Psi_{\alpha\beta\gamma\delta}\, n_\beta n_\gamma n_\delta \qquad (6)$$

This expansion must satisfy relations (5).

For a bidimensional material it has been shown {4} {5} {6} that f_α can be expressed from only four independent variables, noted $\lambda_{\alpha\beta}$:

$$\begin{aligned} \frac{\pi}{2} f_1(\vec{n}) &= n_1 + \lambda_{11}(n_1^2 - 3n_2^2)\, n_1 + \lambda_{12}(n_2^2 - 3n_1^2)\, n_2 \\ \frac{\pi}{2} f_2(\vec{n}) &= n_2 + \lambda_{21}(n_1^2 - 3n_2^2)\, n_1 + \lambda_{22}(n_2^2 - 3n_1^2)\, n_2 \end{aligned} \qquad (7)$$

Variables $\lambda_{\alpha\beta}$ characterize the internal state of the material. It can be easily shown that for a loading without reorientation of principal axes : $\lambda_{12} = \lambda_{21} = 0$, the internal state is then only defined by : λ_{11} and λ_{22}.

3.3. Physical meanings of the state variables

The global stress tensor applied is decomposed as follows :

$$\sigma = \sigma^{(n)} + \sigma^{(t)}$$

$\sigma^{(n)}$: part of σ due to normal components of the average contact forces.

$\sigma^{(t)}$: part of σ due to tangential components of the average contact forces.

To define $\sigma_{\alpha\beta}^{(n)}$ the average contact forces are expressed from relations (4) and (7).

The normal component $\overline{N}(\vec{n})$ is then calculated. Finally relation (3) allows to define $\sigma_{\alpha\beta}^{(n)}$ from $\overline{N}_\alpha(\vec{n})$, which can be written in the principal axes system of $\sigma_{\alpha\beta}$ as :

$$\begin{aligned} 4\,\sigma_{11}^{(n)} &= \sigma_I\,(3 + \lambda_{11}) + \sigma_{II}\,(1 - \lambda_{22}) \\ 4\,\sigma_{22}^{(n)} &= \sigma_{II}\,(3 + \lambda_{22}) + \sigma_I\,(1 - \lambda_{11}) \\ 4\,\sigma_{12}^{(n)} &= -\,(\sigma_I\,\lambda_{12} + \sigma_{II}\,\lambda_{21}) \end{aligned} \qquad (8)$$

For a loading without reorientation of principal axes $\lambda_{12} = \lambda_{21} = 0$, so the principal axes of $\sigma_{\alpha\beta}^{(n)}$ and of $\sigma_{\alpha\beta}$ are identical.

In this case :

$$\sigma_{11}^{(n)} + \sigma_{22}^{(n)} = \sigma_I + \sigma_{II}$$

$$\sigma_{11}^{(n)} - \sigma_{22}^{(n)} = \frac{(1+\lambda_m)}{2}(\sigma_I - \sigma_{II}) + \frac{1}{2}\lambda_d(\sigma_I + \sigma_{II})$$

with

$$\lambda_m = (1/2)(\lambda_{11} + \lambda_{22})$$

$$\lambda_d = (1/2)(\lambda_{11} - \lambda_{22})$$

For a bidimensional material the mobilized angle of internal friction can be measured by s :

$$s = \frac{\sigma_I - \sigma_{II}}{\sigma_I + \sigma_{II}} = s^{(n)} + s^{(t)} = \frac{\sigma_{11}^{(n)} - \sigma_{22}^{(n)}}{\sigma_I + \sigma_{II}} + \frac{\sigma_{11}^{(t)} - \sigma_{22}^{(t)}}{\sigma_I + \sigma_{II}}$$

Then $s^{(n)}$ can be written as :

(9) $$s^{(n)} = \frac{1}{2} s + \frac{1}{2}\lambda_m \, s + \frac{1}{2}\lambda_d$$

From (9) it can be noted that

$$\frac{\partial s^{(n)}}{\partial \lambda_m} < \frac{\partial s^{(n)}}{\partial \lambda_d} \qquad \forall s$$

Thus λ_d characterizes the better yielding mechanism for any value of s . So we assume that λ_m does not change from its initial value ($\lambda_m = 0$) and thus λ_d remains the only internal state variable (for irrotational loadings).

Then relation (9) becomes :

(10) $$s^{(n)} = \frac{1}{2} s + \frac{1}{2}\lambda_d$$

This relation shows that the evolution of λ_d is related to changes in the structure in such a way that deviatoric stresses would be borne in a greater part by the normal components of contact forces which are stabilizing elements on every contact point.

This conclusion is in agreement with the results proposed by Thornton and Barnes {15} (Fig.3).

For loadings with reorientations of principal axes, parameters λ_{12} and λ_{21} are not equal to zero.

Relation (8) shows that $\sigma_{12}^{(n)} \neq 0$ so axes 1 and 2 are not principal for tensor $\sigma^{(n)}$.

The evolution of these parameters can be considered in two different cases :

a) When a reorientation of principal axes occurs.

b) When λ_{12} and λ_{21} are different from zero for a loading increment without reorientation of principal axes.

Case a) can be analyzed as a change of axis system for parameters λ_{ij} .

In a previous work {6}, it has been shown that the rule of axis change can be expressed by :

(11) $$\begin{vmatrix} \lambda'_{11} \\ \lambda'_{12} \\ \lambda'_{21} \\ \lambda'_{22} \end{vmatrix} = \begin{vmatrix} A & B & C & D \\ -B & A & -D & C \\ -C & -D & A & B \\ D & -C & -B & A \end{vmatrix} \begin{vmatrix} \lambda_{11} \\ \lambda_{12} \\ \lambda_{21} \\ \lambda_{22} \end{vmatrix}$$

It is easy to demonstrate that $\lambda_{12} = \lambda_{21}$ implies that $\lambda'_{12} = \lambda'_{21}$

This relation suitable for the "*virgin state*" , will remain true for any change of axis system related to reorientation of principal axes.

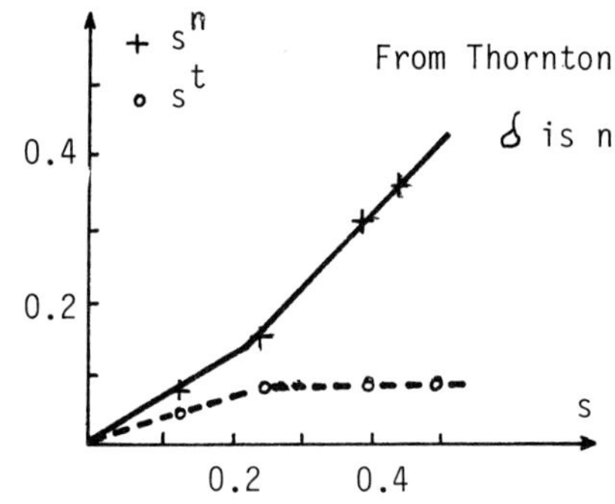

Fig. 3 : Contribution of normal and tangential components of contact forces to the mobilised internal angle of friction .

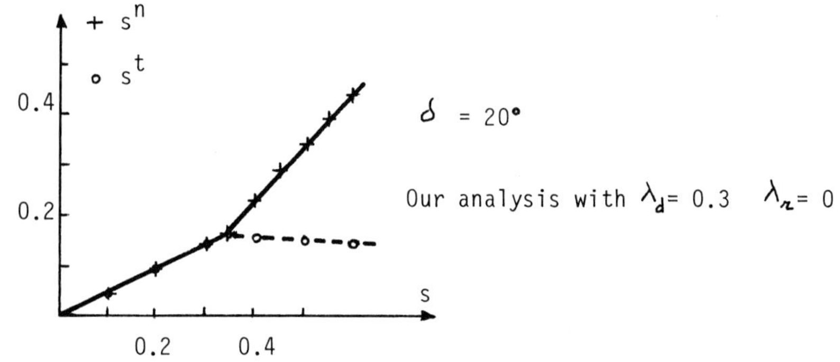

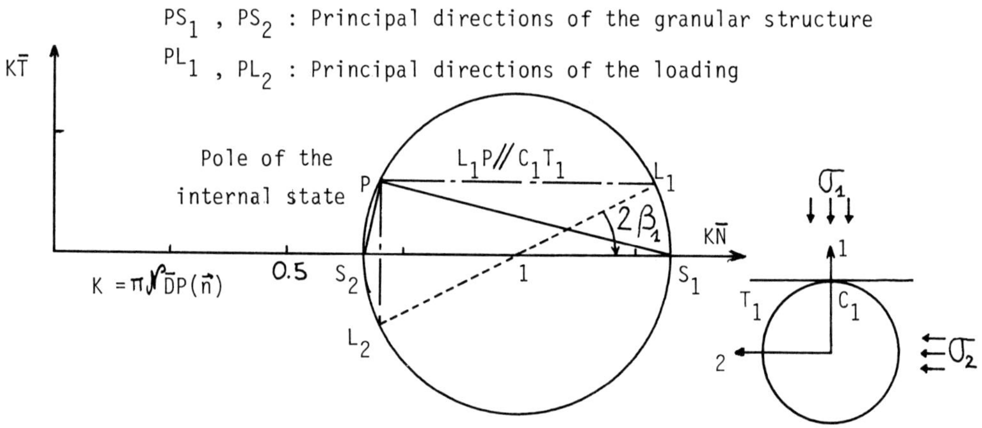

Fig. 4 : Representation of the internal state (λ_r=0.15 ,λ_d=0.30) by a circle (isotropic loading σ = 1)

In case b) $\sigma_{12}^{(n)}$ can be expressed by :

(12) $$\sigma_{12}^{(n)} = -\frac{1}{4}\left[\frac{\sigma_I + \sigma_{II}}{2}(\lambda_{12}+\lambda_{21}) + \frac{\sigma_I - \sigma_{II}}{2}(\lambda_{12}-\lambda_{21})\right]$$

$\sigma_{12}^{(n)}$ is a measure of the difference of the principal orientations of $\sigma_{\alpha\beta}^{(n)}$ and $\sigma_{\alpha\beta}$. It is assumed that the evolution of the internal state occurs in order to minimise this difference.

Relation (12) shows that (λ_{12} + λ_{21}) is the most representative parameter of the evolution of $\sigma_{12}^{(n)}$. Thus it is assumed that (λ_{12} - λ_{21}) = 0. So it can be written :

$$\lambda_{12} = \lambda_{21} = \lambda_r$$

So It can be considered that for every case, parameter λ_r characterizes the evolution of orientation of principal axes.
Thus expansion (7) can be written as :

(13)
$$(\pi/2)\, f_1(\vec{n}) = n_1 + \lambda_d (n_1^2 - 3n_2^2)n_1 + \lambda_r (n_2^2 - 3n_1^2)n_2$$
$$(\pi/2)\, f_2(\vec{n}) = n_2 - \lambda_d (n_2^2 - 3n_1^2)n_2 + \lambda_r (n_1^2 - 3n_2^2)n_1$$

The angle α between the two principal axis systems of σ and $\sigma^{(n)}$ can be calculated :

$$tg\, 2\alpha = \frac{\lambda_r}{\Delta + \lambda_d}$$

Taking into account the only two parameters λ_d and λ_r , the rule of changing axis system (11) becomes simpler and can be written :

(14) $$\begin{vmatrix} \lambda'_d \\ \lambda'_r \end{vmatrix} = \begin{vmatrix} \cos 2\theta & -\sin 2\theta \\ \sin 2\theta & \cos 2\theta \end{vmatrix} \begin{vmatrix} \lambda_d \\ \lambda_r \end{vmatrix}$$

θ : angle between the two axis systems.

3.4. Representation of the material internal state.

For a given loading, internal state variables λ_d and λ_r define the distribution of vectors $P(\vec{n})$ $\overline{F_\alpha(\vec{n})}$.

A convenient representation of this distribution can be provided in axes $\vec{n}$, $\vec{t}$ (Fig.1).

Normal and tangential components can be defined by :

$$P(\vec{n})\overline{N} = n_1 P(\vec{n})\overline{F_1(\vec{n})} + n_2 P(\vec{n})\overline{F_2(\vec{n})}$$
$$P(\vec{n})\overline{T} = -n_2 P(\vec{n})\overline{F_1(\vec{n})} + n_1 P(\vec{n})\overline{F_2(\vec{n})}$$

In these axes the evolution of the internal state can be defined by the curve described by the end of vectors $P(\vec{n})$ $\overline{F_\alpha(n)}$.

For a virgin state, it is easy to demonstrate that this curve is a circle (Fig.5a).

After yielding without reorientation of principal axes, this curve is no longer a circle but remains symmetrical about axis $\vec{n}$ (Fig.5a).

a)

— — — — — $\lambda_d = 0$ $\lambda_\alpha = 0$ (circle)

- - - - - - $\lambda_d = 0.3$ $\lambda_\alpha = 0$ (curve symmetrical about $\vec{KN}$)

· · · · · · · · $\lambda_d = 0$ $\lambda_\alpha = 0.15$

— · — · — $\lambda_d = 0.3$ $\lambda_\alpha = 0.15$

L_1 is the representative point of the mean contact force in the major principal direction of the loading

b)

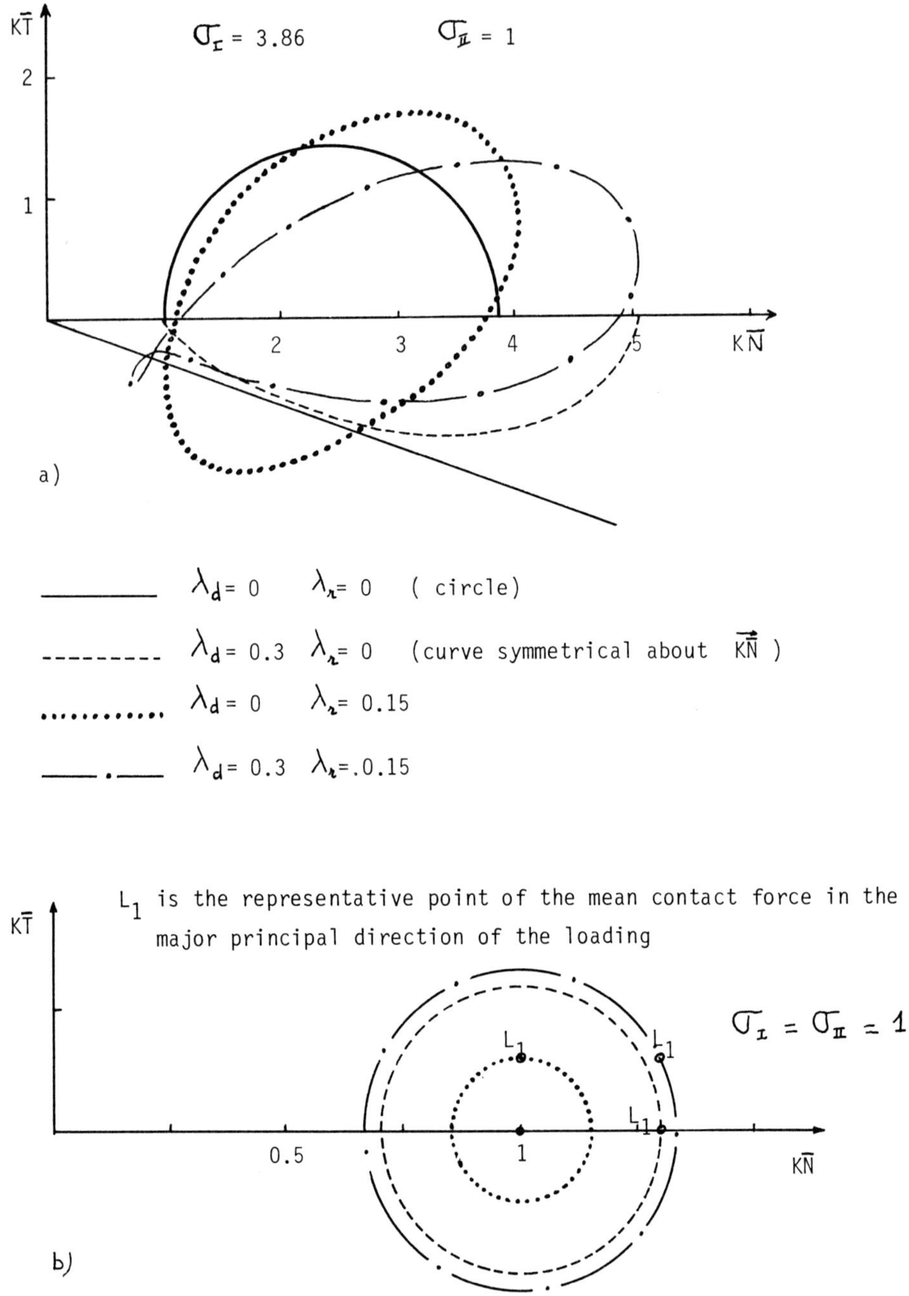

Fig. 5 : Representation of the state of the material by the curve described by the end of vectors $\overline{P(\vec{n})F}(\vec{n})$
  a) For the actual loading
  b) for an isotropic loading ($\sigma = 1$)

After yielding with reorientation of principal axes, this curve is no longer symmetrical about axis $\vec{n}$ (Fig.5a). In fact, this curve gives two informations one about internal state, the other about applied loading.

To define a convenient representation of the only internal state, an isotropic loading ($\sigma = 1$) is considered.

In this case :

$$(15)\qquad \begin{aligned} P(\vec{n})\,\overline{N} &= (1/\pi \mathcal{N} \overline{D})\,(1 + \lambda_d \cos 2\beta - \lambda_r \sin 2\beta) \\ P(\vec{n})\,\overline{T} &= (1/\pi \mathcal{N} \overline{D})\,(\lambda_d \sin 2\beta + \lambda_r \cos 2\beta) \end{aligned}$$

It can be easily demonstrated that the end of vectors $\overrightarrow{P(\vec{n})\, F(\vec{n})}$ describes a circle with a radius equal to $\sqrt{\lambda_d^2 + \lambda_r^2}$ (Fig.4).

The rule of changing axes (14) allows us to make sure that this quantity is independent of the axis system that is used.

To define the internal state completely, it is necessary to know one point of the circle corresponding to one contact orientation. For this purpose, it seems convenient to define pole P of this representation (Fig.4).

The distributions of mean values of contact forces under isotropic loading is then perfectly defined by this representation (circle and pole P).

The principal directions of the structure can be defined :

Contact directions for which mean values of contact forces are normal to the contact plane, for an isotropic loading.

Two principal orthogonal directions can be defined. It is easy to drawn them on the geometrical representation (Fig.4).

In this representation, it is possible to define the major principal direction from equation (15) by angle β_1 (point L_1 on Fig.4) :

$$\operatorname{tg} 2\beta_1 = -(\lambda_r / \lambda_d)$$

It can be noted that if L_1 does not coincide with S_1 (Fig.4) the principal directions of the structure differ from the principal directions of the loading. We can see the representations of various internal states on Fig.5b.

At first glance, this representation seems very similar to the Mohr representation of a stress tensor. In fact, it is quite different, in particular the direction of evolution of the representative point on the circle is opposite. This representation cannot be expressed from a second rank tensor.

Then the internal state of a granular material can be defined by the two parameters λ_d and λ_r with the rule of changing axes (14), the geometrical representation of this state is a circle in axes $\vec{n}$, $\vec{t}$.
The principal directions of the structure can be defined.

It can be noted that for a loading without reorientations of principal axes, the principal axes of tensor σ and of the structure are identical, for a loading with reorientation of principal axes, it is not the case.

3.5. Sliding criterion.

Relation (1) gives the sliding condition at contact point k.

By summation on all contacts defined for a given orientation β, mean values of contact forces must satisfy a similar condition :

$$(16)\qquad |\overline{T(\vec{n})}| \leqslant \overline{N(\vec{n})}\,\mu$$

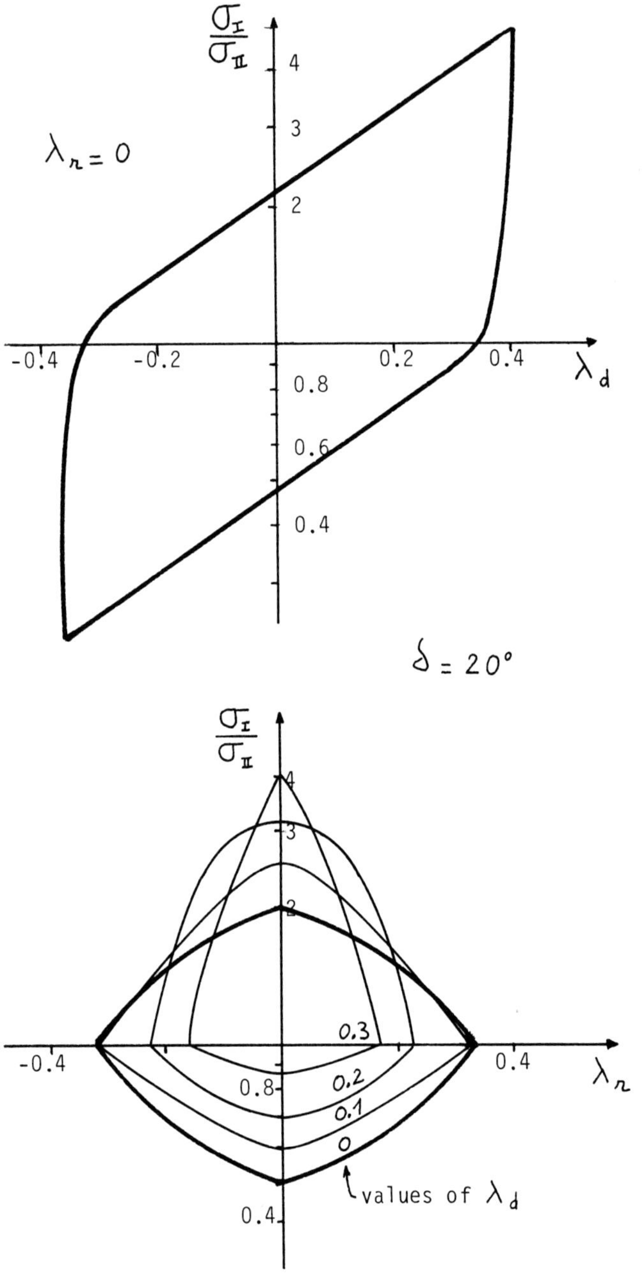

Fig.6 : Representation of the sliding criterion for different values of λ_d and λ_r (yielding mechanism n° 1)

It will be noticed that condition (16) is necessary to have an admissible state, but it is not sufficient. In fact, it is the only one which can be written because only mean values of contact forces are defined.
To go further, it would be necessary to consider a more accurate statistical description of the distributions of contact forces.

In the principal stress space, the sliding criterion (16) can be written :

$$(17) \qquad \mathrm{Sup}_\beta \left\{ \frac{(n_1 - n_2\mu)\, f_2}{(n_2 + n_1\mu)\, f_1} \right\} \leqslant \frac{\sigma_I}{\sigma_{II}} \leqslant \mathrm{Inf}_\beta \left\{ \frac{(n_1 + n_2\mu)\, f_2}{(n_2 - n_1\mu)\, f_1} \right\}$$

It will be noticed that this criterion depends on the friction at contact point (μ) but also on the internal state of the material ($\frac{f_1}{f_2}$ or λ_d and λ_r).

For $\lambda_d = \lambda_r = 0$ criterion (17) is a Mohr-Coulomb criterion with an internal friction angle equal to δ ($\mu = \tan\delta$).

Fig.6 shows criterion (17) depending on parameters λ_d and λ_r .

It can be noticed that for $\lambda_r = 0$ (loading without reorientation of principal axes)the yielding mechanism is of a kinematic type. For $\lambda_d = 0$ the yielding mechanism is of a softening isotropic type. In the general case, the yielding mechanism is quite complicated (kinematic + isotropic).

In the principal stress space, criterion (17) is represented by straight lines (Fig.7) the orientations of which depend on parameters λ .
Within the framework of an elastoplastic analysis, these lines represent the plastic-potential functions depending on parameters λ_d and λ_r .

3.6. Flow rule .

To define the flow rule of this material, it is expressed that the dissipation due to plastic strains is equal to the work of forces at sliding contact points.

$$(15) \qquad \sigma_{\alpha\beta}\, d\varepsilon^{p}_{\alpha\beta} = \Sigma\, \vec{T}^{k}\, \vec{dg}^{k}$$

$\vec{dg}^k$: sliding at contact k ($\vec{T}^k\, \vec{dg}^k > 0$)

quantity $d\tilde{g}(\vec{n})$ is defined by :

$$d\tilde{g}(\vec{n}) = \frac{\overline{T(\vec{n})\, dg(\vec{n})}}{\overline{T(\vec{n})}}$$

From relation (15) expressed in the principal axis system, it is easy to define $d\varepsilon_\Delta$:

$$(18) \qquad \begin{aligned} d\varepsilon_I &= (-1/\bar{D}) \int_0^\pi f_I(\vec{n})\, n_2\, d\tilde{g}(\vec{n})\, d\beta \\ d\varepsilon_{II} &= (1/\bar{D}) \int_0^\pi f_{II}(\vec{n})\, n_1\, d\tilde{g}(\vec{n})\, d\beta \end{aligned}$$

From (16) it has been shown in {5} that volumetric strain can be expressed by :

$$(17) \qquad -(d\varepsilon_I + d\varepsilon_{II}) = -2\, G\, \lambda_d$$

with $$G = \frac{2}{\pi \bar{D}} \int_0^\pi n_1\, n_2\, d\tilde{g}(\vec{n})\, d\beta$$

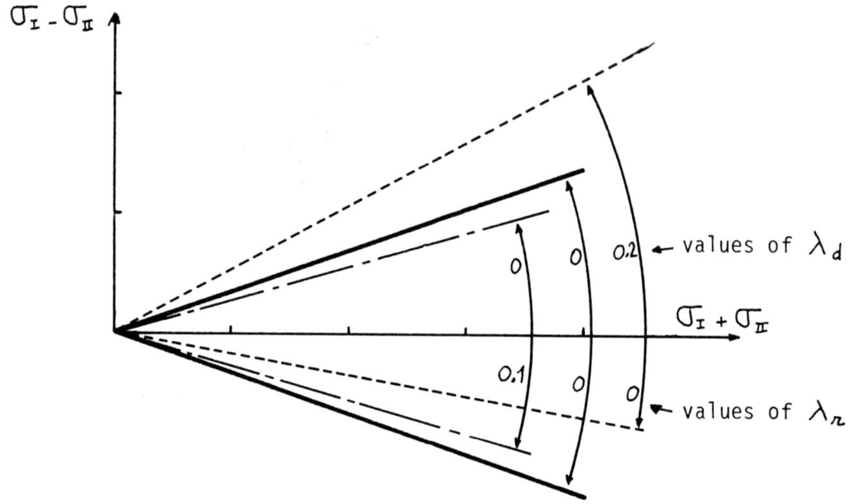

Fig.7 : Representation of the sliding criterion (yielding mechanism n° 1)

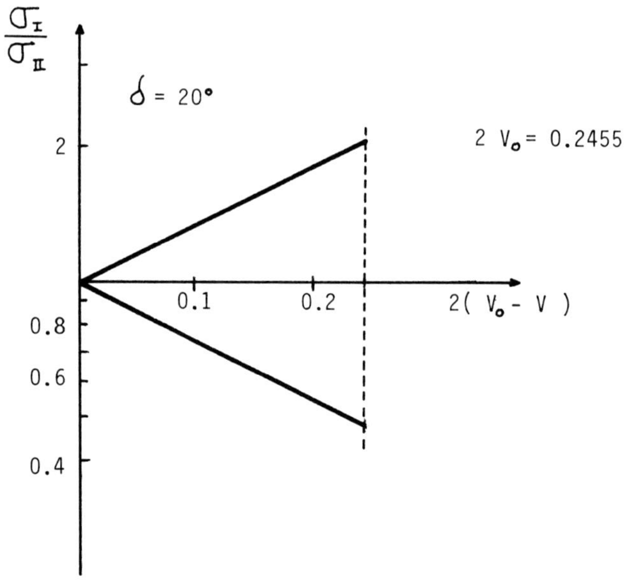

Fig.8 : Representation of the sliding criterion (yielding mechanism n° 2)

G is a quantity with a sign identical to that of $(\sigma_{II} - \sigma_I)$
For a monotically increasing loading path, λ_d has the same sign as $(\sigma_I - \sigma_{II})$, then the volumetric strains are necessarily extensive.

With an additional simplifying assuption it can be demonstrated that fulfilling equations (17) and (18) leads to the stress-dilatancy relation defined by Rowe {11} .

For loading with reorientation of principal axes the volumetric strains depend on another term including λ_r:

$$-(d\varepsilon_I + d\varepsilon_{II}) = -2G\lambda_d - (2/\pi\bar{D})\int_0^\pi \lambda_r \cos 2\beta \, d\tilde{g}(\vec{n})\, d\beta$$

In this case, the stress-dilatancy relation is no longer accurate.

4. Second order statistical description (Yielding mechanism n° 2)

The distributions of random variables $(P(\vec{n})F_\alpha^k)$ are described more accurately in this paragraph by the mean values $P(\vec{n})\ \overline{F_\alpha}(\vec{n})$ and the standard deviations $P(\vec{n})\ \Delta_\alpha(\vec{n})$.

4.1. Discretization of the second order moments $\Delta_\alpha(\vec{n})$. :

The following relation, similar to (4) is proposed :

$$P(\vec{n})\ \Delta_\alpha(\vec{n}) = (1/\rho\bar{D})\ \sigma_\alpha\ v_\alpha(\vec{n}) \tag{19}$$

without summation on subscript α.

Function $v_\alpha(\vec{n})$ characterizes the second order material state.

It would be possible to define an expansion of v_α similar to the one proposed for f_α.

To simplify the analysis we only consider the first order expansion :

$$v_\alpha(\vec{n}) = V_\alpha \qquad (V_\alpha > 0)$$

At virgin state, the material is isotropic so :

$$v_\alpha(\vec{n}) = V_0$$

4.2. Deviatoric loadings.

Distributions of contact forces are assumed to be symmetrical and bounded by $P(\vec{n})[\overline{F_\alpha} \pm 2\Delta_\alpha]$

4.2.1. Sliding criterion :

The sliding criterion can be written

$$\mathrm{Sup}_\beta\left\{\frac{(n_1 - n_2\mu)(f_2 + 2V_2)}{(n_2 + n_1\mu)(f_1 - 2V_1)}\right\} \leq \frac{\sigma_I}{\sigma_{II}} \leq \mathrm{Inf}_\beta\left\{\frac{(n_1 + n_2\mu)(f_2 - 2V_2)}{(n_2 - n_1\mu)(f_1 + 2V_1)}\right\} \tag{20}$$

When a virgin granular material is subjected to an isotropic loading, irrecoverable strains occur, so criterion (20) is satisfied with $\sigma_I = \sigma_{II}$ $f_\alpha = n_\alpha$, $V_\alpha = V_0$.

Then V_0 can be defined :

$$2V_0 = \sqrt{2}\sin(\delta/2).$$

After a numerical analysis of relation (20) it has been concluded that the two parameters V_α have a very similar effect. So it is assumed that :

$$V_\alpha = V$$

Then V is the only parameter of an isotropic yielding mechanism. Criterion (20) can be written with $V_\alpha = V_o$ (Fig.8).

4.2.2. Flow rule :

It is assumed that only the second yielding mechanism is actived.

$$d\tilde{g}(\vec{n}) = \frac{T(\vec{n})\ dg(\vec{n})}{T(\vec{n}) \mp 2\ \Delta_T(\vec{n})}$$

(with sign + if $T(\vec{n}) > 0$ and sign - is $T(\vec{n}) < 0$).

Similarly to 3.5. plastic strains can be defined, and particularly volumetric plastic strains :

$$-(d\varepsilon_I + d\varepsilon_{II}) = \pm \frac{2V}{\bar{D}} \left\{ \int_0^\pi -n_1\, dg(\vec{n})\, d\beta + \int_0^\pi n_2\, dg(\vec{n})\, d\beta \right\} \tag{21}$$

with sign + for $T(\vec{n}) > 0$, $\sigma_{II} > \sigma_I$

- for $T(\vec{n}) < 0$, $\sigma_I > \sigma_{II}$

For a loading without reorientation of principal axes, it is easy to demonstrate that $-(d\varepsilon_I + d\varepsilon_{II})$ takes the sign of $-(\sigma_{II} - \sigma_I)^2$, then volumetric strains are compressive.

With an additional simplifying assumption it can be demonstrated that fulfilling equations (20) and (21) leads to the stress-dilatancy relation defined by Rowe {11} .

4.3. Loading stress path with σ_I/σ_{II} = constant

The previous analysis points out sliding criterions defined by σ_I/σ_{II} . In such a hypothesis, loadings with σ_I/σ_{II} = ct . cannot provide irrecoverable strains, which does not correspond to the observed experimental behavior of granular materials.

To have a better adequacy of the model, it is necessary to modify hypothesis (4) and (19) . In fact, the irrecoverable strains provide an evolution of the number of contact points, which is not proportionnal to the stresses applied.

To take into account this phenomena term $\frac{\sigma_\alpha^{(1+a)}}{P_o^a}$ can be substituted to σ_α in relation (4) and (19).

Term a is a parameter to be defined, P_o equal to the unity of pressure makes it possible to preserve the homogeneity of the formulae.

The analyses presented in section 3 and 4 can be developped with the new assumption {5}.

In this case, the yielding lines of the two mechanisms are slightly curved.

5. Elasto-plastic model.

The previous microstructural analysis leads, in the general framework, of elasto-plastic theory, to a model with two yielding mechanisms (Fig.9).

. The first one defined in section 3 characterizes the evolution of mean contact forces. For a bidimensional material this mechanism depends on two scalar parameters :

- λ_d which is a kinematic hardening parameter, is linked to the evolution of the deviatoric stresses.

- λ_r which is an isotropic softening parameter, is linked to the re-orientation of the principal axes.

For a monotonical loading the evolution of this mechanism leads to extensive volumetric strains. Yielding surfaces are defined in section 3.

. The second one defined in section 4 characterizes the width of the distributions of contact forces.

For a bidimensional material, this mechanism depends only on one scalar parameter which is an isotropic softening parameter. For a monotonical loading the evolution of this mechanism leads to compressive volumetric strains. Yielding surfaces are defined in section 4.

For a first loading path, the limit between the two domains of activation of the two mechanisms is given by (Fig.9) :

$$\sigma_I / \sigma_{II} = tg^2(\pi/4 \pm \delta/2)$$

After complex loading this limit can change. For real loading it can be assumed that the transition from one mechanism to the other occurs progressively around the defined bounds.

6. Predictions of the model and experimental results .

- Isotropic loading :

Hypothesis taken into account in section 4.3. allows to describe irrecoverable strains under an isotropic loading.

- First monotically deviatoric loading:

From experimental data Habib and Luong {9} have shown, that in a stress space, a " *characteristic*" line can be defined. This line is the limit between volumetric compressive strains and volumetric extensive strains, this limit does not depend on the density of the granular material (Fig.10).

These experimental results are in agreement with the analysis proposed herein. The characteristic line is in our analysis the boundary of the two yielding mechanisms.

It is possible to determine experimentally the yielding surfaces of a granular material. It can be shown on Fig.11 from Tatsuoka and al {12} that these surfaces are very similar to the yielding surfaces defined in this work (Fig.7).

Fig.12 shows that at the beginning of a deviatoric stress path the hardening observed is isotropic and kinematic and for greater value of deviatoric stresses it is only kinematic. This kind of behavior is in keeping with the previous analysis .

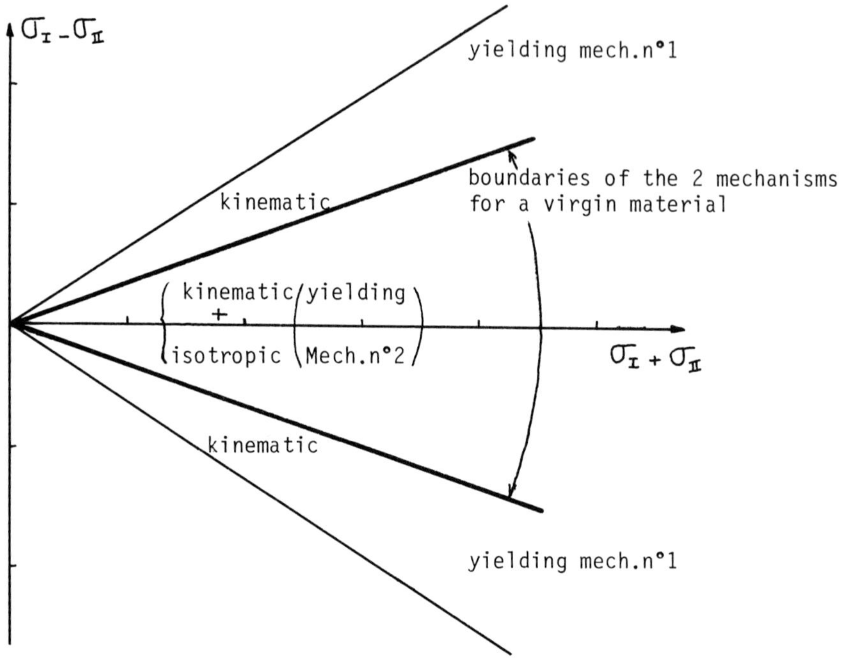

Fig. 9 : Yielding mechanisms.

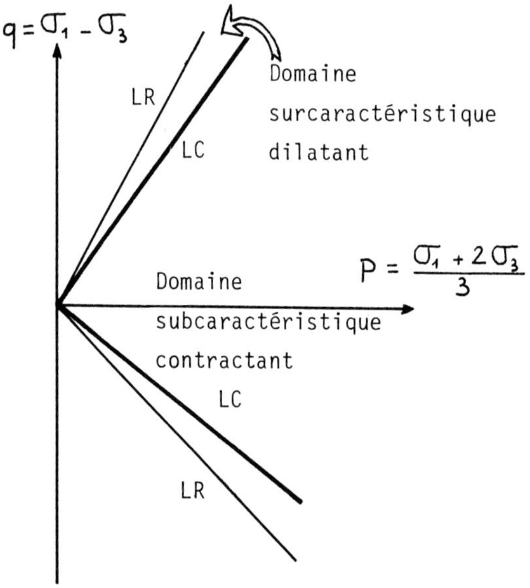

Fig.10 : Definition of the "characteristic" line.

From Luong {9}

Fig.11 : Yielding surfaces
Experimental results from
Tatsuoka and Ishihara {12}

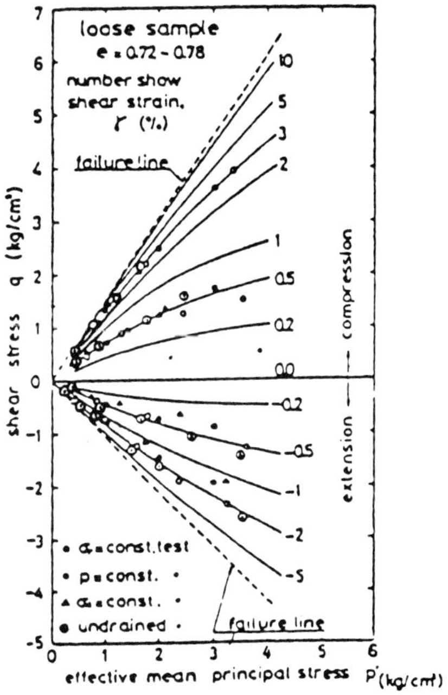

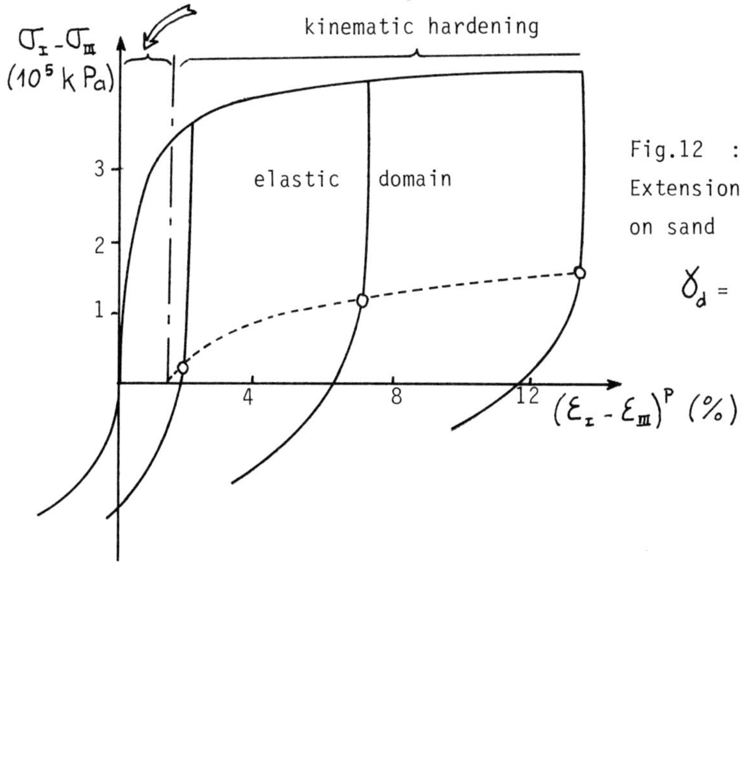

Fig.12 : Compression -
Extension triaxial tests
on sand

γ_d = 15,4 kN/m^3

- cyclic loading :

For cyclic loading with large stress reversal, volume changes show a general tendency to compaction even for dense materials (Fig.13 from Thanopoulos {13}). This can be explained in the framework of this study because, after large stress reversal, it is easy to demonstrate that G and λ_d have the same sign in equation (17), then the volumetric strains are necessarily compressive.

6. Conclusion

The microstructural analysis proposed in this work has allowed us to define the yielding mechanisms of an elasto-plastic model.
This model seems to be in agreement with phenomena observed in experiments on granular media.

Only bidimensional media were considered here, the analysis of three dimensional material does not present any theoritical difficulty but requires taking into account a greater number of yielding parameters.

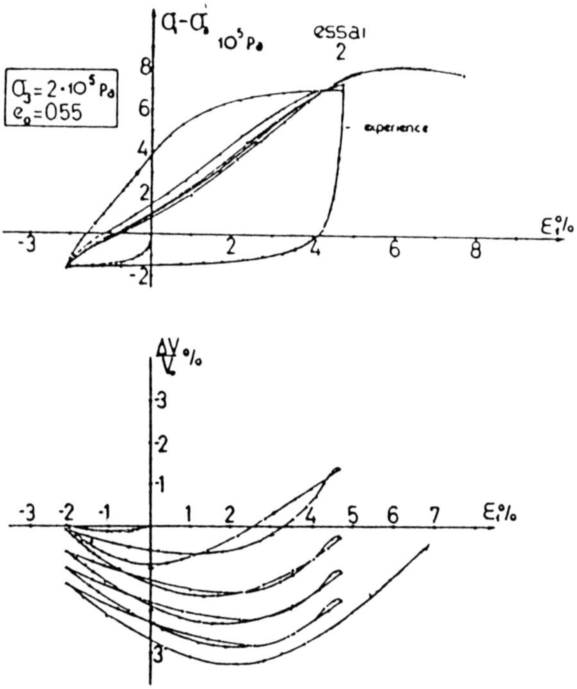

Fig. 13 : Cyclic loading on dense sand.
(From Thanopoulos {13})

REFERENCES

1 CAMBOU B. (1982) " Orientational distributions of contact forces as memory parameters in a granular material".
IUTAM Symposium proc. "Deformation and failure of granular Material" Delft.

2 CAMBOU B. (1984) " Microscopic aspects of hardening in granular material".
International CHISA - Prague.

3 CAMBOU B. - SIDOROFF F. (1983) " Failure criteria for granular material based on statical Microstructural variables ".
Villard de Lans - Juin 1983. CNRS. Symposium Proc.

4 CAMBOU B. - SIDOROFF F. (1984) " Distributions orientées dans un milieu granulaire et leurs représentations".
"Proc. of Journées de Mécanique Aléatoire appliquée à la construction" - Paris.

5 CAMBOU B. " Les mécanismes de déformations plastiques dans un sol granulaire".
Revue Française de Géotechnique n° 31.

6 CAMBOU B. - SIDOROFF F. (1985) " Description de l'état d'un matériau granulaire par variables internes statiques à partir d'une approche discrète".
Journal de Mécanique Théorique et Appliquée. Vol.4, N° 2 ,pp.223/242.

7 CHRISTOFFERSEN J. - MEHRABADI M. - NEMAT-NASSER S. - " a Micromechanical description of granular material behaviour".
J. of. Appl. Mech. Vol.48 , pp. 339.344.

8 LECKIE F.A. - ONAT E.T. (1981) " Tensorial nature of damage measuring internal variables in physical non linearities in structural analysis."
Ed. J. Hult - J.Lemaitre - Springer Berlin, pp. 140/155.

9 LUONG M.P. (1980) " Phénomènes cycliques dans les sols pulvérulents"
Revue Française de Géotechnique n° 10. pp.39/53.

10 MEHRABADI M. - NEMAT-NASSER S. - ODA M. " On statistical description of stress and fabric in granular materials".
Int. J. Num. Anal. Meth. Geom. 6. 1982, pp. 95/108.

11 ROWE P.W. (1969) "The relation between the shear strenght of sands in triaxial compression, plane strain and direct shear".
Géotechnique 19 - Vol.1. pp. 75/86.

12 TATSUOKA F. - ISHIHARA K. (1974) " Yielding of sand in triaxial compression ".
Soils and Foundations, 14, 2 pp. 63/76.

13 THANOPOULOS I. (1981) " Contribution à l'étude du comportement cyclique des milieux pulvérulents".
Thèse D.I. Grenoble.

14 THORNTON C. - BARNES D.J. " On the mechanics of granular material". C.R. IUTAM Symposium "Deformation and Failure of granular material". Delft, Balkema 1982 , pp. 69/77.

15 THORNTON C. - BARNES D.J. (1984) " The relationship between stress and microstructure in particulate media".
C.R. Du Congrès International CHISA - Prague.

16 WEBER J. (1966) " Recherche concernant les contraintes intergranulaires dans les milieux pulvérulents".
Bulletin de Liaison des Ponts et Chaussées n° 20 , pp. 3.1/3.20.

A MATHEMATICAL MODEL FOR THE LIQUEFACTION OF SOILS

Lucia Drăguşin
Polytechnical Institute, Bucharest, Romania

Abstract. A unitary model for mechanical soil behaviour is set forth. By using the particular cases of the material constants occurring in the constitutive equation, the model is able to interpret the behaviour of both granular and cohesive soils, either normally consolidated or overconsolidated under monotonic loading. It accounts for the occurrence of dilatancy in cohesionless soils and of liquefaction in certain cohesionless soils undergoing quasistatic cyclic loading.

1. INTRODUCTION

The work supplies a mathematical model for the mechanical behaviour of saturated soils under monotonic loading. In order to describe the behaviour of this bi-phase material, the model starts from three constitutive equations of the rate-type, an equation for the loading processes, one for the unloading processes and another one for the reloading processes. The stress paths described by these processes join up in the stress space, the initial stress state of a process being equal to the final stress state of the previous process.

Strain and stress will be taken positive in compression.

The material properties will be considered independent of time, i.e. viscous and rate effects will not be covered by the model. A superposed dot will indicate rate, but in the latter, time is intended only as an ordering parameter in the sense generally used in the theory of plasticity. Monotonic and cyclic loading will be considered as quasistatic.

The model may describe the behaviour of two large classes of materials: cohesive soils and cohesionless soils.

Relying on constitutive equations of the hypoelastic type, the model highlights the way in which the initial state of stress and strain has to be formulated.

The dilatancy phenomenon of cohesionless soils in evinced by the existence of a value of maximum density ρ (for certain loading paths).

The occurrence of cohesionless soils liquefaction undergoing undrained cyclic loading is entailed on account of the lack of stability in the constant density curves. The condition ρ = constant will establish the effective stress paths featured by undrained tests where the pore pressure is zero.

2. A CLASS OF HYPOELASTIC MATERIALS OF GRADE THREE

The mathematical model set forth starts from the following hypotheses :

i. soils have memory, their behaviour depending on the initial state and the stress history;

ii. for a certain stress history, the dilatancy of soils may occur;

iii. because for drained soils the contribution of the non-mechanical energy can not be ignored, we suppose that the work done by the stress related to the initial configuration depends on the stress history $T = T(\tau)$, $\tau \in [t_o, t]$

$$[w(T(\cdot))](t) = \int_{t_o}^{t} \left(\int_{V} \mathbf{tr}(TD)\,dV\right)d\tau = \int_{V_o}\left(\int_{t_o}^{t} \frac{(\rho_s)_o}{\rho_s}\,\mathbf{tr}(TD)\,d\tau\right)dV_o,$$

where T is the Cauchy stress tensor (defined by means of the internal normal at body surface); $D = \frac{1}{2}(L + L^T)$ is the rate of deformation tensor ($L = -\mathbf{grad}\,\dot{x}$ being the spatial gradient of the velocity of a particle); $V_o = V(t_o)$ is the initial volume V; $(\rho_s)_o$ is the initial relative mass density of solid phase (mass of the solid phase/total volume) (see Drăguşin [1]).

iv. for a closed (undrained) incompressible system ($\rho = \rho_o$), the work w is a conservative function, namely there is a function ψ so that

$$dw = \mathbf{tr}(TD)\,d\tau = \dot{\psi}\,d\tau.$$

PROPOSITION 1. *The constitutive equation*

$$\begin{aligned}\overset{\circ}{T} = {} & (\alpha_1 x^2 + \beta_o x^3 + \frac{\alpha_7}{6} y + \beta_1 xy - \frac{\beta_4}{3} z) I_D I + (\alpha_2 x^2 + \beta_2 x^3 - \frac{\alpha_7}{6} y + \beta_3 xy + \beta_4 z) D + \\ & + (\alpha_3 x + \beta_5 x^2) I_D T + (\alpha_4 x + \beta_6 x^2 + \frac{\beta_4 - \beta_7}{3} y)\mathbf{tr}(TD) I + (\alpha_5 x - \frac{2\beta_9}{3} x^2 + \beta_7 y)\frac{TD+DT}{2} + \\ & + (-\frac{\alpha_7}{3} + \frac{\beta_4 - \beta_9}{3} x) I_D T^2 + (\alpha_6 + \beta_8 x)\mathbf{tr}(TD) T + (-\frac{\alpha_7}{3} + \frac{\beta_7 - \beta_9}{3} x)\mathbf{tr}(T^2 D) I + \\ & + (\alpha_7 + \beta_9 x)\frac{T^2 D + DT^2}{2} - \beta_4 \mathbf{tr}(TD) T^2 - \beta_7 \mathbf{tr}(T^2 D) T,\end{aligned} \tag{1}$$

where I *is the unit tensor;* I_D = **tr** D *is the trace of* D; $\mathring{T} = \dot{T} + WT - TW$ *is the Jaumann-Noll stress rate,* W *being the spin tensor;* $T^* = T - \frac{1}{3}I_T I$ *is the deviatoric stress tensor;* $x = \mathbf{tr}\, T$; $y = \mathbf{tr}(T^*)^2$; $z = \mathbf{tr}(T^*)^3$, *and*

$$\alpha_1 = -\frac{81\alpha_2^2+\alpha_7(9\alpha_2+9\alpha_5+4\alpha_7)}{27(9\alpha_2+3\alpha_5+\alpha_7)}, \quad \alpha_9 = \frac{-27\alpha_2\alpha_5+\alpha_7(9\alpha_2+12\alpha_5+5\alpha_7)}{9(9\alpha_2+3\alpha_5+\alpha_7)}$$

$$\beta_1 = -\frac{1}{18}(3\beta_4+27\beta_6+2\beta_7+3\beta_8-11\beta_9), \quad \beta_3 = \frac{1}{6}(\beta_4+27\beta_6+2\beta_7+3\beta_8-9\beta_9),$$

$$\beta_5 = -\frac{1}{9}(81\beta_o+27\beta_2+27\beta_6+2\beta_7+3\beta_8-5\beta_9),$$

implies the existence of two potentials ϕ,ψ *so that*

$$\dot{\phi}(x,y,z) = \frac{x}{3}I_D, \quad \dot{\psi}(x,y,z) = \mathbf{tr}(TD).$$

They are

$$\phi(x,y,z)-\phi(x_o,y_o,z_o) = \frac{1}{b^2d}\,\mathbf{ln}\left[\left|\frac{x}{x_o}\right|^{b^2-3ac}\left|\frac{6a^3x^3-3ab^2xy+2b^3z}{6a^3x_o^3-3ab^2x_oy_o+2b^3z_o}\right|^{ac}\right]+$$

$$+\frac{1}{2AD}\left(\frac{2Ax_o^2-Cy_o}{x_o^3}-\frac{2Ax^2-Cy}{x^3}\right), \tag{2}$$

$$\psi(x,y,z)-\psi(x_o,y_o,z_o)=\frac{1}{b^2d}\,\mathbf{ln}\left[\left|\frac{x}{x_o}\right|^{b^2+3a(d-c)}\left|\frac{6a^3x_o^3-3ab^2x_oy_o+2b^3z_o}{6a^3x^3-3ab^2xy+2b^3z}\right|^{a(d-c)}\right]+$$

$$+\frac{1}{2AD}\left[\frac{2Ax_o^2-(C+D)y_0}{x_o^3}-\frac{2Ax^2-(C+D)y}{x^3}\right]$$

where

$$a=\alpha_2+\frac{1}{3}\alpha_5+\frac{1}{9}\alpha_7; \quad b=\alpha_5+\frac{2}{3}\alpha_7; \quad c=3\alpha_4+\alpha_5+\alpha_6;$$

$$d=9\alpha_1+3\alpha_2+3\alpha_3+3\alpha_4+\alpha_5+\alpha_6-\frac{1}{3}\alpha_7; \quad A=\beta_2-\frac{1}{9}\beta_9;$$

$$B=\frac{1}{6}(\beta_4+27\beta_6+4\beta_7+3\beta_8-8\beta_9); \quad C=\frac{1}{3}(\beta_4-9\beta_6-3\beta_8+2\beta_9);$$

$$D=-(18\beta_o+6\beta_2+6\beta_6+\frac{2}{3}(\beta_7-\beta_9)); \quad E=\beta_7-\beta_4,$$

and $a,b,d,A,B,D,E, > 0$, $d-c > 0$, $C+D > 0$, $b-3a > 0$, $b^2-3ac > 0$.

PROOF. From the constitutive equation (1), because

$$\dot{x} = \mathbf{tr}\,\mathring{T}; \quad \frac{1}{2}\dot{y} = \mathbf{tr}(T\mathring{T}) - \frac{x\dot{x}}{3}; \quad \frac{1}{3}\dot{z} = \mathbf{tr}(T^2\mathring{T}) - \frac{2x}{3}\mathbf{tr}(T\mathring{T})+(\frac{x^2}{3} - y)\frac{\dot{x}}{3}$$

we obtain

$$\dot{x}=[\frac{d-c}{3}x^2+\frac{C+D}{3}x^3]I_D+(cx-Cx^2)\mathbf{tr}(TD)$$

$$\frac{1}{2}\dot{y}=[-\frac{3a-b}{9}x^3+\frac{3a(d-c-b)+b^2}{9a}xy-\frac{A}{3}x^4+\frac{C+D}{2}x^2y]I_D+[\frac{3a-2b}{3}x^2-$$
$$-\frac{b^2-3ac}{3a}y+Ax^3-\frac{3C}{2}xy]\mathbf{tr}(TD)+(bx)\mathbf{tr}(T^2D) \tag{3}$$

$$\frac{1}{3}\dot{z}=[\frac{a}{9}x^4-\frac{6a-b}{18}x^2y+\frac{3a(d-c)+b^2}{9a}xz+\frac{A}{9}x^5-\frac{3A-B}{9}x^3y-\frac{6B+E}{18}xy^2+$$
$$+\frac{6B+9(C+D)-2E}{18}x^2z+\frac{E}{3}yz]I_D+[-\frac{2a}{3}x^3+\frac{b}{6}xy-\frac{b^2-3ac}{3a}z-\frac{2A}{3}x^4-$$
$$-\frac{2B}{3}x^2y-\frac{6B+9C-4E}{6}xz]\mathbf{tr}(TD)+(ax^2+Ax^3+Bxy-Ez)\mathbf{tr}(T^2D).$$

Then

$$I_D=3\frac{[6a^3x^3+3a(2ac-b^2)xy+2b(b^2-3ac)z]\dot{x}-3a^2cx^2\dot{y}+2abcx\dot{z}}{dx^2(6a^3x^3-3ab^2xy+2b^3z)}+3\frac{(2Ax^2-3Cy)\dot{x}+Cx\dot{y}}{2ADx^5},$$

$$\mathbf{tr}(TD)=\frac{\{6a^3x^3-3a[2a(d-c)+b^2]xy+2b[b^2+3(d-c)]z\}\dot{x}+3a^2(d-c)x^2\dot{y}}{dx(6a^3x^3-3ab^2xy+2b^3z)}-$$
$$-\frac{2ab(d-c)x\dot{z}}{dx(6a^3x^3-3ab^2xy+2b^3z)}+\frac{[2Ax^2-3(C+D)y]\dot{x}+(C+D)x\dot{y}}{2ADx^4} \tag{4}$$

$$\mathbf{tr}(T^2D)=\left[\frac{2Ax^4+(6A-C-2D)x^2y-9Cy^2-9Dxz}{6ADx^5}+E\frac{2Ax^3y^2-3Cxy^3-18Dxz^2-36Cy^2z}{12ADx^5(Ax^3+Bxy-Ez)}\right]\dot{x}+$$
$$+\left[\frac{(C+2D)x^2+3Cy}{6ADx^4}+\frac{CExy^2-6CEyz+6BDx^2z}{12ADx^4(Ax^3+Bxy-Ez)}\right]\dot{y}+\frac{\dot{z}}{3(Ax^3+Bxy-Ez)},$$

if $x \neq 0$, $6a^3x^3-3ab^2xy+2b^3z \neq 0$, $Ax^3+Bxy-Ez \neq 0$.

The differential forms

$$\dot{\phi}=\frac{x}{3}I_D \quad \text{and} \quad \dot{\psi}=\mathbf{tr}(TD)$$

are exact. By integrating these relations we get the expressions (2).

REMARK. In the axial-symmetrical case, the stress tensor T and the rate of deformation tensor D have the form

$$T=\begin{bmatrix} T_1 & 0 & 0 \\ 0 & T_1 & 0 \\ 0 & 0 & T_3 \end{bmatrix} \quad D=\begin{bmatrix} D_1 & 0 & 0 \\ 0 & D_1 & 0 \\ 0 & 0 & D_3 \end{bmatrix}$$

Then $x = 3p, \quad y = \frac{2}{3}q^2, \quad z = \frac{2}{9}q^3$ and from (2), (4) we have

$$\dot{\phi}(p,q)= \frac{[(9ap+2bq)(9ap-bq)+6acq^2]\dot{p}-6acpq\dot{q}}{dp(9ap+2bq)(9ap-bq)} + \frac{3(9Ap^2-Cq^2)\dot{p}+2Cpq\dot{q}}{81ADp^4} ,$$

$$\dot{\psi}(p,q)= \frac{[(9ap+2bq)(9ap-bq)-6a(d-c)q^2]\dot{p}+6a(d-c)pq\dot{q}}{dp(9ap+2bq)(9ap-bq)} + \tag{5}$$

$$+ \frac{3[9Ap^2-(C+D)q^2]\dot{p}+2(C+D)pq\dot{q}}{81ADp^4}$$

$$\phi(p,q) - \phi(p_o,q_o) = \frac{1}{b^2d}\,\mathbf{ln}\left[\left|\frac{p}{p_o}\right|^{b^2-3ac} \cdot \right.$$

$$\left. \cdot \left|\frac{9^3a^3p^3-27ab^2pq^2+2b^3q^3}{9^3a^3p_o^3-27ab^2p_oq_o^2+2b^3q_o^3}\right|^{ac}\right] + \frac{1}{81AD}\left(\frac{27Ap_o^2-Cq_o^2}{p_0^3} - \frac{27Ap^2-Cq^2}{p^3}\right)$$

$$\tag{6}$$

$$\psi(p,q) - \psi(p_o,q_o) = \frac{1}{b^2d}\,\mathbf{ln}\left[\left|\frac{p}{p_0}\right|^{b^2+3a(d-c)} \cdot \right.$$

$$\left. \cdot \left|\frac{9^3a^3p_o^3-27ab^2p_oq_o^2+2b^3q_o^3}{9^3a^3p^3-27ab^2pq^2+2b^3q^3}\right|^{a(d-c)}\right] + \frac{1}{81AD}\left[\frac{27Ap_o^2-(C+D)q_o^2}{p_o^3} - \frac{27Ap^2-(C+D)q^2}{p^3}\right].$$

In this case, the system (3) may be inverted if

$$p \neq 0, \quad 9ap+2bq \neq 0, \quad 9ap-bq \neq 0, \quad 27Ap^3+2Bpq^2-\frac{2}{9}Eq^3 \neq 0.$$

We consider the set

$$D = \{(p,q)\,|p > 0, \quad 9ap + 2bq > 0, \quad 9ap - bq > 0, \quad \frac{q^3}{p} - \frac{9B}{E}\frac{q^2}{p} - \frac{243A}{2E} > 0\}.$$

3. THE STABILITY WITH RESPECT TO THE INITIAL STRESS STATE

DEFINITION 1. Let $\mathbf{D} \subset \mathbf{R}^2 \times \mathbf{R}$ be the set in which the equation $F(p,q,k) = 0$ implicitly defines the function $p = f(q,k)$ and let $q \to f_k(q)$ be the partial function. We say that q^* is a *critical point* for f_k if $\left(\frac{df_k}{dq}\right)_{q=q^*} = 0$. It is a *nondegenerate critical point* if the condition $\left(\frac{d^2f_k}{dq^2}\right)_{q=q^*} \neq 0$ is also satisfied; if $\left(\frac{d^2f_k}{dq^2}\right)_{q=q^*} = 0$ it is a *degenerate critical point*.

DEFINITION 2. Let k^* be fixed and q^* a solution of the equation $\frac{\partial f}{\partial q}(q,k^*) = 0$. We say that in the point q^* *the function* f_{q^*} *is stable with respect to the parameter* k, if for any $\varepsilon > 0$ there is an $\delta(\varepsilon) > 0$ and a neighborhood $V_{q^*}(\varepsilon)$ of q^* so that for any k, for which $|k^* - k| < \delta(\varepsilon)$, and for which the equation $\frac{\partial f}{\partial q}(q,k) = 0$ has at least a solution $q_\varepsilon \in V_{q^*}(\varepsilon)$, we have

$$\left|\frac{d^2 f_{k^*}}{dq^2}(q^*) - \frac{d^2 f_k}{dq^2}(q_\varepsilon)\right| < \varepsilon .$$

In Cristescu, Drăgușin [2] the following proposition has been proved: let k^* be fixed and q^* a critical point for which $\frac{\partial f}{\partial q}(q^*,k^*) = 0$. The function f_{q^*} is stable with respect to the parameter k in its nondegenerate critical points.

We shall further study from this point of view the curves $\Phi(p,q) = \Phi(p_o,q_o)$ and $\Psi(p,q) = \Psi(p_o,q_o)$ with respect to the initial stress state (p_o,q_o).

From of the continuity equation $\frac{\dot\rho}{\rho} = I_D$ one has $\frac{\dot\rho}{\rho} = \frac{1}{p}\dot\Phi$.

This means that the relation $\Phi(p,q) = k$, where $k = \Phi(p_o,q_o)$ represents the stress paths for which total mass density ρ remains constant thus characterizing the behaviour of a material under undrained tests with pore pressure equal to zero.

PROPOSITION 2. *If equation* $\Phi(p,q) = k$ *implicitly defines the function* $p = p(q,k)$ *then the critical degenerated points of the function* p *pertaining to the domain* D *are* $(p_1 = \frac{adC}{3cAD},\ q_1 = 0)$ *and* $(p_2 = \frac{3adC}{8cAD},\ q_2 = \frac{27a^2dC}{32bcAD})$.

PROOF. The critical degenerated points of function $p = p(q,k)$ satisfy the relations

$$\frac{dp}{dq} = 0, \quad \frac{d^2p}{dq^2} = 0.$$

We shall rearch for the solutions of this system which belongs to the domain *D*. From the relation (5_1), we obtain for $\dot\Phi(p,q) = 0$

$$\begin{cases} q[243acADp^3 - dC(9ap+2bq)(9ap-bq)] = 0 \\ 243acADp^3 - dC[(9ap+2bq)(9ap-bq)+bq(9ap-4bq)] = 0. \end{cases}$$

This system is equivalent to

$$\begin{cases} q = 0 \\ p = \dfrac{adC}{3cAD} \end{cases} \quad \text{and} \quad \begin{cases} q = \dfrac{9p}{4b}[a \pm\sqrt{3a(3a-8\dfrac{cAD}{dC}p)}] \\ q = \dfrac{9a}{4b}p \end{cases} \tag{7}$$

yielding the solutions (p_1,q_1) and (p_2,q_2).

As $\left(\dfrac{d^2p}{dq^2}\right)_{q=0} = \dfrac{2}{27aAp}\,\dfrac{-adC+3cADp}{3Dp+d}$,

$$\left(\frac{d^2p}{dq^2}\right)_{q = \frac{9p}{4b}[a\pm\sqrt{3a(3a-8\frac{cAD}{dC}p)}]} =$$

$$= \pm\frac{8b^2d^2C^2[a\pm\sqrt{3a(3a-8\frac{cAD}{dC}p)}]\,\sqrt{3a(3a-8\frac{cAD}{dC}p)}}{81acADp^2[16b^2A(3Dp+d)+5dC(a\pm\sqrt{3a(3a-8\frac{cAD}{dC}p)})]},$$

if $c,C > 0$ for $p_o > \frac{adC}{3cAD}$ then there exist minimum points along axis $q = 0$; for $\frac{adC}{3cAD} < p_o < \frac{3adC}{8cAD}$ there are minimum points on the curve $q = \frac{9p}{4b}[a + \sqrt{3a(3a - 8\frac{cAD}{dC}p)}]$, maximum points on curve $q = \frac{9p}{4b}[a - \sqrt{3a(3a - 8\frac{cAD}{dC}p)}]$ and minimum points on the axis $q = 0$; for $p_o < \frac{adC}{3cAD}$ there are minimum points on the curves $q = \frac{9p}{4b}[a \pm\sqrt{3a(3a - 8\frac{cAD}{dC}p)}]$ and maximum points on the axis $q = 0$.

If, on the other hand $c,C < 0$, then for $p_o > \frac{adC}{3cAD}$ there are maximum points on the axis $q = 0$; for $\frac{adC}{3cAD} < p_o < \frac{3adC}{8cAD}$ the maximum points are on the curve $q = \frac{9p}{4b}[a+\sqrt{3a(3a-8\frac{cAD}{dC}p)}]$, the minimum points on the curve $q = \frac{9p}{4b}[a-\sqrt{3a(3a-8\frac{cAD}{dC}p)}]$ and the maximum points on the axis $q = 0$; for $p_o < \frac{adC}{3cAD}$ there are maximum points on the curves $q = \frac{9p}{4b}[a \pm \sqrt{3a(3a-8\frac{cAD}{dC}p)}]$ and minimum points on the axis $q = 0$.

From the relation (6) we find that the first instability point (p_1,q_1) occurs for the initial stress state (p_o,q_o) which verifies the relation $\phi(p_1,q_1) = \phi(p_o,q_o)$ viz.

$$\ln p_o + \frac{ac}{b^2}\ln\frac{9a^3p_o^3}{9a^3p_o^3-27ab^2p_oq_o^2+2b^3q_o^3} + d\,\frac{27Ap_o^2-Cq_o^2}{81ADp_o^3} = \ln\frac{adC}{3cAD} - \frac{aC}{cA}$$

while the second instability point (p_2,q_2), for the initial stress state which verifies the relation

$$\ln p_o + \frac{ac}{b^2}\ln\frac{9a^3p_o^3}{9a^3p_o^3-27ab^2p_oq_o^2+2b^3q_o^3} + d\,\frac{27Ap_o^2-Cq_0^2}{81ADp_0^3} = \ln\frac{3adC}{8cAD} - \frac{ac}{b^2}\ln\frac{32}{27} +$$

$$+ c\,\frac{15a^2C-8b^2A}{18ab^2C} .$$

PROPOSITION 3. *If the function* $q = q(p,k)$ *is implicitly defined by the equation* $\phi(p,q) = k$, *then its critical degenerated points from domain D are the solutions of the system*

$$\begin{aligned}&\left(\frac{81a^2A}{b}E_o^2 - E_2F_o\right)^2 + (E_oF_1 - E_1F_o)\left(E_2F_1 - \frac{81a^2A}{b}E_oE_1\right) = 0\\ &q = \frac{p}{E_oF_1-E_1F_o}\left(E_2F_o - \frac{81a^2A}{b}E_o^2\right),\end{aligned} \tag{8}$$

where

$$E_o=-b[540AD^2(b^2-3ac)p^2+24dD(27a^2C+11b^2A-18acA)p+$$
$$+d^2(171a^2C+32b^2A)],$$

$$E_1=9a[108AD^2(3b^2+ac)p^2+9dD(18b^2A+2acA-9a^2C)p+2d^2(10b^2A-9a^2C)],$$

$$E_2=243a^2bA(153D^2p^2+82dDp+11d^2),$$

$$F_o=648A^2D^2(b^2-3ac)^2p^2+36dAD(45a^2b^2C+8b^4A-117a^3cC-24ab^2cC)p +$$
$$+ d^2(729a^4C^2+360a^2b^2AC+32b^4A^2),$$

$$F_1=-27abA[144AD^2(b^2-3ac)p^2+dD(68b^2A^2+81a^2C-108acA)p+$$
$$+2d^2(9a^2C+4b^2A)].$$

PROOF. The critical degenerated points of function $q = q(p,k)$ verify the system

$$\frac{dq}{dp} = 0, \qquad \frac{d^2q}{dp^2} = 0$$

viz.

$$\begin{cases}[9Ap^2(3Dp+d)-dCq^2](9ap+2bq)(9ap-bq)+162acADp^3q^2 = 0\end{cases}$$

$$\left\{ \begin{aligned} & Ap(9Dp+2d)(9ap+2bq)(9ap-bq)+a[9Ap^2(3Dp+d)-dCq^2](18ap+bq) + \\ & \qquad + 54acADp^2q^2 = 0. \end{aligned} \right.$$

By successively eliminating q from these two relations we obtain the relations (8). Relation $(8)_1$ is an equation of degree 8 in p. Hence the functions $q = q(p,k)$ may have 8 or 6 or 4 or 2 or no critical degenerated point.

PROPOSITION 4. *Equation* $\dot{\Phi}(p,q) = 0$ *has no singular points in* D *if* $b^2A - 3a^2C > 0$.

PROOF. In its singular points, equation $\dot{\Phi}(p,q) = 0$ is identically satisfied. Then

$$\left\{ \begin{aligned} & 243acADp^3-dC(9ap+2bq)(9ap-bq) = 0 \\ & 9Ap^2(3Dp+d)-dCq^2](9ap+2bq)(9ap-bq)+162acADp^3q^2 = 0. \end{aligned} \right.$$

This system is equivalent to

$$\left\{ \begin{aligned} & \frac{9A^2D^2}{d^2}(2b^2+3ac)^2p^2 + \frac{3AD}{2d}[(2b^2+3ac)(8b^2A-15a^2C)+9a^3cC]p + \\ & \qquad + (b^2A-3a^2C)(4b^2A-3a^2C) = 0 \\ & q = \pm 3p\sqrt{\frac{3A(3Dp + d)}{dC}} . \end{aligned} \right.$$

The first equation has the roots

$$p_{1,2}=d\,\frac{-\frac{1}{2}[(2b^2+3ac)(8b^2A-15a^2C)+9a^3cC]\pm\sqrt{a^2b^2C^2+4aC(2b^2+3ac)(aC+cA)}}{6AD(2b^2 + 3ac)^2}$$

From the condition that the two roots are negative, we find $b^2A - 3a^2C > 0$.

The curves $\Phi(p,q) = \Phi(p_o,q_o)$ are graphically represented for c, C > 0 in fig.1 and for c, C < 0 in fig.2.

The dotted line represents the functions

$$q = \frac{9p}{4b}[a + \sqrt{3a(3a - 8\frac{cAD}{dC}p)}], \qquad q = \frac{9p}{4b}[a - \sqrt{3a(3a - 8\frac{cAD}{dC}p)}].$$

The slope of the straight line $q = mp$ is the solution of the equation

$$m^3 - \frac{9B}{E}m^2 - \frac{243A}{2E} = 0.$$

If $E < b\,\frac{6a^2B + b^2A}{6a^3}$ then $o < m < \frac{9a}{b}$.

For a stress (p,q) belonging to the straight line q = mp, the soil sample becomes unstable.

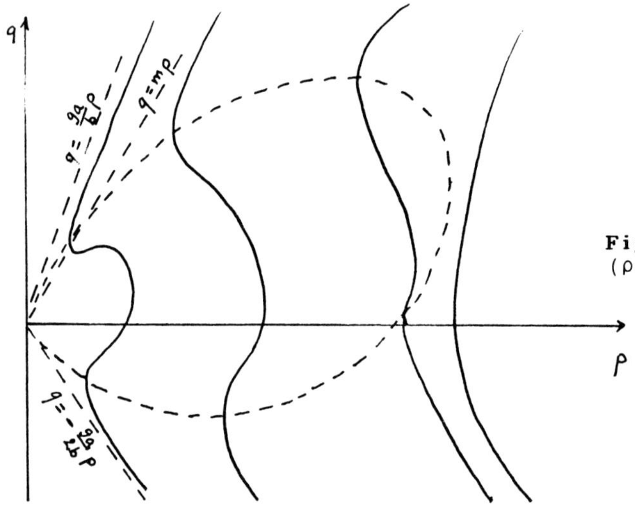

Fig. 1. Undrained tests (ρ = const.) for c, C > 0.

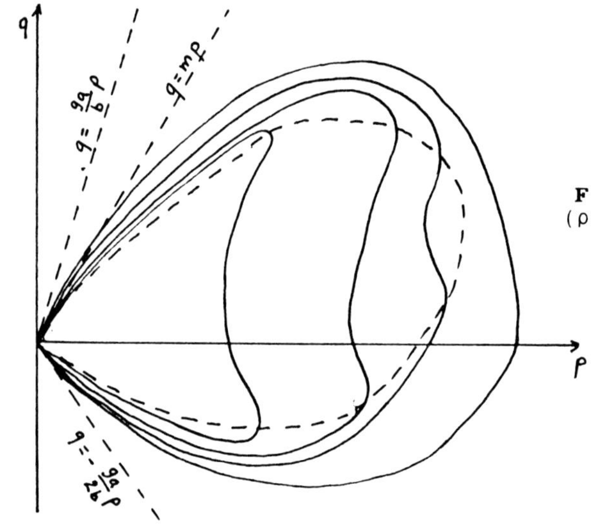

Fig. 2. Undrained tests (ρ = const.) for c, C < 0.

The curves $\psi(p,q) = \psi(p_o,q_o)$ have the equation

$$\ln p+\frac{a(d-c)}{b^2}\ln\frac{p^3}{9^3a^3p^3-27ab^2pq^2+2b^3q^3} - d\,\frac{27Ap^2-(C+D)q^2}{81ADp^3} =$$

$$\ln p_o+\frac{a(d-c)}{b^2}\ln\frac{p_o^3}{9^3a^3p_o^3-27ab^2p_oq_o^2+2b^3q_o^3} - d\,\frac{27Ap_o^2-(C+D)q_o^2}{81ADp_o^3}\ .$$

Their only points of maximum value are on the axis $q = 0$.

Fig. 3 provides a graphical representation of the curves $\psi(p,q) = \psi(p_o,q_o)$.

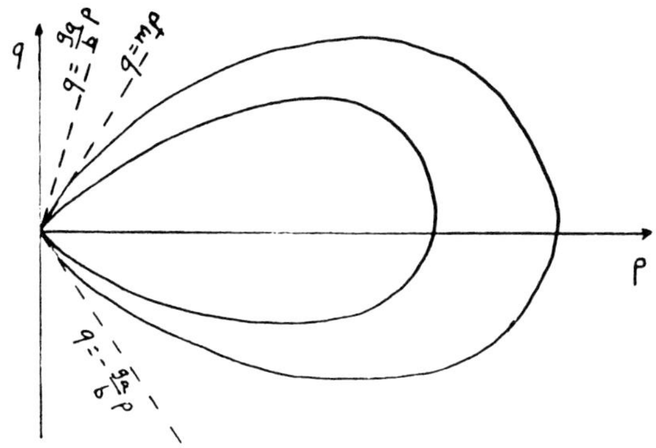

Fig. 3. Curves $\psi(p,q) = \text{const.}$

We shall further consider the response of a material having the constitutive equation (1) in the case of some particular loading.

PROPOSITION 5. *Let* (1) *be the constitutive equation and the stress path* $T_1 = T_1^c = \text{constant}$. *Then, in the domain D, we have*

$$\frac{\rho}{\rho_o} = \left\{ \left|\frac{q_o+3T_1^c}{q+3T_1^c}\right|^{3(2b-3a)} \left|\frac{(b-3a)q-9aT_1^c}{(b-3a)q_o-9aT_1^c}\right|^{4(b-3a)} \cdot \left|\frac{(3a+2b)q+9aT_1^c}{(3a+2b)q_o+9aT_1^c}\right|^{(3a+2b)} \right\}^{\frac{ac}{6b^3dT_1^c}} \cdot$$

$$\cdot \exp\left\{ -\frac{3(b^2-ac)}{b^2d}\left(\frac{1}{q+3T_1^c} - \frac{1}{q_o+3T_1^c}\right) + \frac{3A-C}{2AD}\left(\frac{1}{(q+3T_1^c)^2} - \frac{1}{(q_o+3T_1^c)^2}\right) + \right.$$

$$\left. + \frac{4CT_1^c}{AD}\left(\frac{1}{(q+3T_1^c)^3} - \frac{1}{(q_o+3T_1^c)^3}\right) - \frac{27C(T_1^c)^2}{4AD}\left(\frac{1}{(q+3T_1^c)^4} - \frac{1}{(q_o+3T_1^c)^4}\right)\right\}, \qquad (9)$$

$$\varepsilon = \varepsilon_o + \frac{1}{6b^2T_1^c}\ln\left\{\left|\frac{q_o+3T_1^c}{q+3T_1^c}\right|^{9a} \left|\frac{(b-3aq_o)-9aT_1^c}{(b-3a)q-9aT_1^c}\right|^{2(b-3a)} \left|\frac{(3a+2b)q+9aT_1^c}{(3a+2b)q_o+9aT_1^c}\right|^{(3a+2b)}\right\} +$$

$$+ \frac{1}{4A}\left(\frac{1}{(q+3T_1^c)^2} - \frac{1}{(q_o+3T_1^c)^2}\right) - \frac{3T_1^c}{2A}\left(\frac{1}{(q+3T_1^c)^3} - \frac{1}{(q_o+3T_1^c)^3}\right),$$

where $\dot{\varepsilon} = D_3 - D_1$.

PROOF. For an axial-symmetrical loading we have

$$p = \frac{2T_1^c + T_3}{3}, \quad q = T_3 - T_1^c, \quad I_D = 2D_1 + D_3, \quad \dot{\varepsilon} = D_3 - D_1.$$

Then $T_1^c = p - \frac{q}{3}$, $T_3 = p + \frac{2q}{3}$, $D_1 = \frac{I_D - \dot{\varepsilon}}{3}$, $D_3 = \frac{I_D + 2\dot{\varepsilon}}{3}$, $\dot{\psi} = 2T_1^c D_1 + T_3 D_3 = pI_D + \frac{2}{3}q\dot{\varepsilon}$.

Introducing these relations into (5) we get

$$\frac{\dot{\rho}}{\rho} = \dot{q}\left\{\frac{3}{d}\,\frac{[(3a+2b)q+9aT_1^c][(3a-b)q+9aT_1^c]-6acqT_1^c}{(q+3T_1^c)^2[(3a+2b)q+9aT_1^c][(3a-b)q+9aT_1^c]} + \frac{3A(q+3T_1^c)^2 - Cq(q-6T_1^c)}{AD(q+3T_1^c)^5}\right\} \tag{10}$$

$$\dot{\varepsilon} = \dot{q}\left\{\frac{27aT_1^c}{(q+3T_1^c)[(3a+2b)q+9aT_1^c][(3a-b)q+9aT_1^c]} - \frac{q-6T_1^c}{2A(q+3T_1^c)^4}\right\}$$

By integrating them we find the expressions (9). One finds from the relation (10) that the function $\rho = \rho(q)$ has as its critical points the solutions of the equation of the 5th degree in q : $G_1(q) = 0$, where

$$G_1(q) = [(3a+2b)q+9aT_1^c][(b-3a)q-9aT_1^c][3AD(q+3T_1^c)^3+3dA(q+3T_1^c)^2 - dCq(q-6T_1^c)]+18acADT_1^c q(q+3T_1^c)^3 \tag{11}$$

Since for $c, C > 0$ $\lim_{q\to-\infty} G_1(q) < 0$, $\lim_{q\to q_2} G_1(q) < 0$, $\lim_{q\to 0} G_1(q) < 0$, $\lim_{q\to q_1} G_1(q) > 0$, $\lim_{q\to\infty} G_1(q) > 0$, where $q_1 = \frac{9aT_1^c}{b-3a}$, $q_2 = -\frac{9aT_1^c}{3a+2b}$, it follows that the equation (11) will have at least one real root in the interval $(0,q_1)$, whilst the function $\rho = \rho(q)$ at least one critical point in this interval. If $c, C < 0$, the function $\rho = \rho(q)$ has at least one critical point in the interval $(q_2,0)$.

The curve $\rho = \rho(q)$ has been graphically plotted in fig.4a.

For $c, C > 0$ this curve has the shape (1) if $G_1(q) = 0$ has a real root, (2) if $G_1(q) = 0$ has three roots in the interval $(0,q_1)$, (3) if $G_1(q) = 0$ has one positive and two negative roots in the interval (q_2,q_1), (4) if $G_1(q) = 0$ has three positive and two negative roots in interval (q_2,q_1), (5) if $G_1(q) = 0$ has one positive and four negative roots in interval (q_2,q_1).

For c, C < 0 the curves (1'), (2'), (3'), (4'), (5') have been drawn. The stresses with $G_1(q) = 0$ are related to the material dilatancy.

The relation $(10)_2$ makes obvious that the function $\varepsilon = \varepsilon(q)$ has as critical points the solutions of the third degree equation in q $G_2(q) = 0$, where

$$G_2(q) = 54aAT_1^c(q+3T_1^c)^3+(q-6T_1^c)[(3a+2b)q+9aT_1^c][(b-3a)q-9aT_1^c].$$

The equation $G_2'(q) = 0$ has the positive roots

$$q^*_{1,2} = \frac{T_1^c}{54aAT_1^c+(3a+2b)(b-3a)}\Big[-162aAT_1^c+b(4b-3a) \pm$$

$$\pm\sqrt{-162abAT_1^c(14b-9a)+16b^4-60ab^3-99a^2b^2+405a^3b+27^2a^4}\Big]$$

which occur if

$$T_1^c < \frac{4(b-3a)^3(4b+21a)+9a^2[(5b-15.9a)^2+80.19a^2]}{162abA(14b-9a)} .$$

As $\lim\limits_{q\to q_2} G_2(q) > 0$, $\lim\limits_{q\to 0} G_2(q) > 0$, $\lim\limits_{q\to q_1} G_2(q) > 0$, $G_2''(q_1^*) < 0$, it follows that in the plot of the function $\varepsilon = \varepsilon(q)$ (fig.4b), the curve (1") will occur if $G_2(q_1^*) > 0$, while (2") when $G_2(q_1^*) < 0$.

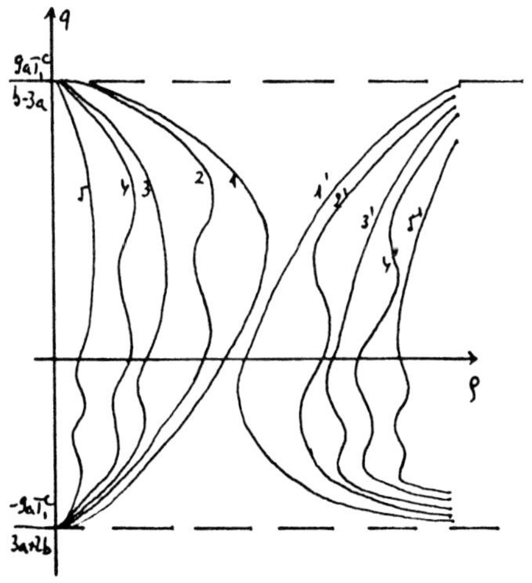

Fig. 4a. Curves $\rho = \rho(q)$ for T_1 = const.

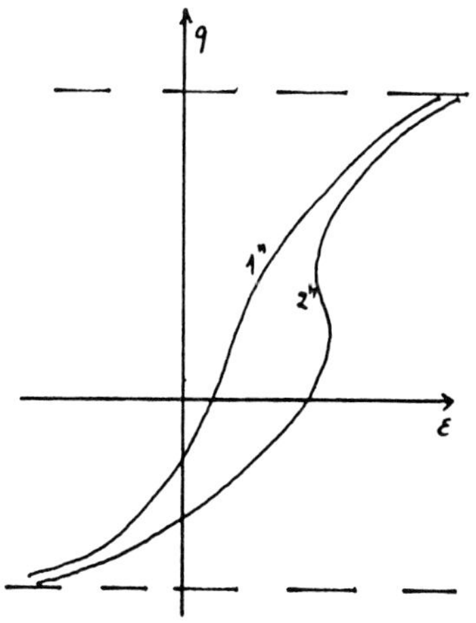

Fig. 4b. Curves $\varepsilon = \varepsilon(q)$ for T_1 = const.

REMARK. The stress path $T_1 = T_1^c =$ constant for $q > 0$ corresponds to the "triaxial al compresion" experiments.

PROPOSITION 6. *Let us take the constitutive equation* (1) *and the stress path* $T_3 = T_3^c =$ *constant. Then, in the domain D we shall have*

$$\frac{\rho}{\rho_o} = \left\{ \left|\frac{2q_o-3T_3^c}{2q-3T_3^c}\right|^{3(3a+b)} \left|\frac{2(b-3a)q+9aT_3^c}{2(b-3a)q_o+9aT_3^c}\right|^{(b-3a)} \cdot \right.$$

$$\left. \cdot \left|\frac{(6a+b)q-9aT_3^c}{(6a+b)q_o-9aT_3^c}\right|^{2(6a+b)} \right\}^{\frac{ac}{b^3 dT_3^c}} \cdot$$

$$\cdot \exp\left\{ \frac{3(b^2-3ac)}{b^2 d}\left(\frac{1}{2q-3T_3^c} - \frac{1}{2q_o-3T_3^c}\right) - \frac{12A-C}{8AD}\left(\frac{1}{(2q-3T_3^c)^2} - \frac{1}{(2q_o-3T_3^c)^2}\right) + \right.$$

$$\left. + \frac{CT_3^c}{AD}\left(\frac{1}{(2q-3T_3^c)^3} - \frac{1}{(2q_o-3T_3^c)^3}\right) + \frac{27C(T_3^c)^2}{16AD}\left(\frac{1}{(2q-3T_3^c)^4} - \frac{1}{(2q_o-3T_3^c)^4}\right) \right\} \tag{12}$$

$$\varepsilon = \varepsilon_o + \frac{1}{3b^2 T_3^c} \ln\left\{ \left|\frac{2q-3T_3^c}{2q_o-3T_3^c}\right|^{9a} \left|\frac{2(b-3a)q+9aT_3^c}{2(b-3a)q_o+9aT_3^c}\right|^{(b-3a)} \cdot \right.$$

$$\left. \cdot \left|\frac{(6a+b)q_o-9aT_3^c}{(6a+b)q-9aT_3^c}\right|^{(6a+b)} \right\} - \frac{1}{8A}\left(\frac{1}{(2q-3T_3^c)^2} - \frac{1}{(2q_o-3T_3^c)^2}\right) -$$

$$- \frac{3T_3^c}{4A}\left(\frac{1}{(2q-3T_3^c)^3} - \frac{1}{(2q_o-3T_3^c)^3}\right).$$

PROOF. If $T_3 = T_3^c$, the equation of the stress path is $p = \frac{3T_3^c - 2q}{3}$. From the relations (5) we get

$$\frac{\dot\rho}{\rho} = \dot q \left\{ \frac{6}{d} \frac{-[(6a+b)q-9aT_3^c][2(b-3a)q+9aT_3^c]+9acT_3^c q}{(2q-3T_3^c)^2[(6a+b)q-9aT_3^c][2(b-3a)q+9aT_3^c]} + \right.$$

$$\left. + 2\,\frac{3A(2q-3T_3^c)^2 - Cq(q+3T_3^c)}{AD(2q-3T_3^c)^5} \right\}, \tag{13}$$

$$\dot\varepsilon = \dot q \left\{ \frac{27aT_3^c}{(2q-3T_3^c)[(6a+b)q-9aT_3^c][2(b-3a)q+9aT_3^c]} + \frac{q+3T_3^c}{A(2q-3T_3^c)^4} \right\}.$$

By integration we obtain (12).

The functions $\rho = \rho(q)$ and $\varepsilon = \varepsilon(q)$ have been plotted in fig. 5a and fig. 5b. One obtains the curves (1), (2), (3), (4), (5) for c, C > 0 and (1'), (2'), (3'), (4'), (5') for c, C < 0.

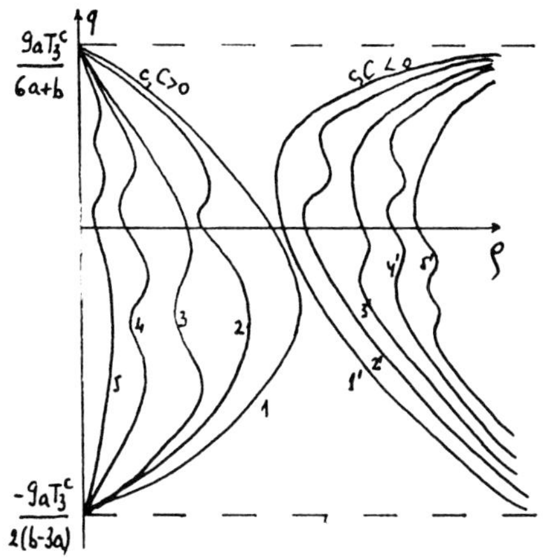

Fig. 5a. Curves $\rho = \rho(q)$ for T_3 = ct.

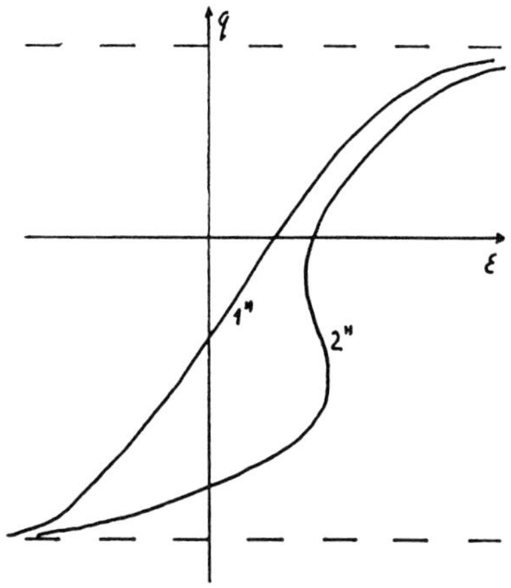

Fig. 5b. Curves $\varepsilon = \varepsilon(q)$ for T_3 = ct.

REMARK 1. The stress path $T_3 = T_3^c$ = const. for q < 0 corresponds to "triaxial extension" experiments.

REMARK 2. For a cyclic loading, these curves will look like the curves shown in fig. 4a, 4b for q > 0 and like those shown in fig. 5a, 5b for q < 0, the joint occurring in point ($p = T_1^c = T_3^c$, q = 0). From

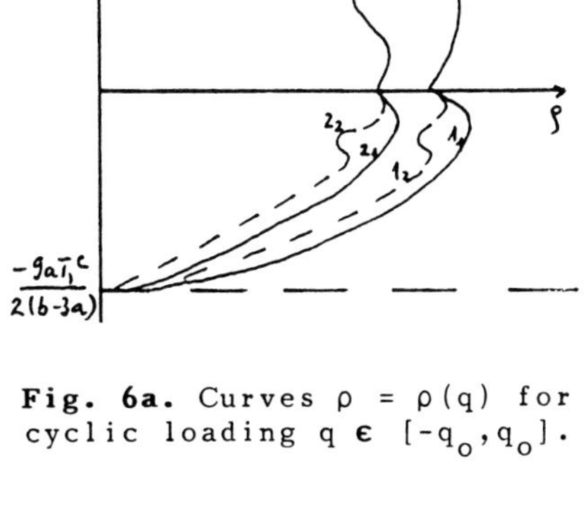

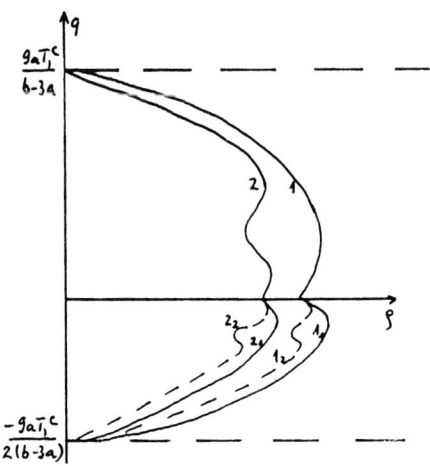

Fig. 6a. Curves $\rho = \rho(q)$ for cyclic loading $q \in [-q_o, q_o]$.

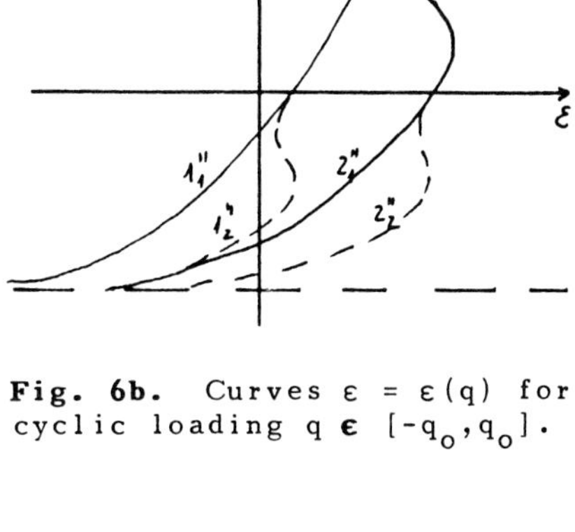

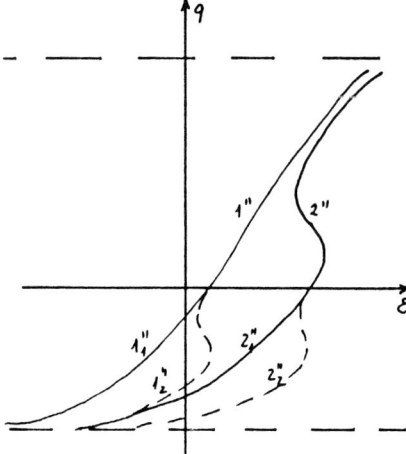

Fig. 6b. Curves $\varepsilon = \varepsilon(q)$ for cyclic loading $q \in [-q_o, q_o]$.

the relations (10) and (13) we notice that this joint is reached with a continuous derivative for the curves $\varepsilon = \varepsilon(q)$ and a discontinuous one for the curves $\rho = \rho(q)$. For a cyclic loading

$$p = \begin{cases} \frac{1}{3}(3T_1^c - 2q), & q \in [-q_o, 0] \\ \frac{1}{3}(q + 3T_1^c), & q \in (0, q_o] \end{cases} \tag{14}$$

we acquire the curves shown in fig. 6a, 6b.

4. MATHEMATICAL MODEL FOR THE MECHANICAL BEHAVIOUR OF SOILS

Let P_1 be the set of unloading processes

$$P_1 = \{(p,q)\,|\,p = p(\tau),\quad q = q(\tau),\quad \tau \in [t_1^i, t_1^f]],\quad \dot{\psi} < 0\},$$

where t_1^i is the initial moment while t_1^f the final moment of the process (for the first unloading one shall consider $(p(t_1^i), q(t_1^i)) = (p_s, q_s)$ viz. the stress state "in situ".

Let P_2 be the set of reloading processes

$$P_2 = \{(p,q)\,|\,p=p(\tau),\ q=q(\tau),\ \tau\in[t_2^i,t_2^f],\ \psi(p,q)<\psi(p(t_1^i),\ q(t_1^i)),\ \dot{\psi} \geq 0,\ t_1^f=t_2^i\}.$$

Let P_3 be the set of loading processes

$$P_3 = \{(p,q)\,|\,p=p(\tau),\ q=q(\tau),\ \tau\in[t_3^i,t_3^f],\ \psi(p,q)\geq\psi(p(t_1^i),\ q(t_1^i)),\ \dot{\psi}\geq 0,\ t_3^i=t_1^f\}.$$

Assuming that the mechanical behaviour of soils is described by the constitutive equation

$$\hat{T} = \sum_{n=1}^{3} \hat{T}^{(n)} \chi(P_n)$$

where

$$\chi(P_n) = \begin{cases} 1, & (p,q) \in P_n \\ 0, & (p,q) \notin P_n \end{cases}$$

is the characteristic function of the process P_n while $\hat{T}$ is the expression resulting from the equation (10) with the coefficients $\alpha_i^{(1)}$, $\beta_i^{(1)}$ for the unloading processes, the coefficients $\alpha_i^{(2)}$, $\beta_i^{(2)}$ for reloading processes and $\alpha_i^{(3)}$, $\beta_i^{(3)}$ for loading processes.

The initial stress state for a process will be equal to the final stress state of the previous process.

For $\psi(p,q) = \psi(p_o,q_o)$ to preserve its shape in various processes we shall assume that

$$\frac{a}{b}=\left(\frac{a}{b}\right)^{(1)}=\left(\frac{a}{b}\right)^{(2)}=\left(\frac{a}{b}\right)^{(3)}; \quad \frac{d-c}{b} = \left(\frac{d-c}{b}\right)^{(1)} = \left(\frac{d-c}{b}\right)^{(2)} = \left(\frac{d-c}{b}\right)^{(3)};$$

$$\frac{d}{D}=\left(\frac{d}{D}\right)^{(1)} = \left(\frac{d}{D}\right)^{(2)} = \left(\frac{d}{D}\right)^{(3)}; \quad \frac{C+D}{A} = \left(\frac{C+D}{A}\right)^{(1)} = \left(\frac{C+D}{A}\right)^{(2)} = \left(\frac{C+D}{A}\right). \tag{3}$$

REMARK. In drained loading, the stress tensor will be equal to the tensor of effective stress while in undrained loading it is equal to the tensor of total stress (the tensor of effective stress minus pore pressure). If the undrained loading confines itself to the stress path which maintains constant total mass density, then the pore pressure is zero and the tensor of effective stress equals the tensor of total stress.

DEFINITION. We shall say that a sand sample tends to liquefy under undrained cyclic loading if the tensor of effective stress tends to zero, the whole loading being taken over by pore pressure.

The undrained cyclic test of loading, unloading, reloading, unloading,... plotted in fig. 7 will have the equations

$$\phi^{(3)}(p,q) = \phi^{(3)}(p_o,0), \quad q \in [0,q_o]$$
$$\phi^{(1)}(p,q) = \phi^{(1)}(p_1,q_o), \quad q \in [0,q_o], \quad \dot{q} < 0$$
$$\phi^{(2)}(p,q) = \phi^{(2)}(p_2,-q_o), \quad q \in [-q_o,0], \quad \dot{q} < 0$$
$$\phi^{(1)}(p,q) = \phi^{(1)}(p_3,q_o), \quad q \in [-q_o,0], \quad \dot{q} > 0$$
$$\phi^{(2)}(p,q) = \phi^{(2)}(p_4,-q_o), \quad q \in [0,q_o], \quad \dot{q} > 0$$

....

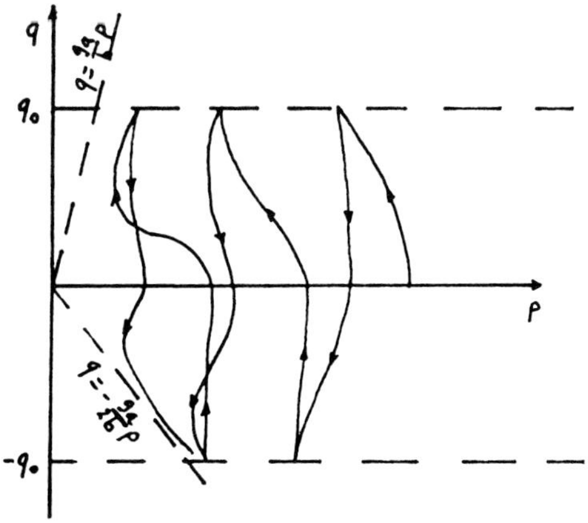

Fig. 7. Stress path for a cyclic undrained test

Owing to the shape change of curves $\phi(p,q)$ = const., one notices a steady decrease of effective pressure p.

5. COMPARISON WITH EXPERIMENTAL DATA

1) Constant density curves plotted in fig. 1. (g for c, C > 0, $p_o < \frac{adC}{3cAD}$) agree with the experimental results for undrained tests on Fuji River sand (Nova, 3, fig. 13.6), dense sand (Thurairajah, 4, fig.3,5,6), loose sand (Thurairajah, 4, fig.4.7).

2) Constant density curves plotted in fig.2 (for c, C < 0) agree with the experimental data for undrained test on a normally consolidated kaolin (Nova, 3, fig.13.11), overconsolidated kaolin (Nova, 3, fig. 13.12) and Weald Clay (Mroz, Norris, 5, fig. 8. 10).

3) The "Yield function" plotted in fig. 3 agrees with the experimental results for a normally consolidated kaolin (Nova, 3, fig. 13.3).

4) For $T_1 = T_1^c$, the curves (1'), (1'') plotted in fig. 4a, 4b (c, C < 0) agree with the experimental results for a constant cell pressure test on a normally consolidated kaolin (Nova, 3, fig. 13.4).

5) For $T_1 = T_1^c$, the curves (2), (1'') (c, C > 0) agree with the experimental results for a constant cell pressure test on a Fuji River sand (Nova, 3, fig.13.7).

6) For $T_1 = T_1^c < T_3$, $T_3 = T_3^c < T_1$ in figs. 6a, 6b (c, C > 0, the curves (1_1), $(1_1'')$ for loading and (1_2), $(2_1'')$ for unloading agree with the experimental data for drained triaxial compression-extension tests on dense sand (Thurairajah, 4, fig.1.2).

CONCLUSION

1) Our model descibes the behaviour of cohesionless soils if c, C > 0 and of cohesive soils if c, C < 0. It involves the occurrence of dilatancy in cohesionless soils.

2) The emphasis falls on the importance of stress history and the initial stress and strain state (p_o, q_o, ρ_o, ε_o respectively).

3) The occurrence of instability points accounts for the change in the shape of unloading and reloading curves with respect to the initial stress state.

4) The sharp change in the shape of unloading and reloading curves for the small values of mean stress p accounts for the occurrence of the liquefaction phenomenon in the case of some cohesionless materials.

REFERENCES

1. **L. Drăguşin**, A hypoelastic model for soils, *Int.J.Enging Sci.* **19**, 511, 1981.
2. **N. Cristescu, L. Drăguşin**, On the stability with respect to the constitutive parameters, *Rev.Roumaine Math.Pures Appl.*, **29**,10, 833,1984.
3. **R. Nova**, A constitutive model for soil under monotonic and cyclic loading, *Soil Mechanics-Transient and Cyclic Loads*, 343, 1982.
4. **A. Thurairajah**, Shear behaviour of sand under stress reversal, *Proc. 8th Int.Conf.Soil.Mech.Found.Enging.* 1.2,439,1973.
5. **Z. Mroz, V.A.Norris**, Elastoplastic and Viscoplastic constitutive models for soils with application to cyclic loading, *Soil Mechanics-Transient and Cyclic Loads*, 173, 1982.

THE KINEMATICS OF SELF-SIMILAR PLANE PENETRATION PROBLEMS IN MOHR-COULOMB GRANULAR MATERIALS

R. Butterfield
University of Southampton, U.K.

1. INTRODUCTION

Throughout this paper we shall assume the material being penetrated to be a rigid-plastic, Mohr-Coulomb continuum, friction angle (ϕ), such that slip lines can develop on planes on which the shear and direct stresses are related by the equation $\tau = C + \sigma\tan\phi$ and the stress characteristics so defined are also the velocity characteristics. Slip can occur independently on either 'slip-line' and the kinematic model is Butterfield and Harkness' (1971) generalisation of Geiringers' (1931) statement : "As successive material points along a slip-line are considered any change in velocity which occurs between one point and the next, relative to the slip line field, is in the direction of the conjugate slip-line".

When a plane, wedge-shaped body is displaced vertically into an extensive bed of material there are three distinct categories of penetration problem of interest.

2. THE INCREMENTAL DISPLACEMENT OF A BURIED WEDGE

Fig.1(a) depicts the simplest case of a slip-line field for a smooth, wedge-shaped body beneath a horizontal, surcharged, 'ground' surface. This is extended in Figs.1(b,c) to cases where the ground surface is uplifted locally at an angle (ρ), for smooth and rough-faced wedges respectively.

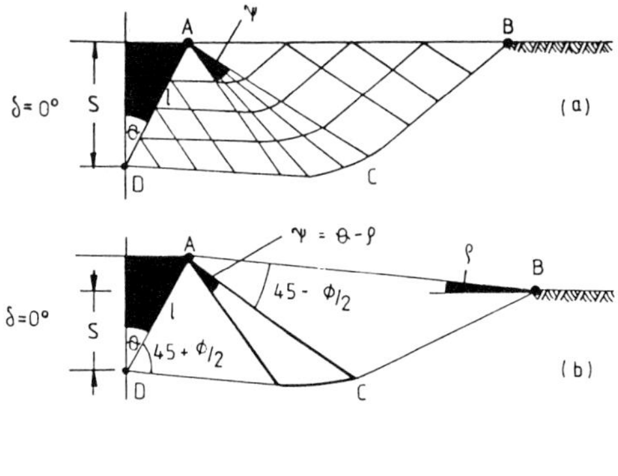

Figure 1. (a), (b).

The general case, in which the wedge-face friction

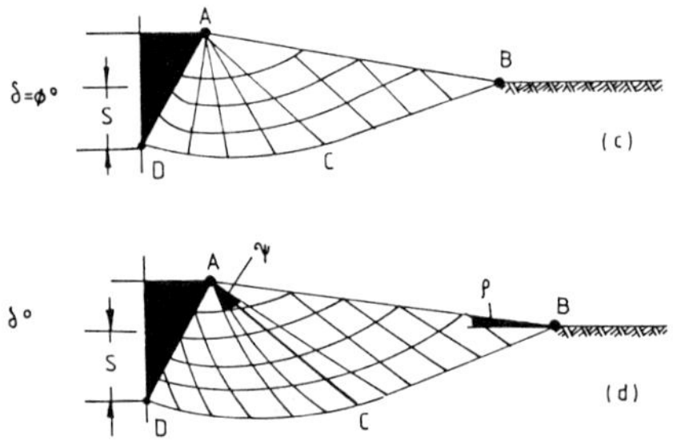

Figure 1. (c), (d).

angle (δ) is $0 \leq \delta \leq \phi$, is shown in Fig.1(d) which serves also to define symbols used elsewhere in the paper.

For a given wedge semi-angle (θ), ϕ, δ and ρ values a statically admissable slip-line field can always be constructed such that B lies on the horizontal surface BB', which is in effect equivalent to ensuring that, when δ = 0

$$\frac{s}{\ell} = \cos\theta - \frac{\cos(45^\circ - \phi/2)}{\cos(45^\circ + \phi/2)} \cdot \sin\rho \cdot \exp(\psi \cdot \tan\phi)\,; \qquad \psi = \theta - \rho \qquad (1)$$

The kinematic theorem quoted above allows us to construct a consistent velocity hodograph for these cases as sketched in Fig.2(a). It immediately follows from this that positive power of dissipation has been ensured and that the incremental volume of material displaced by AD moving vertically downwards is equal to that uplifted by the uniform displacement of AB (Fig.2b). Such consistent hodographs (for identical slip-line fields) can also be drawn for dilating materials in which the dilatancy angle ν > 0 by

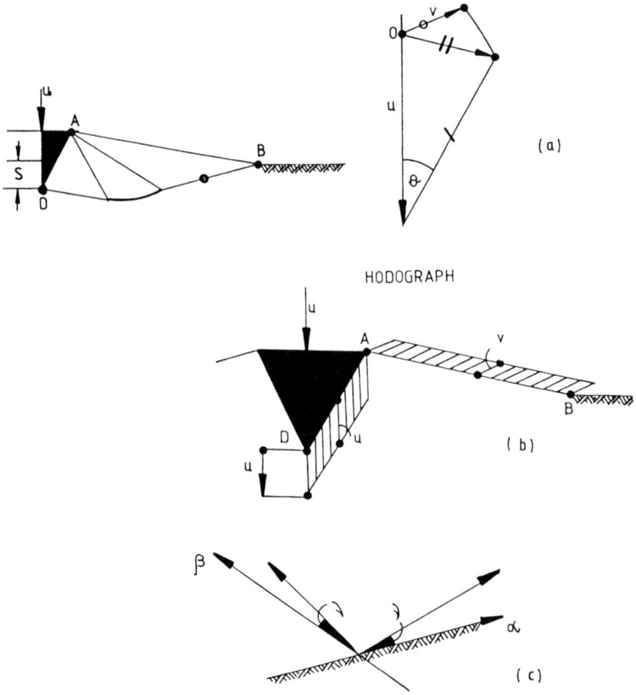

Figure 2.

extending the kinematic statement above to require that all velocity increments should be inclined at ν to the characteristics, Fig.2(c). (Butterfield and Harkness, 1971).

When the material has weight ($\gamma > 0$) both the slip-line field and the hodograph become more complex (Butterfield and Andrawes, 1972). Thus Fig.3 illustrates the full solution for $\theta = 30^\circ$, $\phi = 42^\circ$, $\delta = 12.5^\circ$ and $\rho = 3^\circ$ with the slip-line field in Fig.3(a), the non-dilating material hodograph ($\nu = 0$) in Fig.3(b) and a dilating material hodograph ($\nu = 21^\circ$) in Fig.3(c). All the slip-lines (outside the Rankine zone ABC) become curved leading to an 'expanded' hodograph in which, for example, the velocities of particles along the free surface AB map as points along the curve A'B' in it. Therefore, after the displacement, AB becomes a curved surface and this solution cannot be applicable to a further incremental displacement of the wedge.

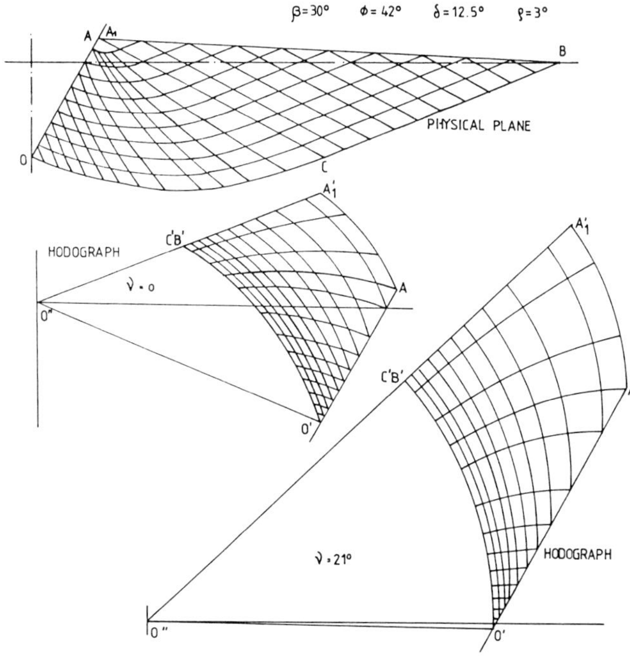

Figure 3.

3. CONTINUOUS DEEP PENETRATION OF A FINITE WEDGE

In this case the wedge (possibly incorporating a suitably shaped tail, see below) is deeply submerged and the depth, for our simple material model, will not be a kinematically relevant parameter.

An obvious question is then whether or not the previous solutions can be extended to encompass this problem, with B located along the 'penetrometer' axis (Fig.4). This in turn, for specified θ, ϕ and δ values on DA and AB (ν has to be zero!), establishes a unique value of the top apex-angle (β) between AB and the penetrometer axis.

For example, when $\delta = 0$, via the relationship,

$$\sin\beta \cdot \exp(\psi \cdot \tan\phi) = \sin\theta \cdot \frac{\sin(45^\circ - \phi/2)}{\sin(45^\circ + \phi/2)}; \qquad \psi = 90^\circ + \beta + \theta \qquad (2)$$

Fig.4(a) shows the slip-line field for such a case (Butterfield and Last, 1983) together with a consistent hodograph which confirms that the volume of material displaced ahead of AD fills the equal-volume void developed behind AB at each displacement increment and that the material crossing AB, as it moves, does so with uniform velocity so that AB remains straight. Additionally, and crucially, XY in the hodograph is parallel to AB, i.e. relative to the penetrometer the material is sliding along AB which can therefore be a solid surface (in this case smooth). It is these latter conditions which preserve the geometry of the void occupied by the solid body and

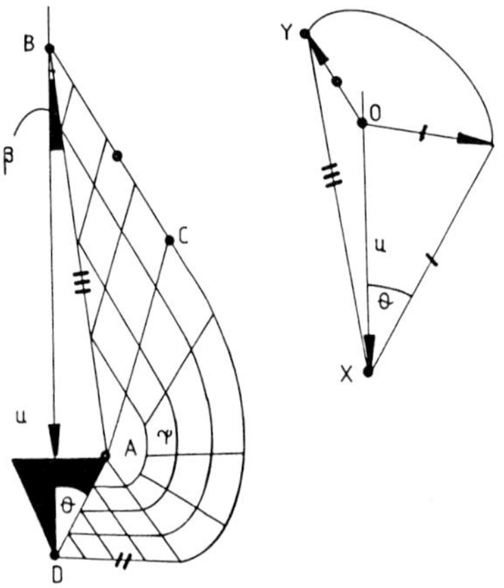

Figure 4.

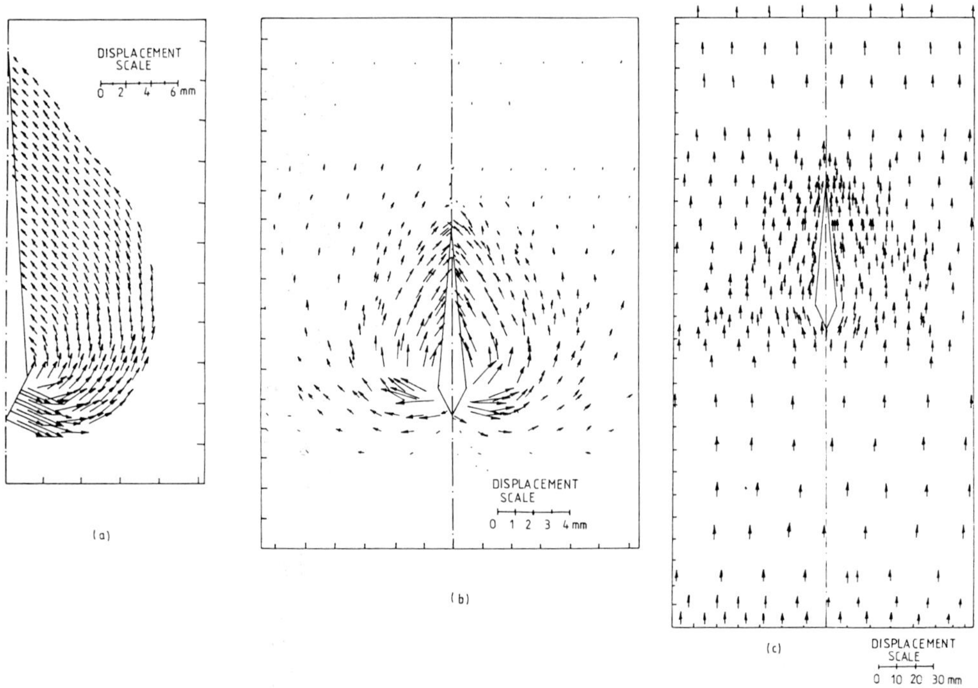

Figure 5. (a) Displacement resultant calculated from the hodograph presented by Butterfield and Last (1983) ($\Phi = 42^\circ$ and PEN = 6.0 mm). (b) Displacement resultant vectors (D/B = 20.0, e = .52 and PEN = 6.0 mm. (c) Displacement resultant vectors. Stationary probe (D/B=20.0, e=0.62 and PEN=6.0 mm).

thereby validate the steady-state solution.

It is also of interest to note that, if we fix the penetrometer by subtracting its velocity (u) from the hodograph (which transfers the velocity origin to X in Fig.4b) the solution becomes that for a stream of frictional fluid flowing past a smooth diamond-shaped pier. $\delta > 0$ can be incorporated into the solution as in the previous examples. The streamlines for this case are displayed in Fig.5(c). Fig.5(a) shows the velocity vectors derived from the Fig.4(b) hodograph ($\phi = 42^{\circ}$) and Fig.5(b) those measured experimentally on a model penetrometer test in a dense bed of sand (Mahmoud, 1985).

4. THE SELF-SIMILAR PENETRATION OF AN 'INFINITE' WEDGE

The third set of solutions concerns the continuous penetration of a wedge-shaped body from the surface of the bed. Such solutions will clearly have something in common with those of Section 2, except that ρ will not now be arbitrary since the geometry of the uplifted surface will have to satisfy, at least, the condition that the TOTAL

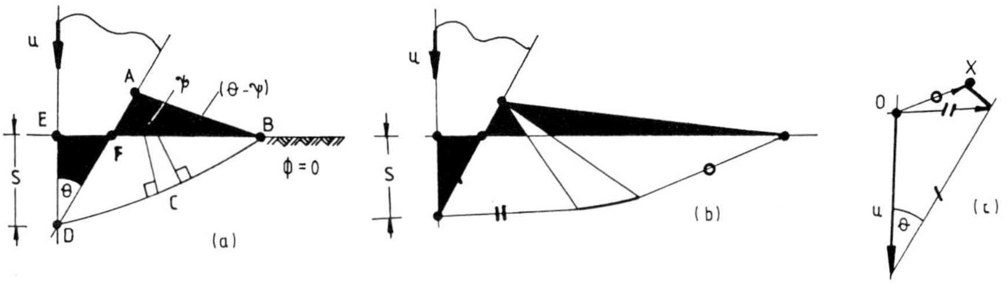

Figure 6.

volume of material displaced by the wedge, from the onset of motion, equals the TOTAL volume of uplifted material (i.e. that the area of triangle DEF = that of triangle ABF, Fig.6a).

Hill *et al.* (1947) studied the $\phi = 0$, smooth wedge case and showed that the above requirement, with B on the material surface, necessitates a unique ρ value such that

$$\cos(\theta + \rho)(1 + \sin\psi) = \cos\psi; \qquad \rho = \theta - \psi . \qquad (3)$$

The equivalent, smooth wedge, non-dilating frictional material condition is that,

$$\cos(\theta+\rho)(1+f\cdot\sin\psi) = \cos\psi - \frac{(1-f^2)}{2f}\cdot\sin 2\theta ; \qquad \rho = \theta-\psi \qquad (4)$$

with

$$f = \frac{\cos(45^{\circ} - \phi/2)}{\cos(45^{\circ} + \phi/2)}\cdot\exp(\psi\tan\phi) .$$

This solution should be distinguished from that published by Shield (1953) which related to a smooth, dilating, associated-flow model of Mohr-Coulomb material.

However, for the self-similar continuous penetration problem we have the additional kinematic requirement (analogous to the invariant geometry of the slip-line field in the steady-state problem of Section 3) that, at all stages of penetration, the slip-line fields shall be exactly scaled replicas of each other (i.e. self-similar). Such a problem (with $\phi = 0$ and weightless material Fig.6a) was first solved by Hill *et al.* (1947) and the following is an extension of his solution to include friction, material weight and dilatancy.

If we consider the general case in which the wedge penetration is s, a typical point in the material is located by $\mathbf{r}$ and moves with a velocity $\mathbf{v}$ then we can define the whole process by a 'unit diagram' (Hill *et al.*, 1947) in which s = 1, our typical point in this diagram is now located by $\boldsymbol{\rho}$ and moves with a suitably determined velocity $\mathbf{u}$. The velocities of all points in such a unit diagram, which represents any penetration stage scaled in the ratio 1:s, must therefore ensure that its geometry remains invariant during the motion.

For everywhere on the wedge, and in the supporting medium, we have,

$$\mathbf{r} = s \cdot \boldsymbol{\rho}$$

and hence

$$\mathbf{v} = \frac{d\mathbf{r}}{dt} = \boldsymbol{\rho} \cdot \frac{ds}{dt} + s\frac{d\boldsymbol{\rho}}{dt} = \boldsymbol{\rho} \cdot \frac{ds}{dt} + s \cdot \mathbf{u}$$

in our unit diagram s = 1 and therefore,

$$\mathbf{u} = \mathbf{v} - \rho \cdot \frac{ds}{dt}$$

We can also, without loss of generality, specify the wedge velocity, $\frac{ds}{dt}$, to be unity whence,

$$\mathbf{u} = \mathbf{v} - \boldsymbol{\rho}; \qquad s = \frac{ds}{dt} = 1 \tag{5}$$

The 'unit hodograph' for the wedge-medium system will then display $\mathbf{v}$ for all points within it. Fig.6c shows such a hodograph for the slip-line field of Fig.6b.

We now superimpose the unit hodograph on the unit diagram, both drawn to the same 'unit' scale, Fig.7a, on which the value of $\mathbf{u}$ for any point in it follows from equation (5) as the difference between vectors $\mathbf{v}$ and $\boldsymbol{\rho}$, Fig.7b. If the geometry of the unit diagram is to be preserved then not only will points within it corresponding to $\mathbf{v} = 0$

move radially towards the coordinate origin (with **u** = −**ρ**) but also all points along the free surface AB must move in the direction of their local tangent, Fig.7c.

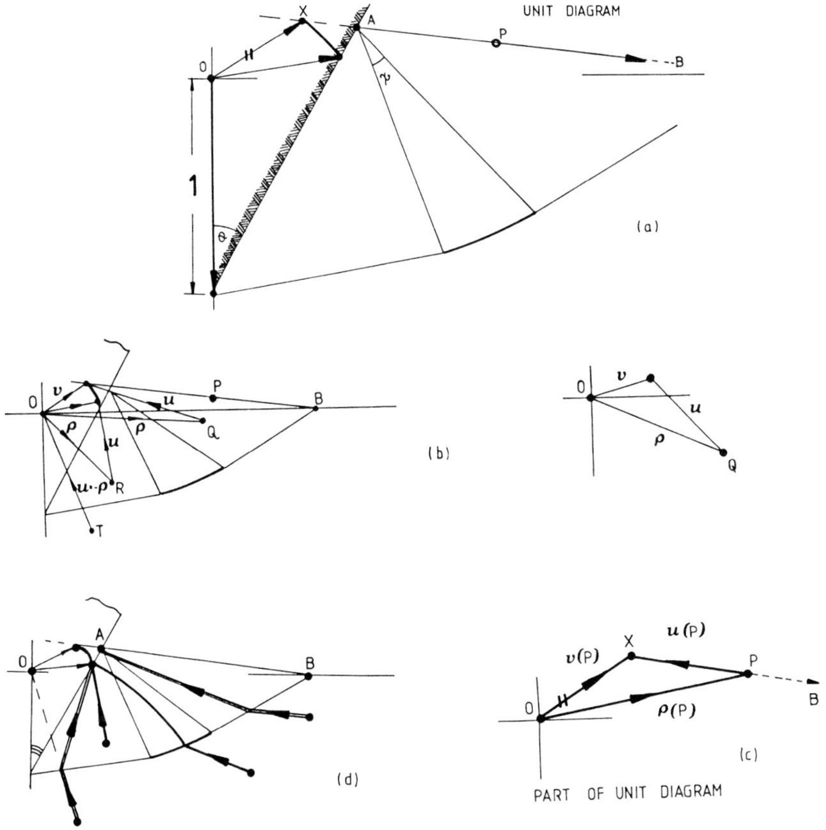

Figure 7.

Since in the present case all points along AB move with the same velocity (OX), AB has to be a straight boundary and X has to lie on AB produced. Whence, from equation (6), **u**(AB) = **OX** − **ρ**(AB). From the geometry of Fig.7a this condition requires that, again θ, ϕ and ψ be interrelated by equation (4) and therefore the unit diagram, hodograph and slip-line fields shown in Figs. 6(a,b) and 7(a) are a correct solution to the continuous penetration, self-similarity problem posed. Typical trajectories of particles in the unit diagram are sketched in Fig.7(d).

It is important to note how such trajectories 'focus' on their corresponding images in the unit hodograph.

We now extend this solution to include the case in which the wedge is no longer smooth (δ > 0) and the material has weight (γ > 0).

The $\delta > 0$ requirement does not introduce any new ideas, it merely changes the angle at which the slip-lines meet the face of the wedge (Fig.6b). The inclusion of material weight does, however, radically affect the self-similar solution. Whereas it will clearly resemble that shown in Fig.3 self-similarity can only be preserved with such an 'expanded' hodograph if the free surface is no longer plane. The precise form of AB is, of course, not known initially and (although the methods used to generate Fig.3 will equally well produce a solution for any assumed shape of AB) only when the tangent at any point (P) along AB passes through the image of P, at P' in the superimposed unit hodograph, will self-similarity be preserved. This is clearly a trial and error process. The salient features of such a solution are sketched in Fig.8.

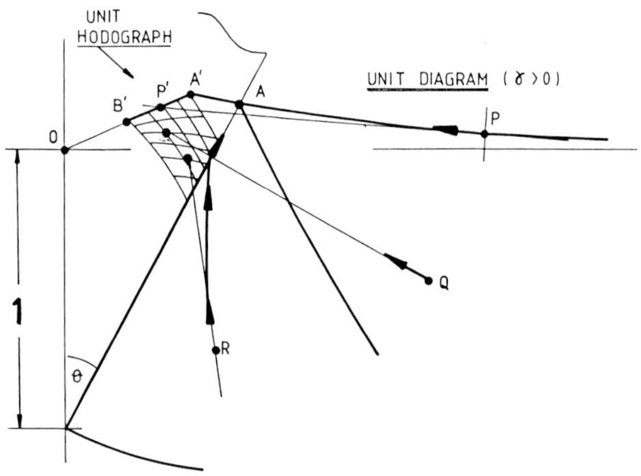

Figure 8.

Fig.9 shows an accurately recorded surface profile obtained by driving a steel wedge ($\theta = 30^\circ$) into a bed of dilatant, brass rods ($\phi = 25^\circ$) Andrawes (1970). It is entirely plausible that tangents to this curve will pass through corresponding points in a unit hodograph of the form described above and a tentative solution is sketched

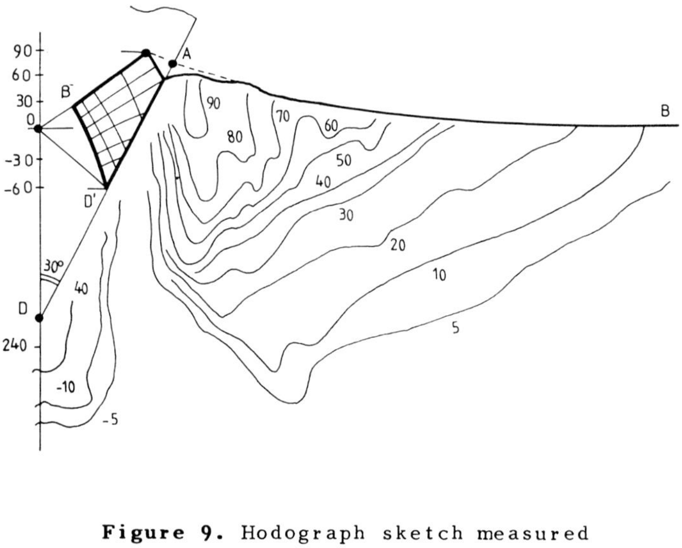

Figure 9. Hodograph sketch measured surface profile and vertical displacement contours.

in the figure.

Measured displacement contours (for a displacement increment of about 7% of the penetration depth) are plotted in the figure. These are seen to agree reasonably with the predictions of even the very crude hodograph.

It is worth noting that different dilatancy angles can be incorporated quite easily in different regions of the system (for example solely around the bounding velocity discontinuity BCD) as explained in Butterfield and Harkness (1973).

More fundamentally, it follows from the kinematic specification of the material that in systems with a 'free-surface', such as depicted in Figs.1 and 3, a single slip-line field can be valid for all admissable values of ν (Butterfield and Harkness, 1971) whereas the steady-state flow solution (Fig.4) is only meaningful for $\nu = 0$. However, because the hodograph and the slip-line field geometries are intimately coupled in self-similar problems (Fig.7) there is a unique consistent slip-line field for each value of ν.

5. CONCLUSIONS

The paper distinguishes between incremental, steady-state and self-similar, planar problems of arbitrarily rough, rigid wedges penetrating rigid-plastic, Mohr-Coulomb materials. It emphasises particularly the kinematic boundary conditions, and hodographs, which establish the validity of any solutions to such problems involving either steady-state flow or self-similarity. In particular a new study of self-similar continuous penetration is presented which can generate fully consistent solutions for dilating, Mohr-Coulomb materials with weight.

6. REFERENCES

Andrawes, K.Z. (1970) *A contribution to Plane Strain Model Testing of Granular Materials*, Ph. D. Thesis, University of Southampton, U.K.

Butterfield, R. and **Andrawes, K.Z.** (1972) A Consistent Analysis of a Soil Cutting Problem, *4th. Intl. Conf. for Terrain Vehicle Systems*, Stockholm.

Butterfield, R. and **Harkness, R.M.** (1971) The Kinematics of Mohr-Coulomb Materials, in *Stress Strain Behaviour of Soils*, Pub. G.T. Foulis.

Butterfield, R. and **Harkness, R.M.** (1973) Idealised Granular Materials, *Symp. Plasticity and Soil Mechanics*, Cambridge, U.K.

Butterfield, R. and **Last, N.C.** (1983) Continuous Penetration Testing in Granular Materials, a New Analytical Solution, *Intl. Symp. on in-situ Testing*, Paris.

Mahmoud, A.M. (1985) *Continuously penetrating bodies in granular media*, Ph. D. Thesis, University of Southampton, U.K.

Geiringer, H. (1931) Beitrag zum Vollstandigen abenen Plastizitas Problem, *Proc. 3rd Intl. Cong. App. Mech. Vol. 2*, Stockholm.

Hill, R., Lee E.H. and **Tupper S.J.** (1947) The Theory of Wedge Indentation of Ductile Materials, *Proc. Roy. Soc. Series A*, Vol.**188**.

Shield, R.T. (1953) Mixed Boundary Value Problems in Soil Mechanics, *Quart. Appl. Math.* Vol. **11**.

SLIP SURFACES IN SOIL MECHANICS

P. Habib
Laboratoire de Mechanique des Solides (Joint Laboratory E.P.-E.N.S.M.P.-E.N.P.C., Associated with CNRS),
Ecole Polytechnique-Palaiseau, France

Abstract. Isolated slip surfaces occur in softening soils, that is to say mainly stiff clays and dense sands. They do not occur for all plastic deformations.

A mechanism of progressive development of discrete slip surfaces in a heterogeneous stress field is first described, in good agreement with actual physical observations. This mechanism makes it possible to analyse mechanical strength tests and to show that the validity of the shear test is especially assured for residual strength. For the triaxial test, the peak value is seen to have physical significance, but the softening slope is shown to be function of sample size, also residual strength is not correct when a localization of strains occurs. A correction is proposed.

The occurence of slip surfaces questions the validity of classical formulae. Approximate practical corrections are proposed for bearing capacity of shallow foundations on sand and for active and passive earth pressures.

I. INTRODUCTION

The localization of quasi-static deformation on one or several slip surfaces is a very common phenomenon in soil or rock mechanics (Fig.1). However, it has not been widely reported and the classical theory of Plasticity disregards heterogeneous deformation (Mandel, 1966). Nevertheless, it remains an important subject but a difficult one, since observations of the physical phenomenon of the actual occurrence of the localization are hard to carry out, soil not being transparent.

In fact, we only have knowledge of slip surfaces when they emerge. Practioners know that research by drilling and trenching on a slip surface in depth, after an accident, is a costly and sometimes disappointing operation, since the interfacing surfaces have a regrettable tendency to rebond perfectly. As for the progression of slip surfaces themselves, the observations are partial and isolated, and it is only very recently that a few data have been collected on the kinetics of the development of slip surfaces. Suemine (1983) indicated rates of progression of deep slip surfaces of around 1 to 100m/h.

Figure 1. Slip lines in granite. The surfaces in contact were opened by weathering.

Figure 2. Two families of slip lines with reject on the first when the second one moved. Note the uncertain parallelism of the second family lines orientation.

Even for an apparently quite elementary and accessible situation, as for example a uniaxial or triaxial compression test, the orientation of one or several slip surfaces, in relation to the direction of main stresses, remains imprecise. Available reports on this subject are sometimes contradictory. Indeed, the identification of the slip lines presents serious difficulties. One reason is that they materialize only on the lateral surface of a test piece after a strain of about 3-10% for soils, so that the reference axes are themselves deformed. We no longer know whether to relate the slip surface to the initial configuration or to the subsequent one.

Moreover, there is a certain scatter in the orientation of slip lines. This is particularly obvious on Figure 2. In an initially homogeneous field there are now two conjugate families of slip surfaces, one of which was obviously active later than the other. This has caused rejects to form.

In addition certain surfaces of the second family show doubtful parallelism, somewhere around 15°. On Figure 3 the identification of the two conjugate families raises complex problems.

Figure 3. Scatter in orientations of slip surfaces in a granite rockmass.

Figure 4. Landslide at Villerville (B.R.G.M. photo).

This dispersion in the orientation is easy to understand. Coulomb defines the orientation of the failure plane by the calculation of the strict minimum of a certain strength along this plane as a function of its angle of incline. It is quite obvious that the neighbouring inclined planes are almost as unfavourable as the critical plane. If there is a slight heterogeneity of the material, a slip surface path will hesitatingly form. In fact, a failure surface is made up of a series of small elementary facettes whose mean direction gives the general orientation.

However, the problems raised by slip surfaces are important. It is well known, for example, that the epicentres of earthquakes are situated on active faults, in other words, on slip surfaces. At a more modest scale, but just as catastrophic, Figure 4 shows the effect of a slip surface on a building. This spectacular accident is particularly remarkable, in as much as the slip displacement is about one metre, whereas the slip emergence (Figure 5) occurred at a distance of about

Figure 5. Landslide at Villerville. Slip surface emergence (B.R.G.M. photo).

250 meters. If the deformation had been spread instead of being localized, there would have been practically no damage to the house.

Finally, the very nature of slip surfaces is debatable (Duthilleul 1983, Desrues 1984). In plane problems, do the slip lines merge with the characteristic stress lines, as was belived over many years, or with the lines characteristic of the rates of deformation (lines of zero extension) as Roscoe (1970) suggested? Although the study of this phenomenon seems very difficult, we shall see how interesting it is and the important consequences it generates.

There is a first source of surprise: in certain cases, soil failure occurs with the formation of a slip surface, but in other cases, the exceeding of the threshold resistance occurs with a continuous deformation and no localization. For example, slip surfaces are observed with certain soils (stiff clays or dense sands), but not with others (compressible materials, muds or loose sands). Sometimes, they occur under moderate mean stresses and disappear under high stresses; this is the case of moderately dense sands. If we consider rocks, the brittle-ductile transition passes through an intermediate state when slip surfaces form. They are well identified for rocks and especially correspond to faults and geological overthrusting (Goguel, 1983).

But for certain stress fields, for materials where slip surfaces usually occur, there are cases where failure surfaces appear, as for example landslines, or on the contrary, do not occur, as under the toe of a pile. All this shows how difficult the problem becomes, since the criterion for slip surface formation must take into account, at one and the same time, the constitutive equation of the material, the mean stress, the type of stress field, and probably boundary conditions.

We shall first examine what happens in a homogeneous stress field before studying the general case.

II. SLIP SURFACES IN A HOMOGENEOUS FIELD

Experience has contibuted to identifying the following phenomenon in a homogeneous stress field: if the consitutive equation of the material indicates hardening, the deformation is homogeneous; if the constitutive equation indicates softening, there is formation of a slip surface. The classical interpretation then follows: the real material is never strictly homogeneous and there is slight scatter of the mechanical characteristics represented by the set of stress-strain curves in Figure 6, which indicate the behaviour of a series of potential slip planes in a test sample.

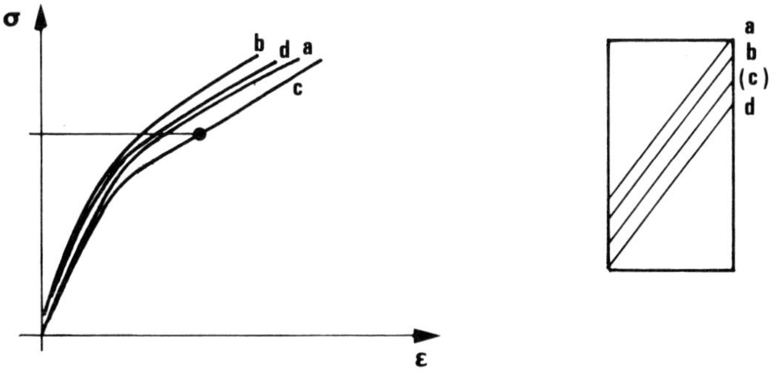

Figure 6. Case of hardening.

Let (c) be the weakest plane. Under stress σ it undergoes a certain slip. An increase in deformation around this plane can only occur through an increase in stress which, in turn, generates the increase in deformation near neighbouring planes. Even if deformation is less there than in the plane which failed first, it is seen to extend and develop throughout the test piece.

On the contrary, if the behaviour curves show a maximum (Figure 7), there exists a plane where the greatest strength is reached before the others.

If the deformation increases, the stress begins to decrease in this plane. This brings about a relaxation in all the test pieces. Young's modulus of a soil is much higher at unloading than the first

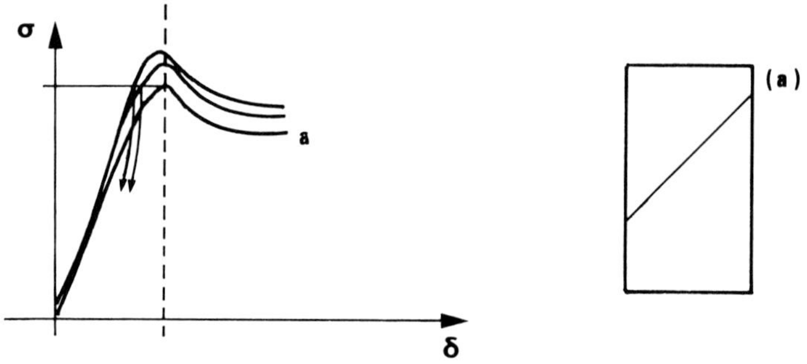

Figure 7. Case of softening.

loading modulus. Then the volumes not affected by large deformations now behave like rigid blocks which slide along the slip planes, so there is localization of deformation.

The stress-strain curve in Figure 7 has no longer any real significance after the maximum because the deformation is not continuous. It must be replaced by a stress-displacement diagram. For a triaxial test on dense sand, the diagram of the volumetric strain $\frac{\Delta V}{V}$ as a function of deformation (Figure 8) loses its significance after the strength maximum and must be replaced by the curve of the variation of volume ΔV as a function of displacement δ.

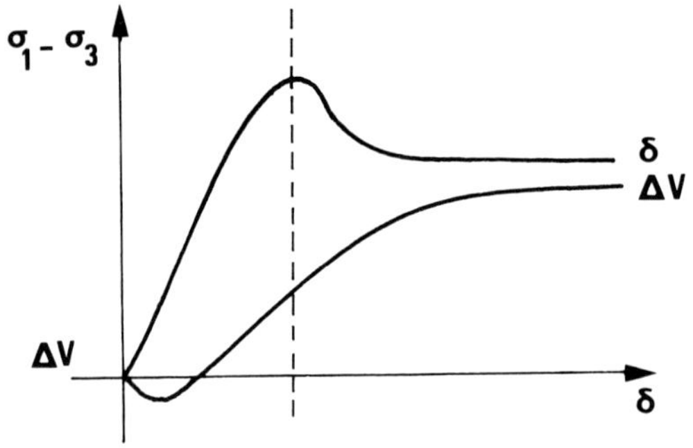

Figure 8. Volume variation of a dense sand during failure (Triaxial test).

This remark is important because, although the formation of a slip surface is perceptible on the stress-strain curve (since it appears and develops physically once the strength maximum is exceeded), on the other hand, it can not be detected on the volume variation curve. Nevertheless, the volume variations are localized at a given time in the slip plane, whereas they ceased in the monolithic blocks as soon as relaxation started.

III. SLIP SURFACES IN A NON-HOMOGENEOUS FIELD

We will now examine what happens in a non-uniform stress field, taking the example of a solid subjected to any loading (Figure 9). When the loading increases, the failure criterion is first reached in a point A, then in a small domain $\mathscr{D}$ surrounding the point A.
If this domain is sufficiently small, the stress field can be considered uniform there; there will be a formation of slip surfaces if the body is subject to softening, and continuous deformation if it undergoes hardening. We will consider the case of Soil Mechanics with a sufficiently slight softening for the classical, elastoplastic solution to be considered as a good approximation of the problem studied *).

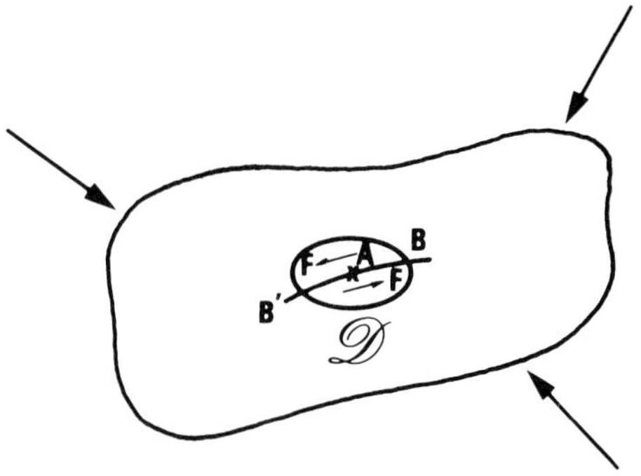

Figure 9. Crossing the plasticity criterion.

Let us examine what is happening on the slip surface AB of Figure 9 and extending in front (Figure 10). At a certain distance, the straight line (a) perpendicular to the crack is practically not deformed. The deformation is greater for the line (b) when approaching B. It is maximum for the line (c) which passes through B. Between A and B a slip occurs between the lips of the crack which crosses the lines (d) and (e). The slip δ increases from B towards A. The displacement $\delta(s)$ and the corresponding shear t(s) are plotted on Figure 10 (assuming $\phi = 0$ to simplify the explanation, but it would be easy to generalize for the case where $\phi \neq 0$).

In a classical elastoplastic solution the shear would be constant along AB; there would be correponding shear force F in each half of the domain $\mathscr{D}$. In the case of softening there is a shear defficiency in the AB crack plane marked by the shaded area on figure 10, whose total value:

$$\Delta F = \int_{AB} (t_{mx} - t(s))\, \overline{ds}$$

an upper bound of which is $(t_{mx} - t_r)\ell$, where t_{mx} is the greatest strength. t_r is the residual shear strenght and ℓ the length of the segment AB. Equilibrium can only be ensured by a displacement of the

*) This would not generally be true in Rock Mechanics where collapse of strength after the maximum is very important.

crack tip towards the right, in fracture mechanics mode II with disturbance of the elastic field beyond this, so that with the approximations of Figure 10, the shaded areas be equal. The displace-

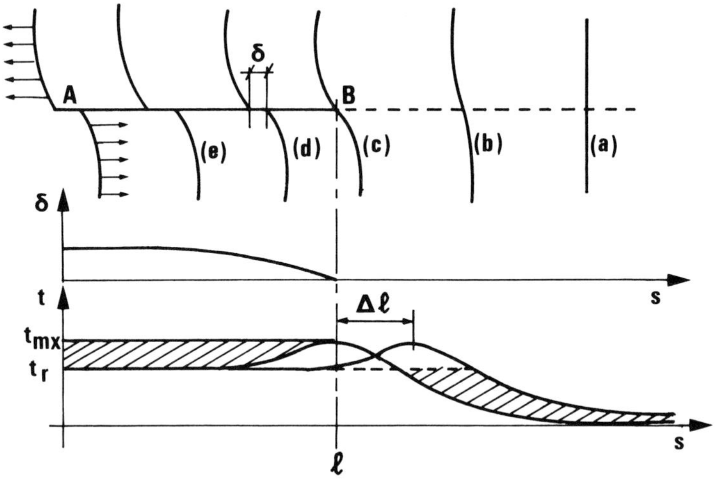

Figure 10. Displacement and shear stress near a slip surface.

ment $\Delta \ell$ of the crack tip allows compensation of the shear deficiency ΔF by bringing additional strength $\Delta \ell \cdot t_r$. An upper bound of the displacement $\Delta \ell$ is therefore:

$$\Delta \ell < \left(\frac{t_{mx}}{t_r} - 1 \right) \ell$$

or again:

$$\frac{\Delta \ell}{\ell} < \frac{t_{mx}}{t_r} - 1 \; .$$

If the softening corresponds to a residual strength of 0.9 t_{mx}, the lengthening of the crack which ensures reestablished equilibrium is 10%. This mechanism is only possible if the "ligament" is capable of resisting the increase of elastic stress, that is to say if the dotted area of Figure 10a is at least equal to the shaded area.

In the contrary case, yielding is total, the slip surface reaches the free boundary and failure occurs. However, if plastic deformation remains restricted, the preceding mechanism blocks the movement of the slip surface by allowing the deformation to concentrate on another slip surface. This may not be in the immediate neighbourhood of the first, since it is difficult to imagine the possibility of resisting t_{mx} there, whereas the first surface can now only resist t_r. The occurence of a multitude of discrete slip surfaces side by side can be expected. They will be difficult to see because they are situated inside the yielded mass, but they lead us back to the concept of quasi-homogeneous deformation. This complies with practical experience.

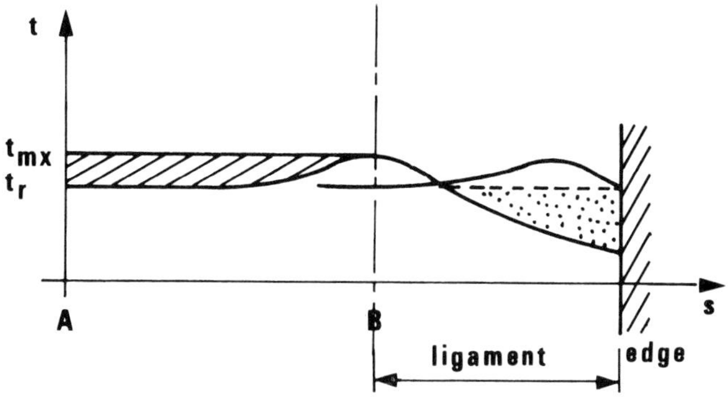

Figure 10a. Ultimate strength of the ligament.

Figure 11 represents a multitude of discrete Lüders lines at the surface of a stamped steel plate, which is a good picture of yielding around a circular section tunnel. Lüders lines are not slip lines, but extensive elongation lines. Contrary to slip lines which are caused by a process of condensation, Lüders lines, which occur in ceratin metals such as aluminium or mild steel, appear immediately after the elastic limitis exceeded. They then progressively widen to completely invade a component in the homogeneous field under the effect of hardening following the plastic stage. Their occurrence in the shape of a discrete double multiplicity on Figure 11 shows that slip lines would even more surely have localized in two distinct families. Other examples are seen later.

We must insist on the fact that progressive elongation of a unique slip line (or of a few slip lines) also correlates with observation, even if this approach is sometimes masked by certain ultimate design intuitions. The slice method, the Fellenius circle, Rendulic logarithmic spirals indeed accept as assumptions the movement between rigid blocks. In fact, the practice of punching foundations or that of very extensive natural slope failures, shows that the moving "blocks" can be subjected to large deformations. For smaller

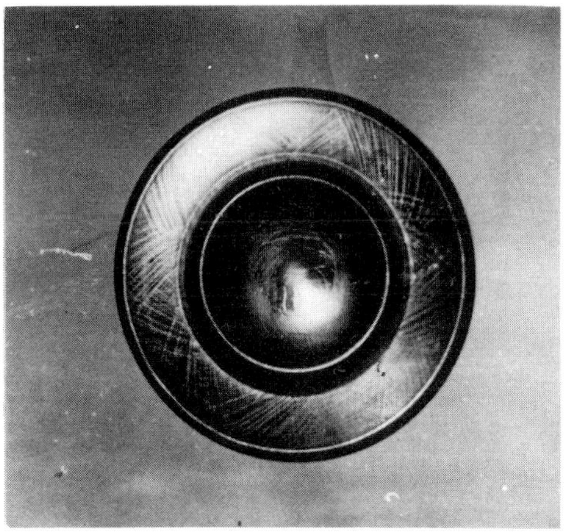

Figure 11. Multitude of discrete Lüders lines in mild stamped steel.

structures such as surface foundations or retaining walls, in active and passive earth pressure, the progression of force with the progression of the slip surface length has been shown and the development of the failure surface has been identified by many research scientists using very different techniques (γ densimeter to follow the development of the slip surface by its dilatancy, radiography of a mesh of lead beads placed in the earth mass, deformation of a mesh in plane deformation behind a pane of glass, and recently by Darve of Grenoble by photographic observation of a mesh traced at the surface of a mass of rollers following the Scheebelli-Dantu method, etc..). Figure 12,

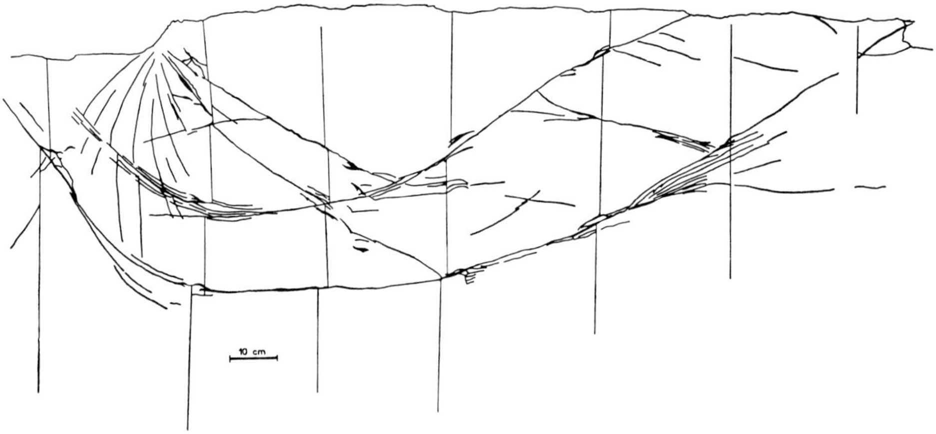

Figure 12. Slip surface under a foundation.

drawn from work by Chazy and Habib (1961) shows the development of a family of slip surfaces under a large punch, where the longest surface emerges just on the surface. It is clearly visible on this figure that the displacement descreases when the free surface is approached. By intuition it is possible to understand that the other slip surfaces have been stopped during development and that they appeared after the principal surface which envelopes them all.

Figure 13a and 13b illustrate some cases where isolated slip surface formation can be observed and others where the deformation remains quasi-homogeneous.

The case of expansion of the circular cavity within an earth mass (Figure 13b) or near the surface of it (Figure 13a) clearly shows the influence of the ligament which can resist or not the increase of stresses in the elastic range, that is to say which allows, or not, the development up to the free surface of an isolated slip surface.

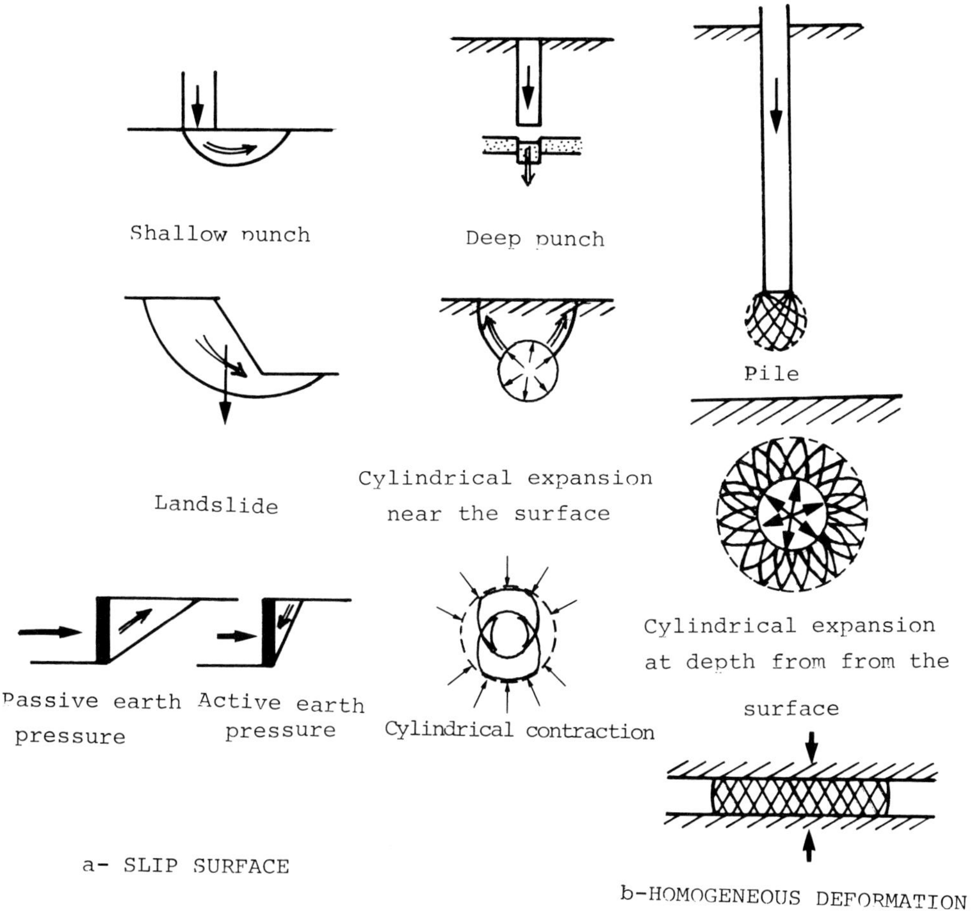

Figure 13. Different cases of deformation (the dark arrows indicate forces, the light arrows displacements).

IV. DIRECT SHEAR TEST

A classical way of measuring the shear strength of a soil is the rectilinear shear test with controlled displacement and particularly the Casagrande box test. It seems that this test method has rather fallen into disuse, without there being any good reasons to explain this. In general, direct shear is criticized for the fact that during the test, the main stresses change direction, whereas they remain fixed in the triaxial test. Also the intermediate principal stress in direct shear is variable during the test. It cannot be mastered and is not known, whereas in the triaxial test it is equal to the smallest of the main stresses, or sometimes to the greatest in a tensile test. The stresses and hence the deformations are perhaps not always homogeneous in the boxes, maybe because of the teeth on the horizontal faces of the half boxes which attach the test piece. This leads to a quasi-impossibility of ensuring the pore pressure during shear, contrary to the triaxial test. Finally, although this reason is not often acknowledged, a triaxial apparatus is undoubtedly less cumbersome and less costly than a shear machine...

However, the first three motives are not very convincing since the situation in the shear box often corresponds to real conditions. At the edge of extensive earthworks during excavation, for example, the direction of main stresses varies as work goes forward. The intermediate main stresses cannot be mastered and the deformation field is not uniform. The shear test (Figure 14) however, raises a

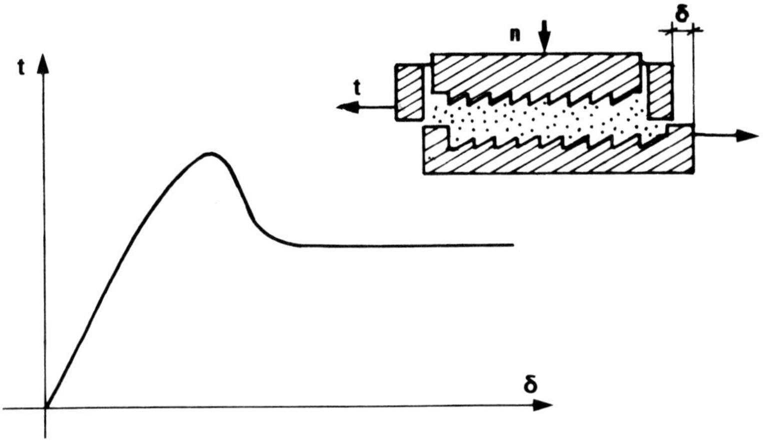

Figure 14. Direct shear test.

certain number of problems. When the curve (shearing force - box displacement) shows hardening, it is difficult to identify a slip surface in the box, and if the test is stopped in the course of experimentation, a lenticular area of deformation is found in place of a failure plane. If the curve presents a maximum, then a residual stress with which ϕ_{mx} and ϕ_r can be defined for a sand (or t_{mx} and t_r for a clay), we find that the strength maximum is reached typically for a displacement of a few millimeters in the classical Casagrande box. The slip plane dilatancy is erratic, sometimes starting at the front, sometimes at the back. This is shown by different swaying movements before obtaining complete dilatancy of the slip plane. Finally, the initial slope of the shear curve is difficult to understand and all the more impossible to interpret in terms of slip modulus since we cannot know to which thickness of slip "surface" it must be related.

Torsion of a cylinder is a much clearer experiment. On Figure 15 we have plotted the results of two torsion tests on a solid cylinder and a solid cylinder with a horizontal notch formed by a crack of zero thickness. The test pieces are made of wax. This is a $\phi = 0$ material with hardening. Consequently, if localization of deformations were to occur, it would take place along a horizontal plane (and not on a helicoid as in the case of $\phi \neq 0$). The comparison of the two failure curves shows behaviour as if we had added together the elastic curve of torsion on the solid cylinder (Figure 15a) and the rigid-plastic curve of the crack of zero thickness (Figure 15b). Moreover, during the tests, the generatrices of the cylindrical test piece coiled around the test piece in spirals. In the case of the notched test piece, the slip rejection both sides of the crack was practically imperceptible before the weakened section yielded.

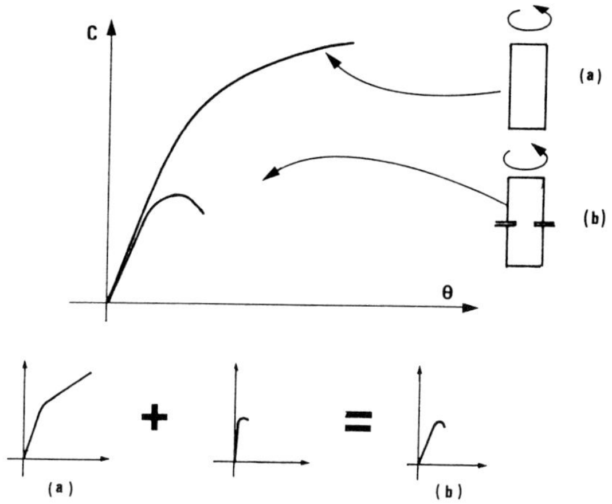

Figure 15. Torsion test on a notched test piece.

There was no rotation between the lips of the crack during the torsion test, which was normal since the thickness of the slip plane was infinitely small. Relative movements occurred only at the immediate approach of failure. The shape of the curve after the yield value very probably corresponds to kinetic hardening generated by the progression of yielding from the periphery of the test piece towards the centre in a deformation made non-uniform by kinematic effects.

Comparison between this torsion test and the shear test is not immediately possible since the progression of the crack in torsion corresponds to mode III propagation, whereas direct shear follows mode II. However, the analogy is obvious and the interpretation of the initial slope of the shear test stress-strain curve using the Casagrande box becomes very simple. A mode II progression of two cracks initiating on the front and back edges of the test piece occurs; they run towards the middle of the test piece. The erratic tiltings of the upper half-box correspond to unequal rates of progression according to the quality of the initial filling of the box. The initial slope of the stress-strain curve corresponds to the irreversible progression of this yielding and not just any elasticity phenomenon. By applying Figure 10 to the Casagrande test, we can produce a diagram of the shear stress distribution in the horizontal plane (Figure 16). The representativeness of this diagram to practical conditions increases with the thickness of the test piece. The teeth of the upper and lower porous stones are obviously aimed at transmitting the slip plane deformation in the most uniform way possible. However, maybe the effect of the attacking edges of the boxes is important.

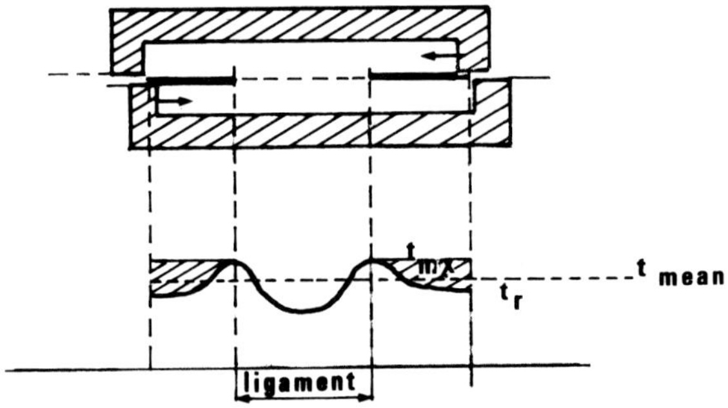

Figure 16. Distribution of shear stress in the failure plane of a direct shear test.

The mean value of the stress is, after all, the only magnitude measured, but it does not correspond to the maximum strength of the soil. There certainly exists a scale effect in this test. The progression of failure in a very large box would define a mean force coming increasingly closer to the residual value. If we assume that the test is undertaken with a series of increasingly thin test pieces giving interlocked stress-strain curves, the sample of zero thickness producing a stress-strain path is obviously the most representative (Figure 17).

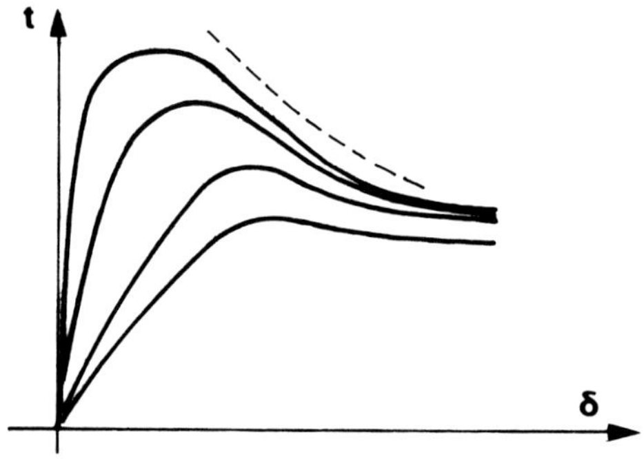

Figure 17. Experimenting to find an idealdirect shear curve

On the other hand, we observe that, after a certain displacement, the linear shear will give the residual strength value t_r. When the two cracks joint and the two humps in the shear diagram in Figure 16 are smoothed out, the value of t_{mean} merges completely with t_r. The linear shear test is perfectly adapted to measuring residual stregth.

V. THE TRIAXIAL TEST

Now we must see if the determination of maximum strength t_{mx} by a triaxial test will draw the same criticism as the linear shear test. If the stress field is uniform, the shear stress on a potential slip plane is constant. Accepting the existence of a weak point in this plane, Figure 10 is adapted to this case in Figure 18 and shows that

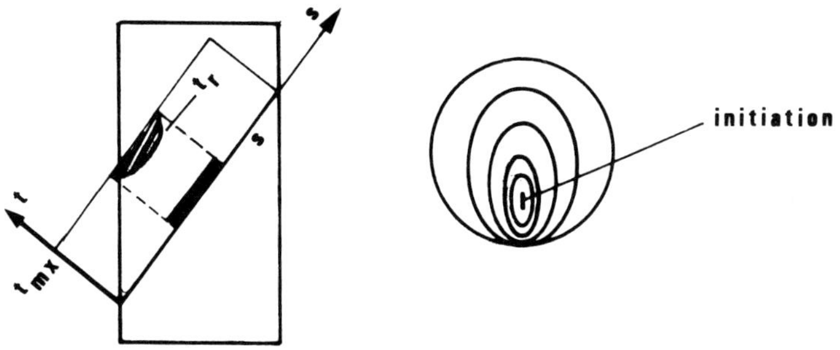

Figure 18. Progression of failure in a triaxial test piece and progressive development of the failure plane.

the deficiency in strength cannot be supported by the ligament. Failure should be of the brittle type as soon as onset occurs, and this is not so.

It is much more logical and in better compliance with experimentation to accept the development of a slip plane starting at a weak point situated near the base singularity. This development probably occurs by successive halos, as for any notch. This explains the formation of one or a limited number of slip planes and not of a failure cone as observed on brittle materials with deterioration, as in rocks or concrete under uniaxial compression. In the triaxial test the maximum value of this strength which develops progressively under compression is therefore prone to the same crticism as that of the shear test. Certainly the tests are perfectly repetitive, but does the maximum value measured in this way really correspond to the strength of the material?

Triaxial test on short test pieces were carried out to make sure. Test pieces with slenderness ratio between 2 and 2.5 were used in a routine triaxial test so that the central part was remote from the base hooping. This also left the possibility of free development of a slip plane (Figure 1). A slenderness ratio lower than 1.5 is acceptable for Paris Sparnacian clay test pieces where the failure plane is inclined at almost 45°. For Sannoisian clay test pieces the failure plane is inclined at about 30° on the test piece axis. If it were possible to impose supports with no friction, nothing would prevent the use of shorter test pieces. We will demonstrate that this statement must be limited to materials susceptible to hardening or at least to the part before the maximum on the stress-strain curve of materials prone to softening.

Many researchers have endeavoured to build friction-free supports for soils, concretes and rocks. The solution we have chosen for sand consists of placing a thin sheet of rubber resting on a film of grease between the test piece and the supporting steel point (Figure 19a). This solution is not perfect, but represents a good

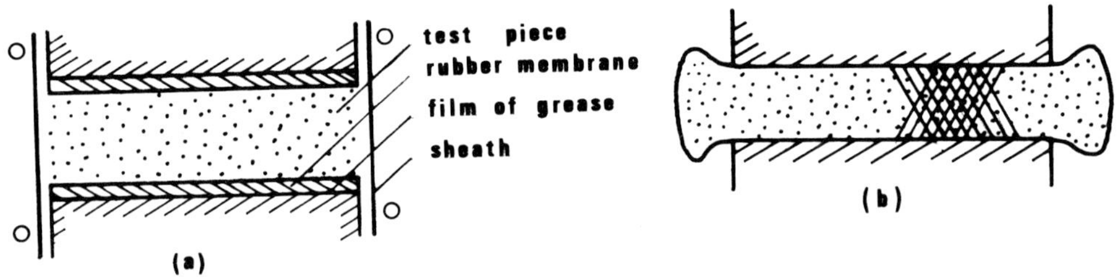

Figure 19. Triaxial test on a short test piece.

approximation of a normal stress at the contact surface.

The results obtained were as follows: during the crushing of a short sand test piece, the material is pushed sideways and a bulge forms round the initial cylinder which probably disturbs the end of the test (Figure 19b), when the deformation exceeds 15 or 20%. In the case of dense sand, a great number of slip lines appear on the bulge.

The stress-strain curves of short samples of dense sand show no softening for extensive deformation, whereas the curves for long test pieces of the same density present a maximum followed by softening, accompanied by an isolated slip plane, in other words with localization of deformation. The particularly interesting aspect of these tests was the following quantitative result: the value of the maximum strength of long test pieces was equal to the stage value for the short test piece (Figure 20).

Crushing tests on short test pieces were also carried out on a clay of low plasticity taken from sampling at great depth. Under uniaxial compression this material relevated localization of deforma-

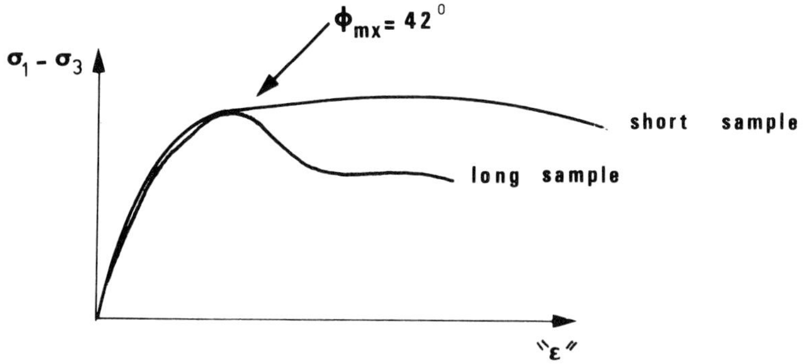

Figure 20. Triaxial tests on short and long sand test pieces.

deformation on a failure plane (Figure 21), but its behaviour was prac-

Figure 21. Localization of deformation in a stiff clay test piece.

Figure 22. Multitude of broken fragments in a short stiff clay test.

tically brittle with a quasicomplete loss of strength after failure. Considering the stiffness of the material, we tried to avoid the presence of a grease film in contact with the clay, likely to be pushed out by the pressure and to generate tension on the planes passing through the test piece axis. Another means of avoiding friction was used on the heads of short samples. It consisted of layers of aluminium confetti, lubricated by grease or molybdenum bisulphide. The clay-confetti contact does not influence tensile strength, so there is no hooping of the test piece base. The lubricant ensures normal stress with the confetti. Failure of short samples occurred with the onset of multiple fractures. It is presented on Figure 22 which clearly shows the anti-hooping device. The result of the tests is indicated on Figure 23. There is still a certain softe-

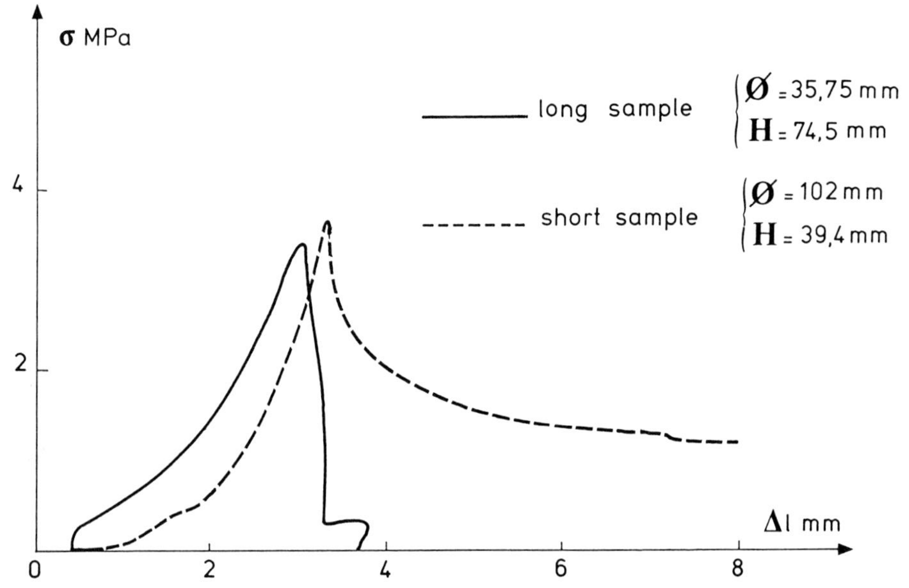

Figure 23. Uniaxial compression test on short and long test pieces of clay.

ning after failure of the short sample, but the value of the maxima is practically the same for the long test pieces and the short one with anti-friction protection.

Van Mier (1984) obtained similar results with triaxial tests on concrete. These results are obviously different from the one which would be obtained with short test pieces and friction at supports. We know that in this case, the strength is much higher (Hudson, Brown, Fairhurst 1971).

For sand and clays, it seems that the value of maximum strength obtained with the triaxial apparatus is independent of the shape of the sample. It therefore corresponds to an intrinsic property of the material. However, the physical significance of the slope of the softening curve remains uncertain.

In the course of this article we have thrown, then lifted,

doubt on the maximum strength value measured in a triaxial test and consequently on the ϕ_{mx} value of a sand. We will now show that the value of ϕ_r in the triaxial test is incorrect and generally calls for correction. In a triaxial test in compression on sand, the angle β of the slip surface with the test piece axis is all the smaller as the angle of internal friction ϕ is wide.
We can also say that β is all the smaller as the density of sand is higher or as the dilatancy increases. Without wishing to take sides as to the physical origin of slip lines and simply to be able to complete the calculation, we can accept the Coulomb equation $\beta = \frac{\pi}{4} - \frac{\phi}{2}$ (which means that the slip line is a line characteristic of the stresses). After softening, when dilatancy is accomplished, the residual angle of friction ϕ_r in the slip plane corresponds, for example, to the critical value ϕ_c, but the slip direction is still β. The Mohr circle corresponding to slip on this plane is therefore larger than in the case of a homogeneous test piece with friction angle ϕ_r. Figure 24 indicates the different magnitudes. The maximum strength on the stress-strain curve defines $\sigma_{max} - \sigma_3$, from which we obtain ϕ_{mx}; the strength corresponding to the test stage defines $\sigma_{1\,min} - \sigma_3$, giving an angle ϕ' which is greater than the residual strength ϕ_r.
A simple, but rather lengthy computation yields the following equation:

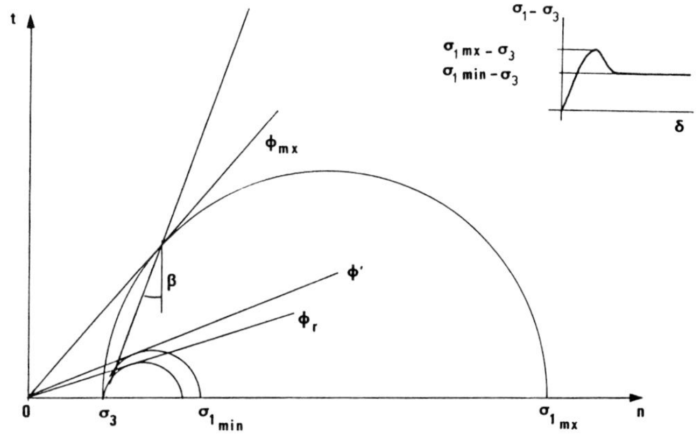

Figure 24. Mohr circles of maximum and residiual friction.

$$\tan^2\left(\frac{\pi}{4} + \frac{\phi'}{2}\right) = \frac{\tan\phi_r\left[\tan^2\left(\frac{\pi}{4} + \frac{\phi}{2}\right) + 1\right]}{\tan\left(\frac{\pi}{4} + \frac{\phi}{2}\right) - \tan\phi_r} + 1 \;.$$

With Figure 25 we can calculate the correction $\phi' - \phi_r$ to make to the ϕ' measured value to determine the value of ϕ_r.

This correction only entails a few degrees in normal cases of Soil Mechanics; it can become quite considerable in Rock Mechanics for wide maximum friction angles.

VI. PRACTICAL APPLICATIONS

We have just stated that in the Soil Mechanics laboratory the correction to make to the triaxial test to determine ϕ_r is small, around 1° or 2°, when a slip plane has settled in the test piece.

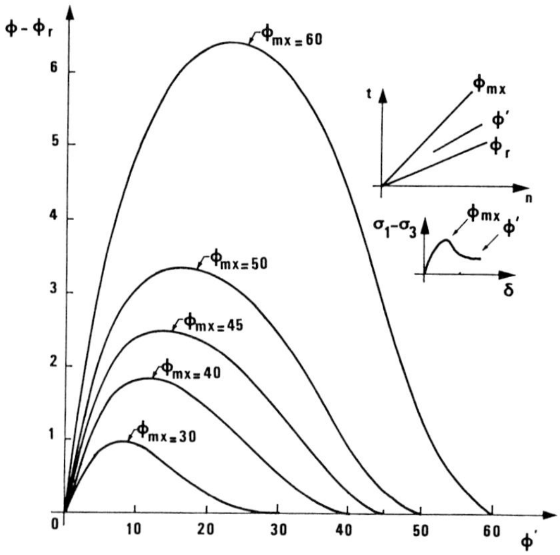

Figure 25. Correction (ϕ' - ϕ_r) for residual friction (triaxial test).

The same phenomenon can be of much greater importance when the slip surfaces are not plane and the stress fields not uniform, for instance the bearing capacity of shallow foundations on non-cohesive materials. Generalized slip occurs on a failure surface where the orientation and the path are functions of ϕ_{mx} (or of the dilatancy of the sand corresponding to ϕ_{max}. However, progressively with localization, the friction approaches critical ϕ, so we can simply say that the limit bearing capacity q_u, calculated by the classical formula of the surface term $\gamma\frac{B}{2}N\gamma$ lies between two values:

$$\gamma\frac{B}{2}N_\gamma(\phi_c) < q_u < \gamma\frac{B}{2}N_\gamma(\phi_{mx}).$$

A numerical example will make this clear : the Caquot-Kerisel tables (1956) give $N\gamma(\phi_c = 32^\circ) = 31$ and $N\gamma(\phi_{mx} = 40^\circ) = 114$, which show the order of magnitude of the theoretical uncertainty linked to the phenomenon of softening. It is obviously extremely difficult to reduce this range by a theoretical approach because of the discontinuous nature of deformation. However, there remains the possibility of proposing approximations.

If we look first at the upper bound, it is quite clear that the angle of internal friction ϕ_c is rapidly reached in the slip surface (or in the slip surfaces) and that ϕ_{mx} is not doubt only representative of the state of rigid blocks. By analogy with the computation used in the slip method, in other words by taking the shear strength of the failure surface, we can try to lower the upper bound value by the coefficient $\frac{\tan\phi_c}{\tan\phi_{mx}}$ to bring friction back to its true value.

The situation is similar for the lower bound. Whereas the angle of friction ϕ_c corresponds to the residual strength, the slip line is too short, so we can consider an increasing coefficient which is a function of the slip surface extension. We must remember that the width L of the side discharge is proportional to the width B of the foundation. Using the classical network of characteristic stress lines, we can formulate:

$$\frac{L}{B} = \tan\left(\frac{\pi}{4} + \frac{\phi}{2}\right) e^{(\pi/2)\tan\phi} .$$

There remains to define the function of L to choose for the increasing coefficient. Again by analogy with the calculation of slip circles, we can consider that L intervenes twice, first to define the length of the resisting arc, secondly to define the stress normal to the slip surface. Naturally, the two slip surfaces are not homothetic but the effective correction function is unboudtedly closer to L^2 than to L, which means that an increasing coefficient of the lower bound can be proposed as follows:

$$\left(\frac{L(\phi_{mx})}{L(\phi_c)}\right)^2 = \frac{\tan^2\left(\frac{\pi}{4} + \frac{\phi_{mx}}{2}\right) e^{\pi\tan\phi_{mx}}}{\tan^2\left(\frac{\pi}{4} + \frac{\phi_c}{2}\right) e^{\pi\tan\phi_c}} .$$

The two corrections proposed are indicated on Figure 26. They provide similar results, which cannot be decisive as a demonstration but they do ensure a certain degree of confidence in the approximation obtained. We must remember that Meyerhof (1961) had given a semi--empirical formula for the surface term. It was devoid of any theoretical pretention but in good correlation with the slip circle calculation, expressed as follows:

$$N_\gamma = (N_q - 1)\tan 1.4\phi = [\tan^2(\frac{\pi}{4} + \frac{\phi}{2})e^{\pi\tan\phi} - 1]\tan 1.4\phi.$$

So it is not surprinsing that the two corrections proposed show satisfactory agreement.

The same approach can be considered for other Soil Mechanics problems to propose corrections to classical formulae for materials prone to softening. We can take the example of active and passive earth pressure normal to a vertical wall retaining a horizontal mass of non-cohesive material. The active and passive pressure coefficients are:

$$K_o = \tan^2\left(\frac{\pi}{4} + \frac{\phi}{2}\right) \quad \text{and} \quad K_a = \tan^2\left(\frac{\pi}{4} - \frac{\phi}{2}\right) .$$

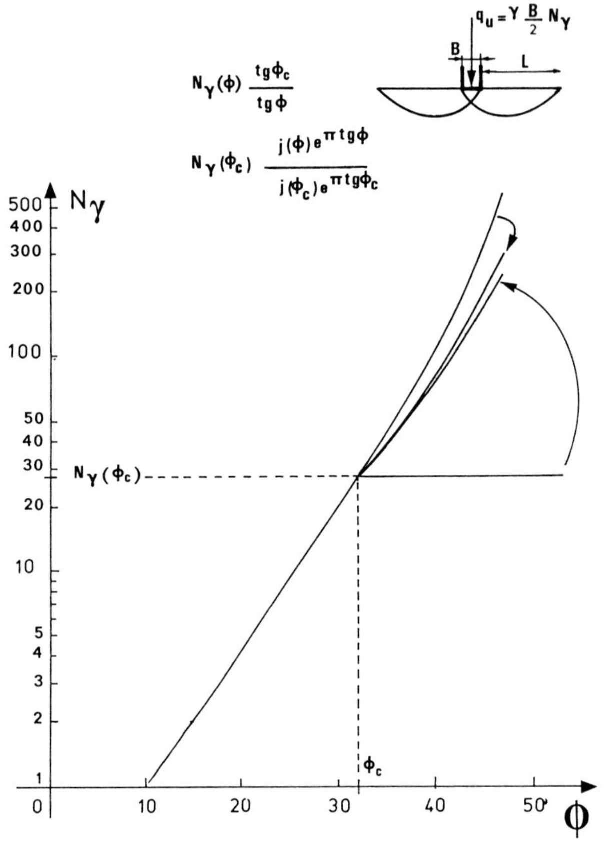

Figure 26. Corrections proposed for bearing capacity of shallow foundation on non-cohesive material with an internal angle of friction ϕ.

With passive earth pressure when the sand is dense enough for a slip surface to occur, the following range can be formulated:

$$\tan^2\left(\frac{\pi}{4} + \frac{\phi_c}{2}\right) < K_p < \tan^2\left(\frac{\pi}{4} + \frac{\phi_{mx}}{2}\right).$$

We can propose the same decreasing coefficient of the upper bound as before, that is :

$$\tan^2\left(\frac{\pi}{4} + \frac{\phi_{mx}}{2}\right) \times \frac{\tan\phi_c}{\tan\phi_{mx}}$$

and for the lower bound a coefficient function of the length $\mathscr{L} = H/\cos\left(\frac{\pi}{4} + \frac{\phi}{2}\right)$ (H being the height of the retaining wall). For this we simply take a linear function of $\mathscr{L}$ since there is no reason for the stress normal to the slip surface to be proportional to $\mathscr{L}$ (the depth of the slip plane is fixed by the foot of the wall). We can except an increasing coefficient expressed as:

$$\frac{\mathscr{L}(\phi_{mx})}{\mathscr{L}(\phi_c)}$$

and as a correction:

$$\tan^2\left(\frac{\pi}{4} + \frac{\phi_c}{2}\right) \times \frac{\cos\left(\frac{\pi}{4} + \frac{\phi_c}{2}\right)}{\cos\left(\frac{\pi}{4} + \frac{\phi_{mx}}{2}\right)} .$$

The two corrections proposed are indicated on Figure 27. They are

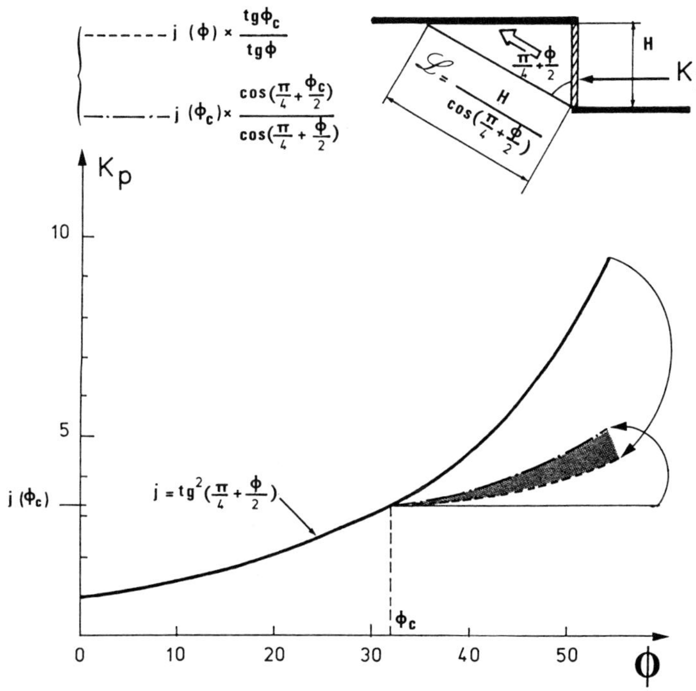

Figure 27. Corrections proposed for normal passive earth pressure on a wall retaining an earth mass of sand with internal angle of friction ϕ .

still close enough to each other to provide sufficient practical information.

The transposition to the case of active earth pressure is obviously immediate by changing $\left(\frac{\pi}{4} + \frac{\phi}{2}\right)$ to $\left(\frac{\pi}{4} - \frac{\phi}{2}\right)$.

VII. CONCLUSIONS

We have shown that the phenomena of deformation localization in one or several slip surfaces during failure in soils have important consequences in the laboratory and also in site practice. A correction is proposed for the determination of the residual shear strength in the triaxial test, but this test is especially interesting for determining maximum strength. In the same way, the direct shear test, either rectilinear or in torsion, is particularly useful and even irreplaceable for the determination of residual strength.

In practice, the occurrence of discontinuities at failure of an earth mass causes extensive discrepancy compared with classical behaviour and modifies the values calculated from the hypothesis of deformation homogeneity. The corrections proposed here for the surface punch and for active earth pressure are obviously rough approximations which call for more precise definition in the cases examined. They can also be extended to other routine applications in foundation engineering.

REFERENCES

Caquot A.; Kerisel, J., (1956), *Traite de Mecanique des Sols*, Gauthier-Villars, Paris, Ch. XVI, p. 389.

Chazy, C.; Habib, P., (1961), Les Piles du Quai de Floride, *5eme Congres Int. de Mec. des Sols*, Paris, juillet 1961, Com. 6/27, p. 669.

Darve, F.; Desrues, J.; Jacquet, M., (1980), *Les Surfaces de Rupture en Mechanique des Sols en tant qu'Irreversibilite de Deformation*, Cahiers du G.F.R., V. 3, janvier 1980.

Desrues, J., (1984), *La Localisation de la Deformation dans les Materiaux Granulaires*, These de Doctorat-es-Sciences, INPG Grenoble, mai 1984.

Duthilleul, B., (1983), *Rupture Progressive:Simulation Physique et Numerique*, These de Docteur-Ingenieur, INPG Grenoble.

Goguel, J., (1983), Etude Mecanique des Deformations Geologiques, B.R.G.M. Orleans, Manuels et Methodes n°6, ch. 6:Rupture Discontinue, Rupture et Glissement, p. 85.

Hudson, J.A.; Brown, E.T.; Fairhurst, C., (1971), Shape of the Complete Stress-Strain Curve for Rock, *13th Symp. on Rock Mechanics*, sept. 71, pp. 773-795.

Mandel, J.. (1966), *Mecanique des Milieux Continus*, Gauthier-Villars, Paris, Deformation Plastique Heterogene, tome II, p. 708.

Meyerhoff, G.G., (1961), Fondations Superficielles: Discussion, *C.R. 5eme Cong. Int. Mecanique des Sols* (Paris), tome III, p. 193.

Roscoe, K.H., (1970), The Influence of Strains in Soil Mechanics, *Geotechnique*, vol. **XX**, 2, june 1970, pp. 129-170.

Suemine, A., (1983), Observation Study on Landslide Mechanism in the Area of Crystalline Schist (part I). An Example of Propagation of Rankine State. *Bull. of the Disaster Prevention Inst.*, sept. 83, vol. 3, Part 3, pp. 105-127, Kyoto University, Japan.

Van Mier, J.G.M., (1984), Complete Stress-Strain Behaviour and Damaging Status of Concrete under Multiaxial Conditions, *Int. Conf. on Concrete under Multiaxial Conditions*, Universite Paul Sabatier, mai 1984, vol. I, p. 79, Toulouse (France).

II. EXPERIMENTS AND APPLICATIONS

UNDRAINED CREEP DEFORMATION OF A STRIP LOAD ON CLAY

A.F.L.Hyde*) and J.J.Burke**)

*) Loughborough University of Technology
**) IBM (United Kingdom) Ltd

SYNOPSIS

Time-dependent creep effects can play an important role in the stress distributions and deformations of foundations. Using a phenomenological model, the analysis of undrained creep behaviour has been introduced into an elasto-plastic finite element programme. The creep behaviour of a strip load on a finite layer of soil has been illustrated and a study has been made of the effects of small changes in the values of the creep parameters on the overall analysis of creep deformation. Time dependent creep deformations of a strip load are also compared with those occurring due to consolidation.

The treatment of creep behaviour has been restricted to the modelling of deviatoric creep. When comparing creep effects on different clays, the shape of the yield surface is an important consideration. Sensitivity analyses on the creep parameters revealed a necessity for their accurate evaluation. Small variations in these parameters caused correspondingly large variations in predicted settlements. The inclusion of creep behaviour in a consolidation and creep analysis resulted in a marked increase in settlements, creep settlements causing heave at points distant from the loading. Consolidation and creep settlements have opposite effects on horizontal displacements below the edge of a strip load.

Notation

a	semi-width of strip foundation
c	coefficient of consolidation (two dimensional)
m	slope of the logarithm of strain rate versus logarithm of time
p'	effective mean normal stress
q'	invariant shear stress
p'_c	preconsolidation pressure
q'_f	invariant shear stress at the point of critical states
t	time
t_1	unit time
Δt	time step size
A	strain rate at time t_1 and $D = 0$ (projected value)
D	deviator stress
$\bar{D}$	ratio of deviator stress to deviator stress at failure
E	Young's Modulus
K_o	coefficient of earth pressure at rest
M	slope of the projection of the critical state line in q', p' space
N	specific volume on normal consolidation line for unit mean normal effective stress
α	value of the slope of the linear portion of a plot of logarithm of strain rate versus deviator stress
$\bar{\alpha}$	αq_f
δ_{ij}	Kronecker delta
ε	direct strain
ε^c	creep strain
$\dot{\varepsilon}^c_{ij}$	creep strain rate tensor

k	swelling index
λ	compression index
ν	Poisson's ratio
$\sigma_{x,y,z}$	direct stresses
σ_{ij}	stress tensor
Γ	specific volume on the critical state line for unit mean normal effective stress

Introduction

Time-dependent creep effects can play an important role in the stress distributions and deformations of foundations and embankments. Kaufman and Weaver (1967) studying deformations which occurred over 10 - 15 years of the Atchafalaya levee on the Mississippi River compared field data with nonlinear elastic and elasto-plastic finite element analyses. The results of their comparison showed that creep effects should be included in these kinds of analyses. Lo et al. (1974) monitoring a test embankment near Ottawa attributed more than half of the settlements to creep behaviour.

Using a phenomenological model for creep behaviour proposed by Singh and Mitchell (1968) the analysis of undrained creep behaviour has been introduced into an elasto-plastic finite element programme. Using this programme the creep behaviour of a strip load on a finite layer of soil has been illustrated and a study has been made of the effects of small changes in the values of the creep parameters on the overall analysis of creep deformation. Creep deformation is likely to be accompanied by consolidation settlements and so the time dependent creep deformations of a strip load are also compared with those occurring during the consolidation process.

Creep Model

Researchers studying creep behaviour of cohesive soils have tended to adopt one of two methods of analysis. Either a rheological model of soil behaviour has been developed followed by the analysis of empirical data to check the applicability of the model (Murayama and Shibata, 1958), or experimental data has been analysed on a phenomenological basis to give predictive equations connecting the various measured parameters (Singh and Mitchell, 1968). Any model of creep behaviour which is adopted must use easily determined parameters preferably obtained from standard soil tests, must be applicable to a reasonable range of creep stresses and must describe the behaviour of a range of soil types. The phenomenological approach, particularly if normalised soil parameters are used, meets these criteria and lends itself to easier use for the prediction of soil behaviour.

Creep tests on many soils such as London Clay (Bishop, 1966), Osaka Alluvial Clay (Murayama and Shibata, 1958) and remoulded illite (Campanella, 1965) show a linear relationship between logarithm of creep strain and logarithm of time (Figure 1(a)) and also between logarithm of creep strain rate and applied deviator stress (Figure 1(b)). Upon analysing experimental data on a number of clays, Singh and Mitchell (1968) derived an equation which was held to be valid irrespective of whether clays are undisturbed, remoulded, normally consolidated or overconsolidated or tested drained or undrained. The equation expresses the strain rate, $\dot{\varepsilon}^c$, as a function of time, t, and sustained deviator stress, D.

$$\dot{\varepsilon}^c = Ae^{\alpha D} \left(\frac{t_1}{t}\right)^m \quad \ldots \quad \ldots \quad (1)$$

where: A = strain rate at time t_1 and $D = 0$ (projected value);

α = value of slope of the linear portion of a plot of logarithm of strain rate versus deviator stress;

t_1 = unit time; and

m = slope of logarithm of strain rate versus logarithm of time.

The authors have chosen to use this model as it needs few parameters to define it and these may be determined by carrying out a small number of creep tests at different stress levels on triaxial samples.

Equation 1 can be written in a more useful form as:

$$\dot{\varepsilon}^c = Ae^{(\bar{\alpha}\,\bar{D})}\left(\frac{t_1}{t}\right)^m \quad \ldots \quad (2)$$

where $\bar{D}$ is the ratio of deviator stress to deviator stress at undrained failure, q'/q'_f;

$\bar{\alpha} = \alpha q'_f$; and

A, t_1, α and m are defined above.

The use of parameters $\bar{\alpha}$ and $\bar{D}$ instead of α and D is more convenient because they are both dimensionless and the value of $\bar{\alpha}$ does not vary greatly with moisture content. Thus predictions of behaviour over a wide range of conditions can be made from a limited number of creep tests.

To use Equation 2 a starting value of elapsed time must be specified. Typical values for practical problems may range between one day and one month. The creep strain rate predicted by equation 2 for a given point in the material under a time varying stress level is shown schematically in Figure 2. Under a level of stress, $\bar{D}_1$, the creep strain rate is initially represented by point 1 and this value gradually decreases until the rate is represented by point 2. At this time an increase in load produces a stress of $\bar{D}_2$ giving rise to a strain rate represented by point 3 in the figure. The predicted strain rate gradually decreases with time as shown as long as no further disturbance is introduced.

The predicted creep rate, $\dot{\varepsilon}^c$, is the vertical creep strain rate of a triaxial sample. Problems arise when attempting to apply this essentially uni-dimensional creep strain rate to situations involving more dimensions (such as plane strain conditions). Chang et al. (1974) overcame these difficulties by making the following assumptions:

(i) no volume change occurs due to creep strains;

(ii) the principal shear strain rates are directly proportional to the corresponding principal shear stresses;

(iii) the principal strain axes do not rotate under deformation; and

(iv) the strains are small.

The flow rule for creep strain rates resulting from these assumptions is

$$\dot{\varepsilon}^c_{ij} = \frac{3\dot{\varepsilon}^c}{2q'}(\sigma'_{ij} - p'\delta_{ij}) \qquad \ldots \qquad (3)$$

where $i,j = x,y,z$ and $\sigma'_{xx} = \sigma'_x$, etc.;

$\dot{\varepsilon}^c$ is given by Equation (2);

δ is the Kronecker delta;

p' is the mean normal effective stress; and

q' is the invariant shear stress.

Proper consideration must be given to the constraints in the out of plane direction (z-direction). For plane strain conditions the total strain in the out of plane direction is zero. The non-plastic stress-strain relationship for a material when creep is continuing can be written as:

$$\varepsilon_x = \frac{1}{E}\{\sigma'_x - \nu(\sigma'_y + \sigma'_z)\} + \varepsilon^c_x \qquad (4(a))$$

$$\varepsilon_y = \frac{1}{E}\{\sigma'_y - \nu(\sigma'_z + \sigma'_x)\} + \varepsilon^c_y \qquad (4(b))$$

$$\varepsilon_z = \frac{1}{E}\{\sigma'_z - \nu(\sigma'_x + \sigma'_y)\} + \varepsilon^c_z \qquad (4(c))$$

For plane strain conditions $\varepsilon_z = 0$ and so from Equation (4(c)):

$$\sigma'_z = \nu(\sigma'_x + \sigma'_y) - E\varepsilon^c_z \qquad (5)$$

and on substituting this in Equations (4(a)) and (4(b)) one obtains:

$$\varepsilon_x = \{\frac{(1 - \nu^2)}{E}\sigma'_x - \nu\frac{(1 + \nu)}{E}\sigma'_y\} + \varepsilon^c_x + \nu\varepsilon^c_z \qquad (6(a))$$

$$\varepsilon_y = \{\frac{(1 - \nu^2)}{E}\sigma'_y - \nu\frac{(1 + \nu)}{E}\sigma'_x\} + \varepsilon^c_y + \nu\varepsilon^c_z \qquad (6(b))$$

The quantities within the brackets of Equations (6) are the elastic strains and so to account for the out of plane creep strains a strain equal to $\nu\varepsilon^c_z$ is added to the strains in the x- and y-directions. Equation (5) is used to calculate σ'_z.

The three creep parameters necessary to define the model of deviatoric creep behaviour can be obtained from a minimum of two identical cylindrical triaxial samples, at the same moisture content and same initial stress conditions. The samples must be subjected to creep tests under different deviator stresses covering a range, say, of 30% to 90% of the maximum deviator stress depending on the stress history. Under these sustained loads, strain is observed with time. Singh and Mitchell (1968) expand on the subject of parameter evaluation in their Appendix I. It should be noted that the parameter, m, is not unique for a given soil and may vary depending on whether the soil sample is on the 'wet' or 'dry' side of critical states. Hyde (1974) has obtained results for Keuper Marl which indicate a value of m of 0.86 on the 'wet' side and 1.00 on the 'dry' side of critical states.

The computer program used was developed by Burke (1983). This program allows nonlinear analyses to be carried out independently of creep analyses and at any stage of a load deformation analysis time dependent behaviour may be introduced. To model creep displacements, an increment of time is allowed to elapse, the creep strains are calculated as described above and then these are converted into a set of equivalent forces. The equivalent loads are added to the external load vector and the solution for the end of a time step involves a re-solution. Because creep response under working load situations is generally a decay process, progressively larger time steps may be used. Equation 2 was modified to include a lower cut-off for creep strains whereby values of $\bar{D}$ lower than 0.3 did not cause creep flow. A flow chart summarising the basic solution algorithm as stated above is shown in Figure 3.

Creep Deformation of a Strip Load on a Finite Layer of Soil

To show the kind of behaviour one is likely to expect from Singh and Mitchell's (1968) creep model, it has been applied to the analyses of a strip load of width 2a underlain by a clay layer of thickness 3a and 6a (Figure 4). The two materials used in the study were San Francisco Bay Mud and Keuper Marl (the material parameters of which are stated in Table 1) and creep deformations under different load intensities were investigated.

The start of the analysis was taken at 7 days after the application of the loads. The time stepping sequence began with a time increment of 1 day and subsequent time step sizes were ever increasing and had a value of 1.5 times the previous value.

Figures 5 and 6 show the creep behaviour for both depths of layer for San Francisco Bay Mud and Keuper Marl, respectively, at various loading pressures. At a loading pressure of 50 kN/m^2 the amount of predicted centreline creep displacement for each material is similar. As the loading increases, however, the creep displacements of San Francisco Bay Mud increase more than those of Keuper Marl. In the case of Keuper Marl doubling the loading pressure from 50 to 100 kN/m^2 has the effect of increasing the centreline creep displacements at 2960 days by a factor of approximately 1.3 and 1.5 for the deep and shallow layers, respectively. For San Francisco Bay Mud the same loading increase causes increases in the creep displacements at 2960 days by a factor of 1.4 and 2.0 for the deep and shallow layers, respectively. In doubling the loading pressure from 100 to 200 kN/m^2 these factors become 2.3 and 2.4.

The above analyses show that San Francisco Bay Mud is more prone to creep than Keuper Marl and this could have been deduced from an inspection of their creep parameters. However, at low stress intensities the amount of creep for each material is shown to be similar. This is in part due to the fact that although the loading is identical Keuper Marl has a lower undrained failure stress for a given stress history (this is illustrated in Figure 7 where the wet side yield locus ellipsi for the two materials are shown). For a given stress intensity, therefore, this would imply that the ratio, $\bar{D}$ (see Equation 2), would be higher for Keuper Marl than it would be for San Francisco Bay Mud resulting in enhanced creep strain rates for Keuper Marl.

Sensitivity Analyses on the Creep Parameters

When using any model of soil behaviour it is worthwhile considering what effect small changes in the values of the material parameters will have on an analysis so that material testing yields values of material parameters to the desired accuracy. Using the parameters for San Francisco Bay Mud such a sensitivity analysis has been carried out on the creep parameters ,$\bar{\alpha}$, and ,m, for the strip load on the deeper layer problem (depth = 6a) at a loading pressure of 200 kN/m^2.

Figure 8 shows the effect of varying the parameter ,$\bar{\alpha}$,. Values of ,$\bar{\alpha}$, have been taken at 10% and 20% above and below the actual value for the material. An increase of 10% and 20% causes an increase in centreline creep displacements at 2960 days of 25% and 54%, respectively. A decrease of 10% and 20% causes - decrease in centreline creep displacements at the same time of 21% and 39%, respectively. Increasing ,$\bar{\alpha}$, therefore has a greater effect on creep displacements than decreasing ,$\bar{\alpha}$. This may also be explained with reference to the equation of creep strain rate (Equation 2). Because ,$\bar{\alpha}$, appears in the equation as an exponent, increasing the value by any amount will have more effect than decreasing it. This partly explains why San Francisco Bay Mud is more prone to creep than Keuper Marl which have values of 5.40 and 1.13, respectively.

Figure 8 shows the effect of varying the parameter ,m . Again, values of the parameter have been taken at 10% and 20% above and below the actual value for the material. An increase of 10% and 20% causes a decrease in centreline creep displacements of 28% and 48%, respectively. A decrease of 10% and 20% causes an increase in centreline creep displacements of 38% and 90%, respectively. Thus decreasing the value of ,m, has a greater influence on creep displacements than a similar increase and this may be explained by the fact that ,m, appears in the creep strain rate equation as an exponent to the reciprocal of time. Figure 9 also shows that creep rupture is associated with low values of ,m .

The above sensitivity analyses show that the creep parameters ,α, and ,m, must be carefully determined because the prediction of creep displacements is sensitive to small changes in their values.

Consolidation and Creep

To examine the effects of creep during the consolidation process the authors have analysed the problem of a semi-infinite layer of San Francisco Bay Mud supporting a flexible, porous strip load of width 20 m. Consolidation settlements were computed using a nonlinear (elasto-plastic) analysis developed by Burke (1983). The material parameters are shown in Table 1. The finite element mesh used to approximate the semi-infinite layer and boundary conditions are shown in Figure 10 and the additional boundary conditions are that free drainage was allowed only along the upper surface boundary, the vertical and lower boundaries being impermeable. The soil had an initial vertical stress of $-150\ kN/m^2$ throughout its depth, the value of the coefficient of earth pressure at rest, K_o, was 0.8 and Poisson's ratio had a value of 0.444. The soil was assumed to be lightly overconsolidated with an overconsolidation ratio of 1.2. The initial values of bulk and shear moduli were 4717 and $455\ kN/m^2$, respectively, and the horizontal and vertical permeabilities were assumed to be 1.15×10^{-5} m/day. The coefficient of consolidation, c, of $5.72 \times 10^{-3}\ m^2/day$ was calculated using the following formula:

$$c = \frac{k}{\gamma_w}(K + \frac{1}{3}G) \quad \ldots \quad \ldots \quad (7)$$

where k is the permeability;

γ_w is the bulk density of water; and

K,G are the bulk and shear moduli, respectively, of the soil skeleton.

Two analyses were carried out assuming:

(a) consolidation of a nonlinear (elasto-plastic) soil skeleton; and

(b) consolidation of a nonlinear soil skeleton with the inclusion of creep effects.

A uniform ramp loading was applied such that the full loading pressure of $100\ kN/m^2$ was obtained after the first ten time steps. The time stepping scheme was as follows:

10 steps of Δt = 10 days;

9 steps of Δt = 10^2 days;

9 steps of Δt = 10^3 days;

9 steps of Δt = 10^4 days; and

9 steps of Δt` = 10^5 days.

The initial time step size of 10 days violated the stability criterion of Vermeer and Verruijt (1981) for the consolidation analysis, however no problems such as oscillating excess pore pressures were encountered with this value of time step size. This may, in part, be attributed to the fact that the criterion strictly only applies to regular finite element meshes for one-dimensional consolidation problems. The example used herein is two-dimensional and the finite element mesh is graded. Another important feature of the analysis is that the loading was not applied suddenly but gradually over the first ten time steps, thus reducing any tendency towards oscillating excess pore pressures. Creep effects were considered from the point in time at which the ramp loading was complete. The time at the end of the analysis corresponds to a value of time factor, T, of 57.2 calculated using the following formula:

$$T = \frac{ct}{a^2} \qquad \ldots \qquad \ldots \qquad (8)$$

where c is the coefficient of consolidation;
t is the elapsed time; and
a is a reference length (e.g. the semi width of a strip load).

Figure 11 shows the development of settlements at the centre of the strip load with time factor for the two analyses performed. A large porportion of the total settlement occurred during the loading period and may be attributed to the fact that dissipation of the excess pore pressures was allowed during the relatively long loading period. Thus at the end of loading the settlements comprise both the 'immediate' settlement due to the load and a contribution due to the consolidation of the underlying soil. The inclusion of creep behaviour in the analysis shows a marked increase in settlement at all times.

Figure 12 shows the dissipation of excess pore pressure with time factor at a point below the strip load (position A in Figure 10). The effect of creep behaviour is to increase the peak value of excess pore pressure and cause this peak to occur at a later time than those shown by the nonlinear consolidation analysis. Also, creep behaviour tends to delay still further the subsequent dissipation of the excess pore pressures.

Figure 13 shows the surface settlement profile at the end of loading and at T = 5.72 for the nonlinear analysis and the nonlinear analysis including creep. The figure shows that settlements due to creep may be significant when compared to consolidation settlements and also that creep may cause heave along part of the surface distant from the strip load.

Figure 14 shows profiles of horizontal movements below the edge of the strip load ($\frac{x}{a} = 1$). It can be seen that the consolidation and creep effects are in opposition, consolidation causing an inward movement and creep causing an outward movement. It is conceivable that at certain times and at other locations in the soil these effects may cancel each other or cause oscillations in horizontal movements. Care may therefore be needed when horizontal movements are monitored in the field if it is known that the underlying material is prone to creep.

Conclusions

The treatment of creep behaviour has been restricted to the modelling of deviatoric creep. Of the two basic approaches to creep analysis, the authors have used the approach utilizing a phenomenological model and in particular that of Singh and Mitchell (1968).

When comparing creep effects on different clays at low stress levels the shape of the yield surface and its effect on the stress ratio $\bar{D}$ is an important consideration.

Sensitivity analyses were carried out on the Singh & Mitchell creep parameters ,$\bar{\alpha}$, and ,m, (because they appear as exponents in Equation (2)) of San Francisco Bay Mud and showed that they must be carefully determined. An over- or under-estimate of the value of the parameter ,$\bar{\alpha}$, by 10% may cause an increase or decrease, respectively, in predicted creep settlements of the order of 25%. An over-estimate of the value of the parameter ,m, by 10% may cause a decrease in predicted creep settlements of the order of 30%; an under-estimate of the value by the same amount may cause an increase in predicted creep settlements of the order of 40%. This last finding is consistent with the fact that creep rupture is associated with low values of m.

The effects of combining creep with a nonlinear (elasto-plastic) consolidation analysis have been studied. The inclusion of creep behaviour resulted in a marked increase in settlement. It also resulted in increased values of peak excess pore pressures (at a later time) and these excess pore pressures took longer to dissipate. When considering vertical movements, creep settlements appear to cause heave along part of the surface distant from the loading. Consolidation and creep have opposite effects on horizontal displacements below the edge of a strip load, consolidation causing an inward movement and creep causing an outward movement. Care may therefore be needed when analysing in-situ horizontal displacement records if it is known that the underlying material is prone to creep.

REFERENCES

BISHOP, A.W. (1966), "The strength of soils as engineering materials", Geotechnique, Vol. 16, pp. 91-128.

BURKE, J.J. (1983), "A non-linear finite element analysis of soil deformation", PhD Thesis, Loughborough University of Technology.

CAMPANELLA, R.G. (1965), "Effect of temperature and stress on time-deformation behaviour in saturation clay", PhD Thesis, University of California, Berkeley.

CHANG, C.-Y., NAIR, K. and SINGH, R.D. (1974), "Finite element methods for the nonlinear and time-dependent analysis of geotechnical problems", Proceedings of the Converence on Analysis and Design in Geotechnical Engineering, Austin, Texas, Vol. II, pp. 269-302.

HYDE, A.F.L. (1974), "Repeated load triaxial testing of soils", PhD Thesis, University of Nottingham.

KAUFMAN, R.I. and WEAVER, F.J. (1967), "Stability of Atchafalaya levees", Proceedings of the Journal of the Soil Mechanics and Foundations Division, ASCE, Vol. 93, No. SM4, pp. 157-176.

LO, K.Y., BOZOZUK, M. and LAW, K.T. (1974), "Settlements resulting from secondary compression", Report RR211, Research and Development Division, Ministry of Transportation and Communications, Ontario, Canada.

MURAYAMA, S. and SHIBATA, T. (1958), "On the rheological characteristics of clay, Part 1", Bulletin No. 26, Disaster/Prevention Research Institute, Kyoto, Japan.

SINGH, A. and MITCHELL, J.K. (1968), "General stress-strain time function for soils", Proceedings of the Journal of the Soil Mechanics and Foundations Division, ASCE, Vol. 94, No. SM1, pp. 21-46.

VERMEER, P.A. and VERRUIJT, A.(1981), "An accuracy condition for consolidation by finite elements", International Journal for Analytical Methods in Geomechanics, Vol. 5, pp. 1-14.

Parameters	Symbol	San Francisco* Bay Mud	Keuper Marl**
Specific volume on normal consolidation line for unit mean normal effective stress (kN/m^2)	N	3.213	2.143
Specific volume on critical state line for unit mean normal effective stress (kN/m^2)	Γ	3.009	2.040
Slope of normal consolidation line in v-ln(p') space	λ	0.326	0.101
Slope of swelling curve in v-ln(p')	k	0.043	0.010
Slope of the projection of the critical state line in q'-p' space	M	1.44	1.25
Singh and Mitchel parameters	A	6.270×10^{-5} day^{-1}	3.734×10^{-4} day^{-1}
	t_1	1 day	1 day
	$\bar{\alpha}$	5.40	1.13
	m	0.73	0.86

* Values taken from Kavazanjian and Mitchell (1980)

** Values taken from Hyde (1974)

Table 1 Material parameter values for different soil types

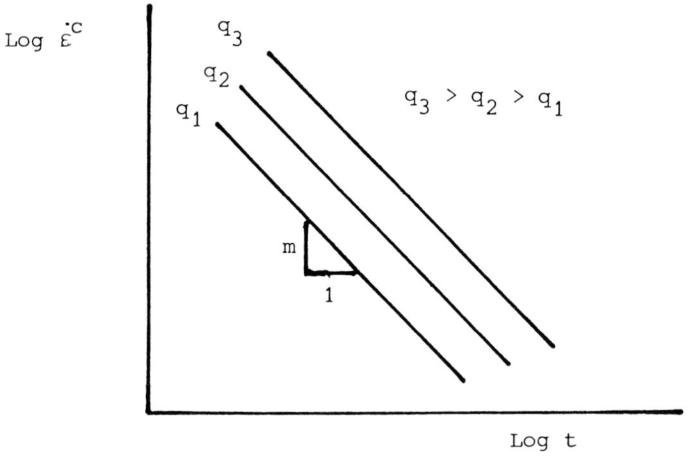

(a) Creep strain rate versus time

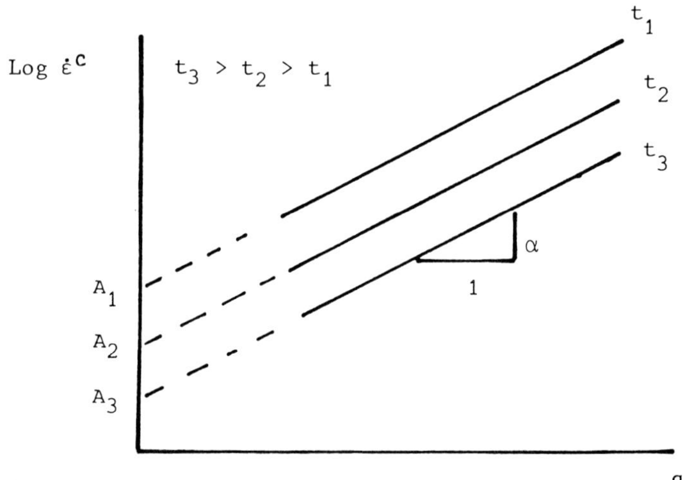

(b) Creep strain rate versus deviator stress

Fig 1. Typical creep strain relationships

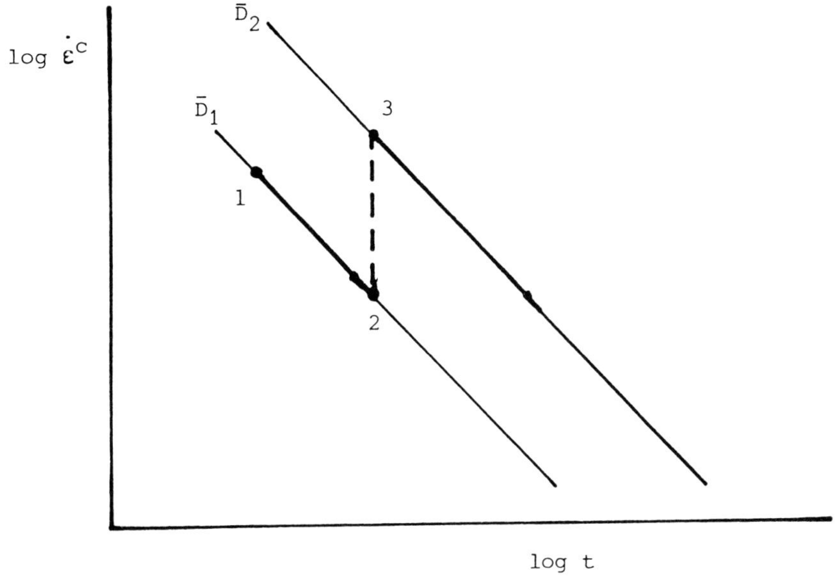

Figure 2 Schematic representation of creep response to a varying stress level

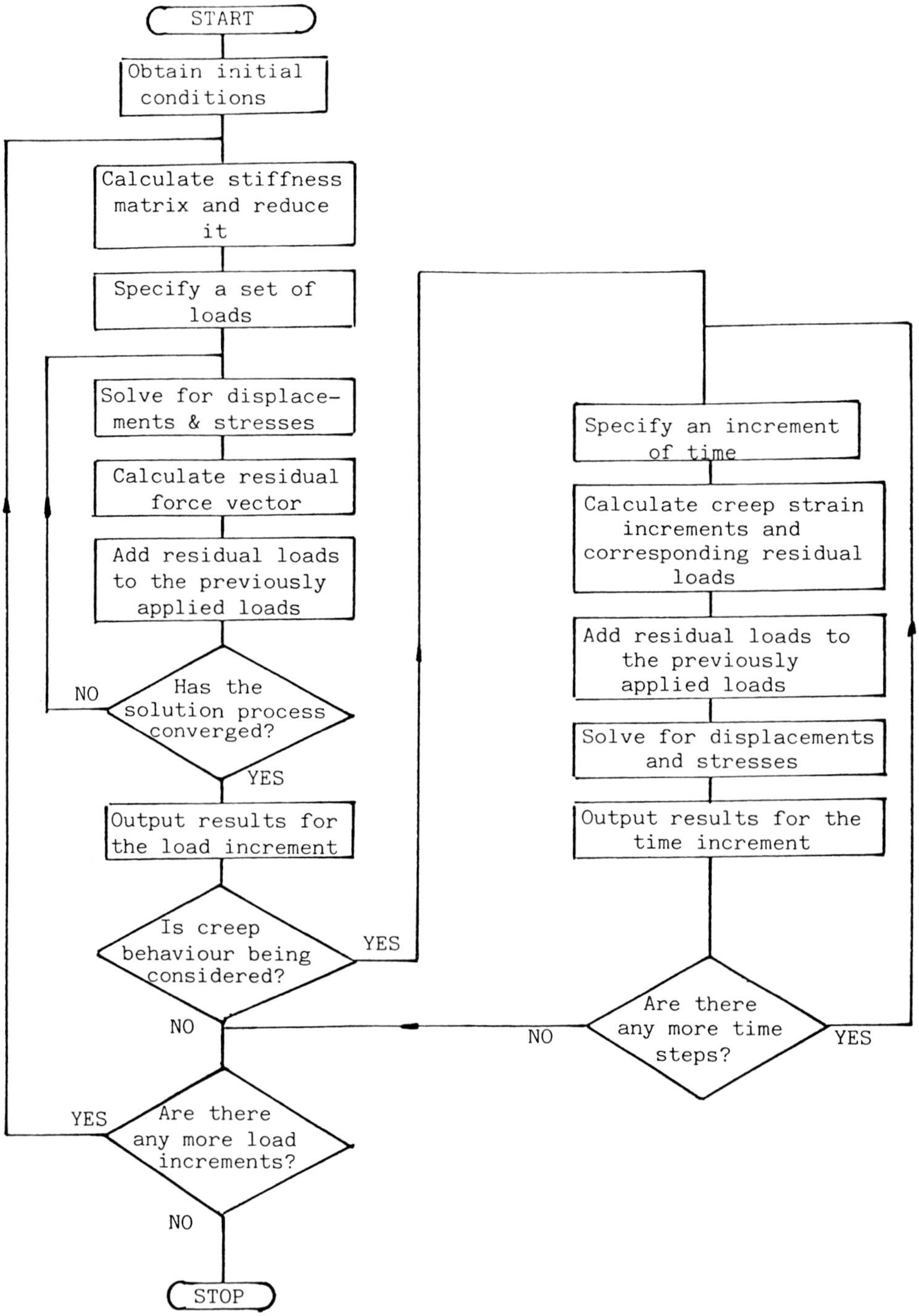

Figure 3 Flowchart of the basic solution procedure for elasto-plastic and creep analyses by finite elements

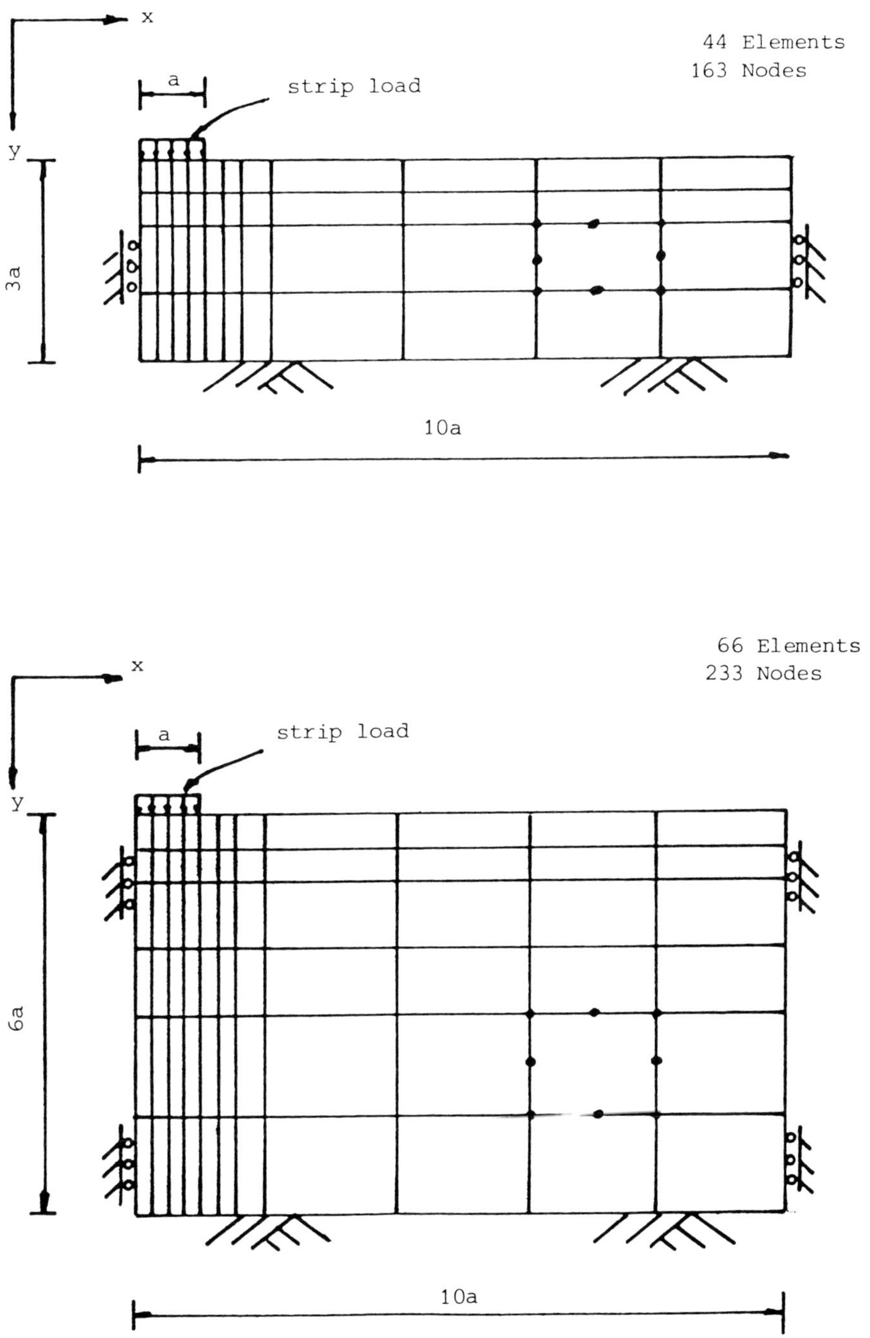

Figure 4 Finite element meshes and boundary conditions used in the analysis of shallow and deep layer problems

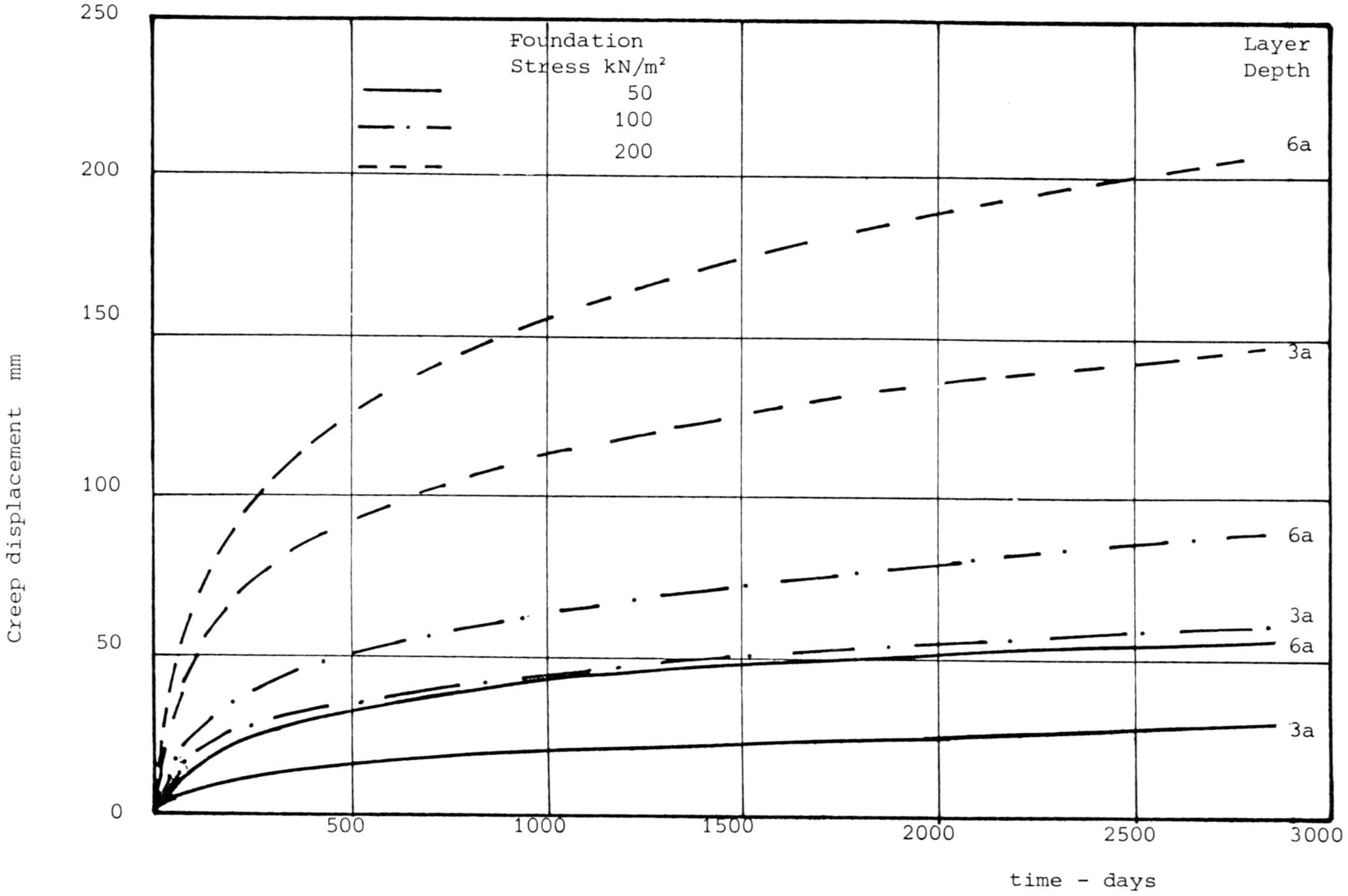

Figure 5 Centreline creep displacements of a strip load underlain by a layer of San Francisco Bay Mud

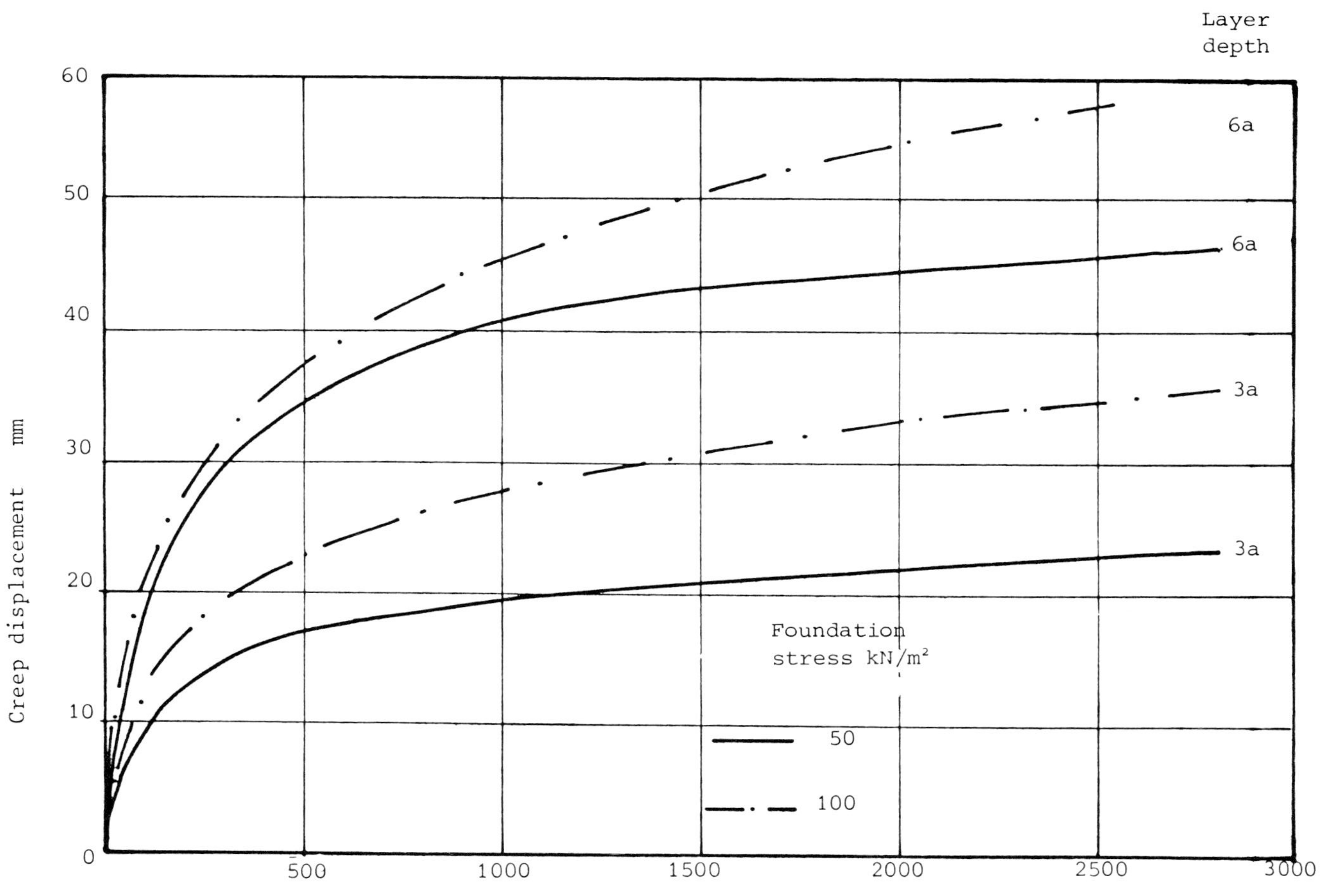

Figure 6 Centreline creep displacements of a strip load underlain by a layer of Keuper Marl

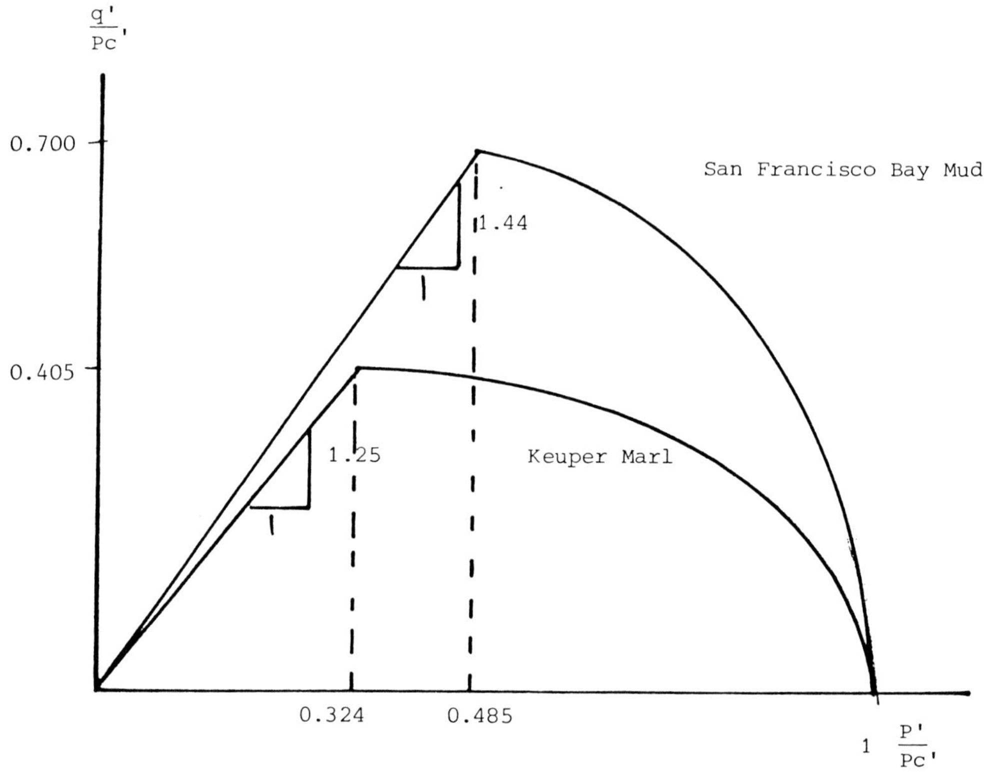

Figure 7 Wet side yield locus ellipsi for Keuper Marl and San Francisco Bay Mud

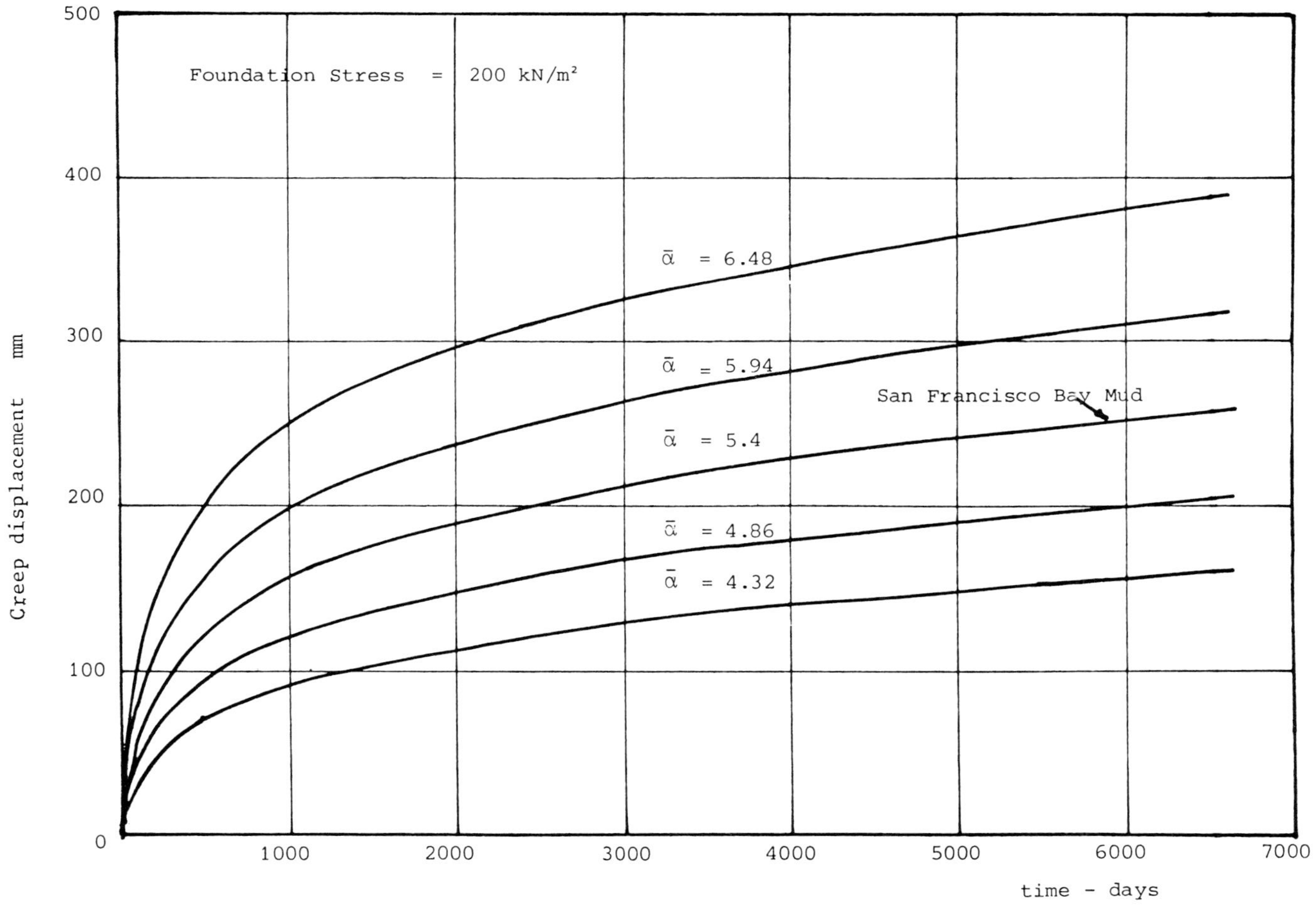

Figure 8 The effect of varying the creep parameter, $\bar{\alpha}$, on the centreline creep displacements of a strip load

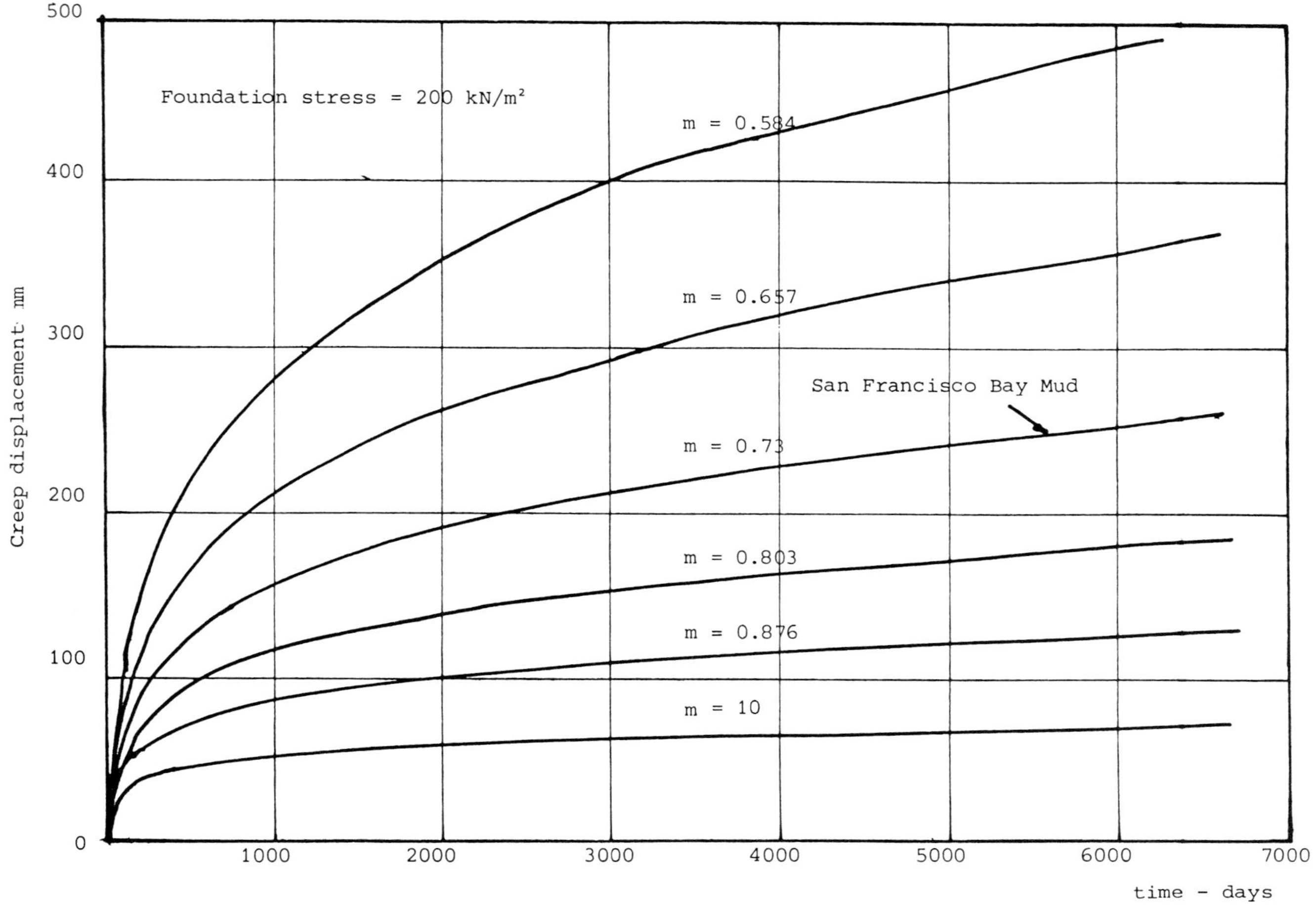

Figure 9 The effect of varying the creep parameter, m, on the centreline creep displacement of a strip load

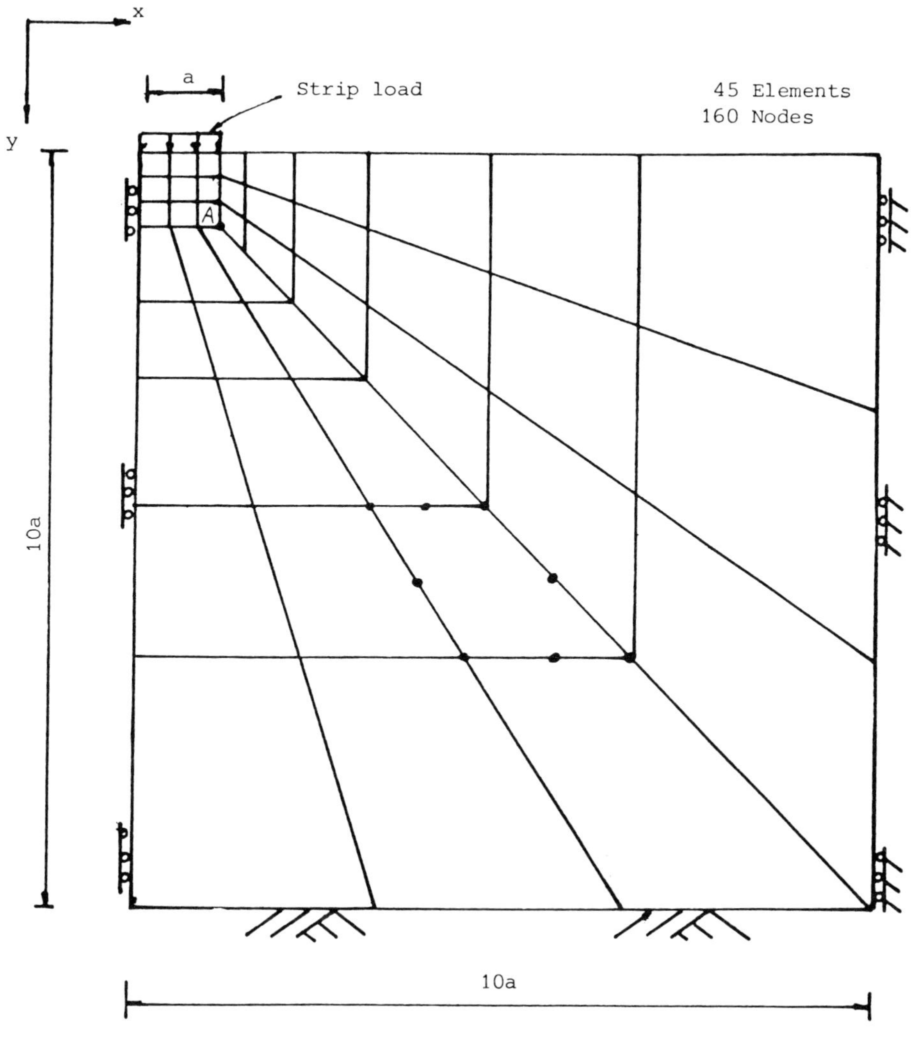

Figure 10 Finite element mesh and boundary conditions for the analysis of a strip load on a semi-infinite layer

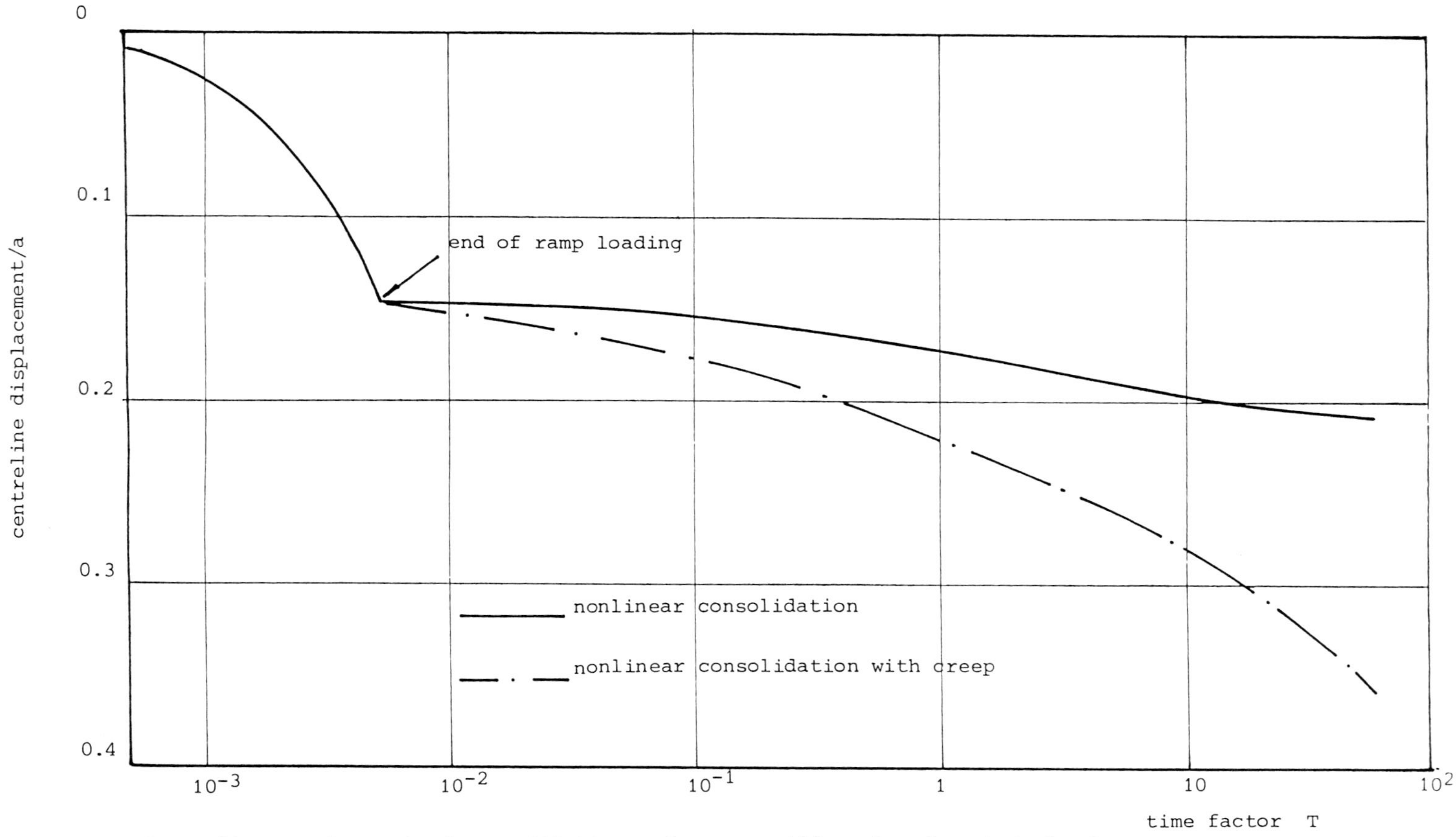

Figure 11 Development of consolidation and creep settlements of a strip load with time factor

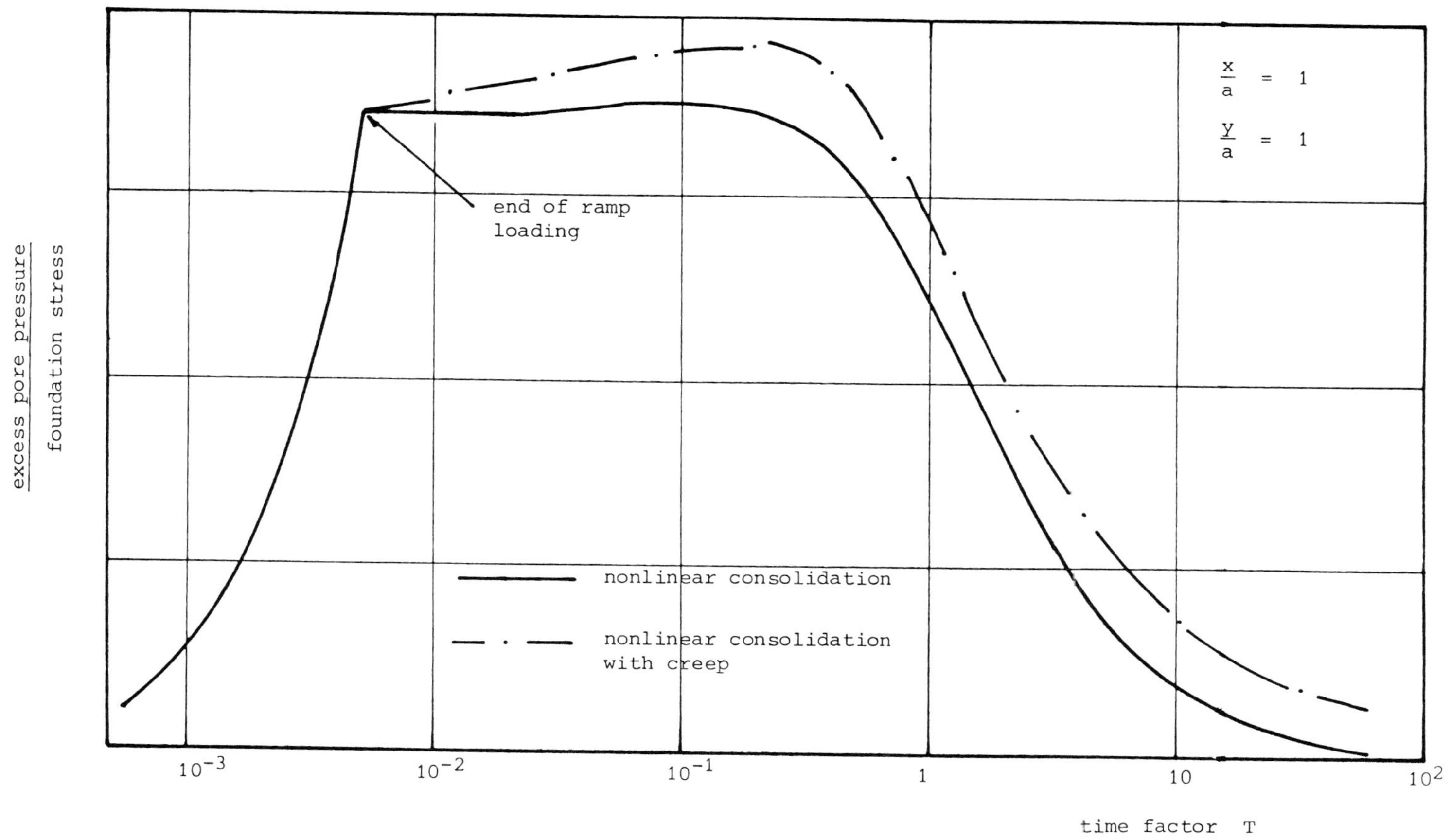

Figure 12 The generation and dissipation of excess pore pressure with time factor below a strip load for nonlinear consolidation and creep analyses

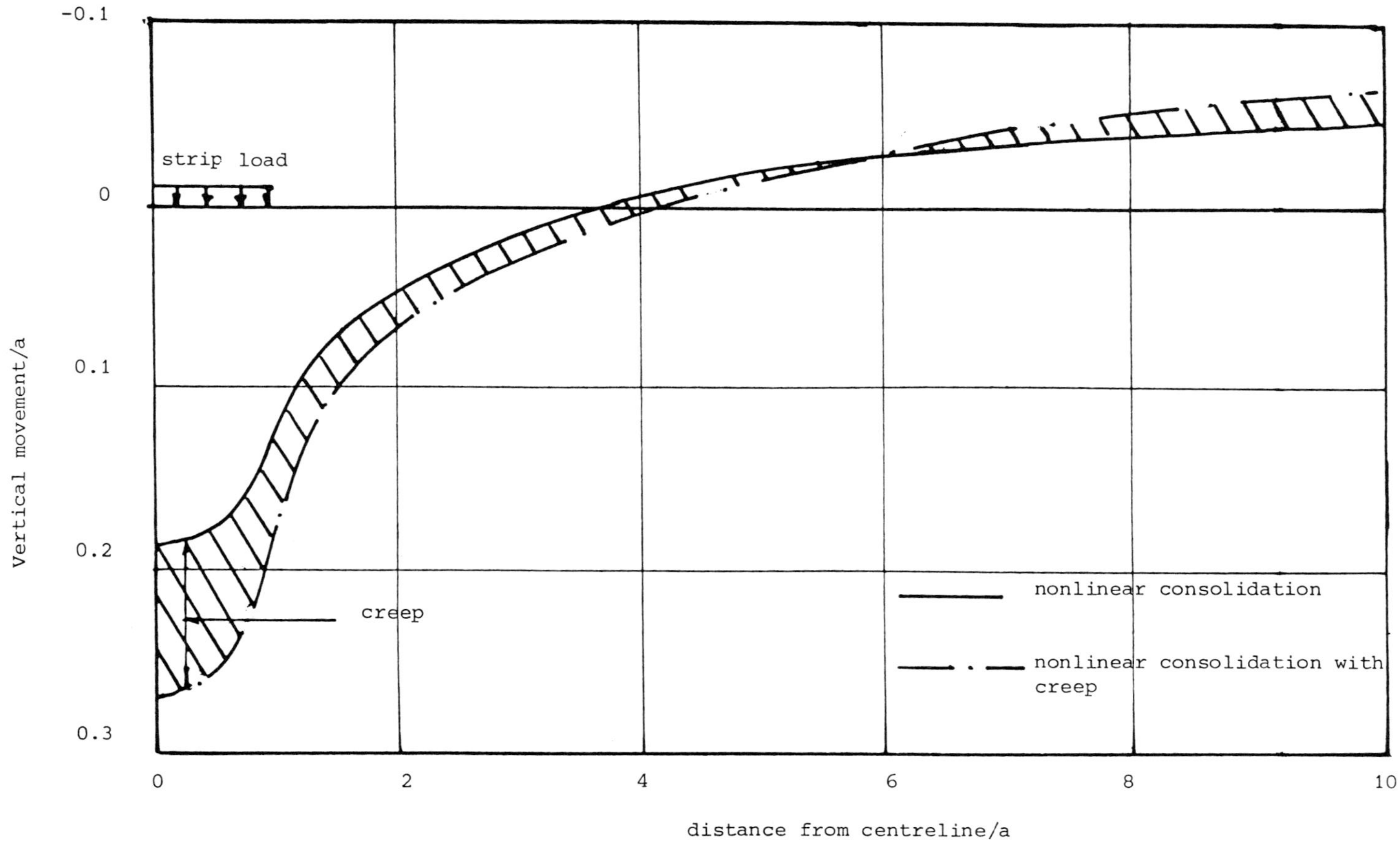

Figure 13 Surface settlement profiles for nonlinear consolidation and creep analyses of a strip load on a semi-infinite layer

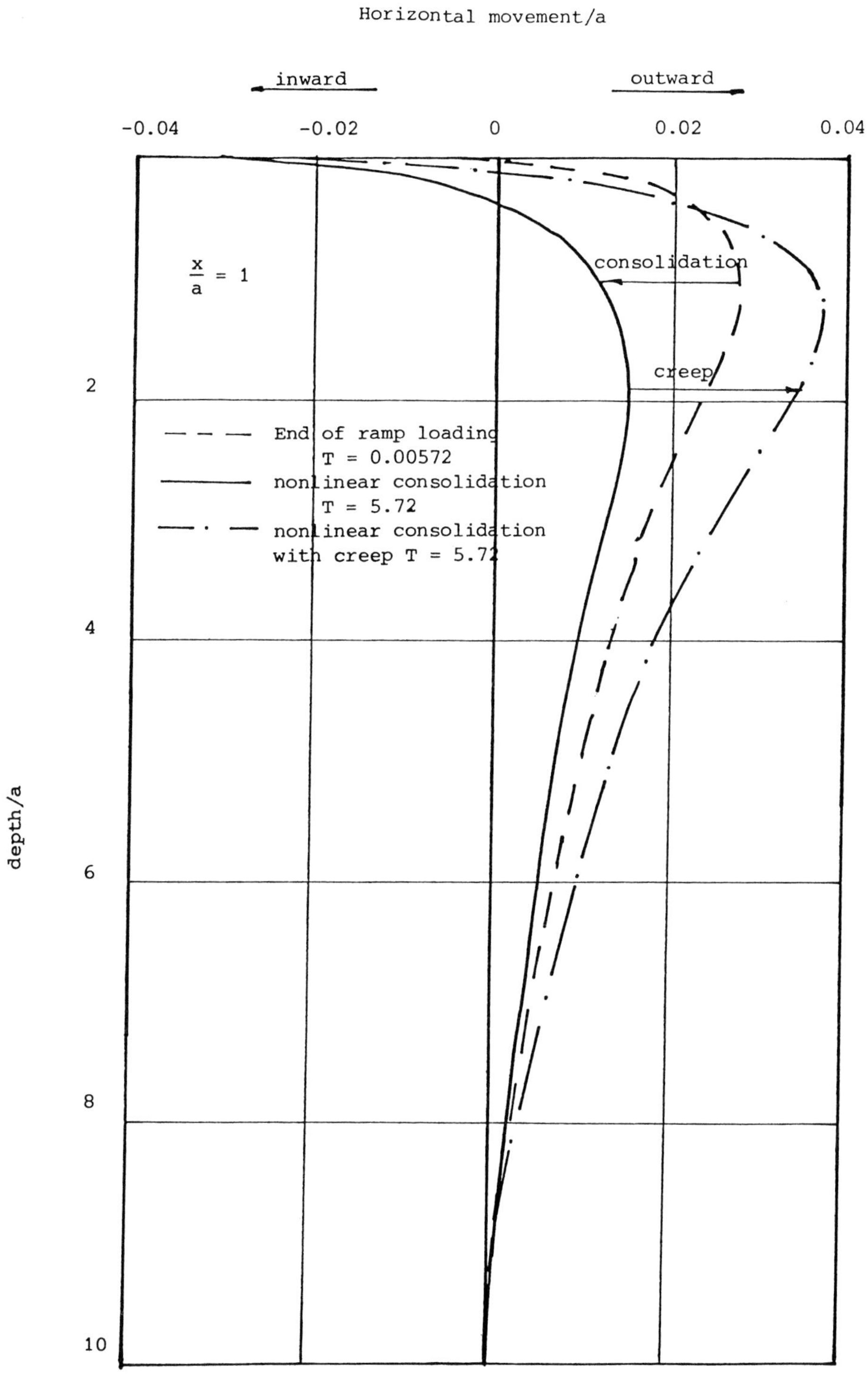

Figure 14 Horizontal movement below the edge of a strip load for nonlinear consolidation and creep analyses

A CLOSED-FORM SOLUTION FOR THE PROBLEM OF A VISCOPLASTIC HOLLOW SPHERE APPLICATION TO UNDERGROUND CAVITIES IN ROCK SALT

P. BEREST
LABORATOIRE DE MECANIQUE DES SOLIDES
ECOLE POLYTECHNIQUE
91128 Palaiseau Cedex - France

Summary :

This paper is divided in two parts. In the first part we give a closed form solution for the problem of a hollow sphere in a elasto-viscoplastic medium, submitted to a varying internal pressure. In the second part, some case stories and actual datas are discussed ; they allow to try to back-calculate the constitutive equation of the rock mass.

Acknowledgment :

Some results of the present paper are based upon a former study of the author and Nguyen Minh Duc, Laboratoire de Mécanique des Solides.

FIRST PART

BEHAVIOUR OF VISCOPLASTIC THICK WALLED SPHERES

INTRODUCTION

The quasi static time dependent behaviour of thick walled spheres and cylinders is of great importance in many areas, including rock mechanics. Indeed, many cavities can be approximatively considered as spherical voids in an infinite medium.

This problem has been considered by Wierzbicki (1963), Aufaure (1975), Tijani (1978), for the special case of a monotonous variation of the applied internal pressure. Madejski (1960) established a solution, adopting a simplified elastic viscoplastic behaviour ; but Ottosen (1985), who gives a solution for a viscoelastic-viscoplastic sphere, considers this solution as erroneous. Bérest and Nguyen M.D. (1983) presented the solution for a spherical cavity in an infinite medium, submitted to a variable internal pressure. A solution for a thick walled sphere will be given in this paper.

We consider a hollow sphere, homogeneous and isotropic as regards its mechanical properties. The internal radius of the sphere is 1, and the external radius is $\rho > 1$.

Initially, the medium is in a natural state (strains and stresses are equal to zero).

The medium exhibits an instantaneous elastic behaviour and a viscoplastic behaviour, with a Tresca's viscoplastic criterion and associated flow rule.

The sphere is loaded by a pressure applied inside the cavity and (or) a pressure applied outside the cavity. However the mechanical behaviour of the sphere only depends upon the difference between the internal pressure and the external pressure, due to the considered rheology of the medium. We can then assume, without loss of generality, that the external pressure is zero.

Only the quasi-static problem and small displacements will be considered.

Under these hypothesis, the constitutive equations can be written as follows :

$$E \frac{\partial \varepsilon}{\partial t} = (1+\nu) \frac{\partial \sigma}{\partial t} - \nu \operatorname{tr}\left(\frac{\partial \sigma}{\partial t}\right) \mathbb{I} + E \frac{\partial \lambda}{\partial t} \frac{\partial F(\sigma)}{\partial \sigma}$$

$$F(\sigma) \equiv \sigma_1 - \sigma_2 - 2\,C$$

$$\frac{\partial \lambda}{\partial t} = \frac{1}{2} F(\sigma) \quad \text{if} \quad F(\sigma) \geqslant 0 \; ; \quad = 0 \quad \text{if} \quad F(\sigma) \leqslant 0 \quad .$$

With the notations :

r, θ, φ	:	spherical coordinates
ε	:	strain tensor with principal components : ε_1, ε_2, ε_3
σ	:	stress tensor with principal components : $\sigma_1 > \sigma_2 > \sigma_3$
E	:	Young's modulus
ν	:	Poisson's ratio
C	:	cohesion of the material
F	:	yield criterion (Tresca)
η	:	viscosity constant
$\frac{\partial \lambda}{\partial t}$	:	absolute value of the equivalent viscoplastic strain rate
$\mathbb{I}$	:	unit tensor

. *Statement of the problem* :

Since we have spherical symmetry, $\sigma_r, \sigma_\theta, \sigma_\varphi$ are the principal stresses ; moreover $\sigma_\theta = \sigma_\varphi$; the displacement is purely radial, $u = u(r,t)$; therefore :

$$\varepsilon_r = \frac{\partial u}{\partial r} \quad \text{and} \quad \varepsilon_\theta = \varepsilon_\varphi = \frac{u}{r} \quad .$$

Let $v = \frac{\partial u}{\partial r}$.

The yield limit is reached simultaneously by two pairs of principal stresses :

$$f_\varphi \equiv \omega(\sigma_\varphi - \sigma_r) - 2\,C = 0$$

$$f_\theta \equiv \omega(\sigma_\theta - \sigma_r) - 2\,C = 0$$

where $\omega = \pm 1$ has the same sign as $\sigma_\varphi - \sigma_r = \sigma_\theta - \sigma_r$. The total plastic strain rate is, in this particular case, the sum of two associated viscoplastic strain rates :

$$\frac{\partial \lambda}{\partial t} = \frac{1}{\eta} \frac{< f_\varphi > + < f_\theta >}{2}$$

$$\dot{\varepsilon}^{vp} = \omega \frac{\partial \lambda}{\partial t} \frac{\partial}{\partial \sigma} \left\{ \frac{< f_\varphi > + < f_\theta >}{2} \right\}$$

Then the constitutive relationship can be written as follows :

$$E \frac{\partial v}{\partial r} = \frac{\partial \sigma_r}{\partial t} - 2 \nu \frac{\partial \sigma_\varphi}{\partial t} - \omega E \frac{\partial \lambda}{\partial t} \qquad (1)$$

$$E \frac{v}{r} = (1 - \nu) \frac{\partial \sigma_\varphi}{\partial t} - \nu \frac{\partial \sigma_r}{\partial t} + \frac{1}{2} \omega E \frac{\partial \lambda}{\partial t} \qquad (2)$$

$$\frac{\partial \lambda}{\partial t} = \frac{1}{\eta} < \omega (\sigma_\varphi - \sigma_r) - 2 C >$$

Since at t = 0, the state is natural, (1) and (2) can be integrated with respect to time between t = 0 and any other instant t. If we take

$$\varepsilon^{vp}(r,t) = \int_0^t \omega(r,t) \frac{\partial \lambda}{\partial t} (r,t) \, dt ,$$

where $\varepsilon^{vp}(r,t)$ is the equivalent viscoplastic strain, whose sign can be positive or negative, we then obtain :

$$E \frac{\partial u}{\partial r} = \sigma_r - 2 \nu \sigma_\varphi - E \varepsilon^{vp} \qquad (1')$$

$$E \frac{u}{r} = (1 - \nu) \sigma_\varphi - \nu \sigma_r + \frac{1}{2} E \varepsilon^{vp} \qquad (2')$$

with :

$$\frac{\partial \varepsilon^{vp}}{\partial t} = \frac{1}{\eta} \left\{ \sigma_\varphi - \sigma_r - 2 \omega C \right\} \quad \text{if} \quad | \sigma_\varphi - \sigma_r | > 2 C .$$

Moreover, the equation of equilibrium reduces to :

$$\frac{\partial \sigma_r}{\partial r} = 2 \frac{\sigma_\varphi - \sigma_r}{r} \qquad (3).$$

NOTATIONS

Once the viscoplastic criterion has been exceeded, a viscoplastic zone will develop from the cavity inside the medium (this will be proved later). Later on, this region may regress and eventually disappear, leaving residual strains in the volumes it has reached. Then, three types of zones must be distinguished :

- Elastic zone : $\varepsilon^{vp} = 0$
- Viscoplastic zone : $\partial \varepsilon^{vp}/\partial t \neq 0$
- Zone with residual strains : $\partial \varepsilon^{vp}/\partial t = 0$, $\varepsilon^{vp} \neq 0$ (or "residual zone").

These different kinds of zones may exist, depending on the evolution of the loading parameter $\sigma_i = \sigma_i(t)$. More precisely, 8 cases must be considered ; they are determined by the evolution of the viscoplastic boarder x = x(t) dividing :

- the viscoplastic zone $1 \leqslant r \leqslant x$, where $\frac{\partial \varepsilon^{vp}}{\partial t} \neq 0$
- the non viscoplastic zone $x \leqslant r \leqslant \rho$, where $\frac{\partial \varepsilon^{vp}}{\partial t} = 0$

The following example shows the eight cases (fig. 1).

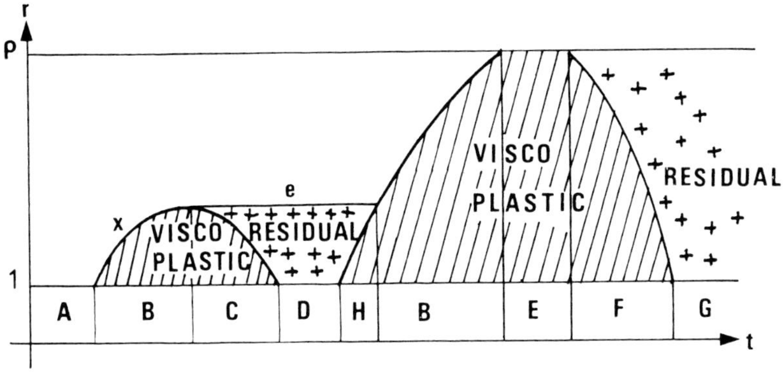

Figure 1 : Evolution of the viscoplastic boarder.

A. Fully elastic sphere
B. Viscoplastic zone $1 \leqslant r \leqslant x$, elastic zone $x \leqslant r \leqslant \rho$
C. Viscoplastic zone $1 \leqslant r \leqslant x$, residual zone $x \leqslant r \leqslant e$, elastic zone $e \leqslant r \leqslant \rho$, $(\dot{x} < 0)$
D. Residual zone $1 \leqslant r \leqslant e$, elastic zone $e \leqslant r \leqslant \rho$
E. Fully viscoplastic sphere
F. Viscoplastic zone $1 \leqslant r \leqslant x$, residual zone $x \leqslant r \leqslant \rho$, $(\dot{x} < 0)$
G. Fully residual sphere
H. Viscoplastic zone, $1 \leqslant r \leqslant x$, residual zone $x \leqslant r \leqslant e$, elastic zone $e \leqslant r \leqslant \rho$, $(\dot{x} > 0)$.

. *General relationships* :

a- the elimination of σ_φ and ε^{vp} between (1') (2') (3) leads to

$$E \left(\frac{\partial u}{\partial r} + 2 \frac{u}{r} \right) = (1 - 2 \nu) \left(3 \sigma_r + r \frac{\partial \sigma_r}{\partial r} \right)$$

which can be integrated with respect to r :

$$E \frac{u}{r} - (1 - 2 \nu) \sigma_r = - \frac{3}{2} (1 - \nu) \frac{A(t)}{r^3}$$

A(t) being a variable of integration. The same function A(t) can be considered for the whole sphere, as for u and σ_r must be continuous.

b- The elimination of σ_φ between (2') and (3) leads to :

$$E \frac{u}{r} - (1-2\nu)\,\sigma_r = \frac{1-\nu}{2}\, r\, \frac{\partial \sigma_r}{\partial r} + \frac{1}{2} E\, \varepsilon^{vp} \quad .$$

c- Then :

$$r \frac{\partial \sigma_r}{\partial r} + 3 \frac{A(t)}{r^3} + \frac{E}{1-\nu}\, \varepsilon^{vp} = 0 \qquad (4).$$

Note that (4) is true everywhere in the medium ; the flow law has not yet been used.

d- The flow law can be written as follows, once σ_φ has been eliminated :

. In a viscoplastic zone :

$$\frac{\partial \varepsilon^{vp}}{\partial t} = \frac{1}{\eta}\left(\frac{r}{2} \frac{\partial \sigma_r}{\partial r} - 2\,\omega\, C \right) \qquad (5).$$

. In an elastic zone :

$$\varepsilon^{vp} = 0 \qquad (5')$$

. In a residual zone :

$$\frac{\partial \varepsilon^{vp}}{\partial t} = 0 \quad \text{and} \quad \varepsilon^{vp} \neq 0 \qquad (5'')$$

. *Viscoplastic relationships* :

In a viscoplastic zone, (4) and (5) can be combined :

a- the elimination of $\partial\sigma_r/\partial r$ between (4) and (5) leads to

$$\frac{\partial \varepsilon^{vp}}{\partial t} + \alpha\, \varepsilon^{vp} + \frac{1}{2\eta}\left(\frac{3\,A(t)}{r^3} + 4\,\omega\, C \right) = 0 \qquad (6)$$

Setting : $\alpha = E/(2\eta\,(1-\nu))$.

b- The elimination of ε^{vp} between (4) and (5) leads to

$$\alpha \left(\frac{\partial \sigma_r}{\partial r} - \frac{4\,\omega\, C}{r} \right) + \frac{\partial^2 \sigma_r}{\partial r \partial t} + 3 \frac{\dot{A}}{r^4} = 0$$

The boarders of the viscoplastic zone are $r = 1$ and $r = x$. Then integration with respect to r between $r = 1$ and $r = x$ leads to :

$$\alpha\, \sigma_x + \frac{\partial \sigma_x}{\partial t} - \alpha \frac{4\,\omega\, C}{3} \operatorname{Log} x^3 - \dot{A}(t) \left(\frac{1}{\rho^3} - 1 \right) = \alpha\, \sigma_i(t) + \dot{\sigma}_i(t) \qquad (7).$$

Setting : $\sigma_x = \sigma_r\,[\, x(t), t\,]$.

SOLUTION

In the following, we will find a first order differential equation governing the evolution of the unknown quantity A(t) (and, if necessary, the evolution of the viscoplastic boarder x(t)). The other unknown quantities can then easily be deduced.

Let ξ be any particular radius (ξ can be 1 or ρ or any value between 1 and ρ).

Let $t_c(\xi)$ an instant when becomes viscoplastic :

$$t = t_c(\xi) \quad \Leftrightarrow \quad x(t) = \xi \;, \quad \dot{x}(t) > 0 \quad .$$

An $t_d(\xi)$ an instant when undergoes a viscoplastic unloading :

$$t = t_d(\xi) \quad \Leftrightarrow \quad x(t) = \xi \;, \quad \dot{x}(t) < 0 \quad .$$

I - FULLY ELASTIC SPHERE (A)

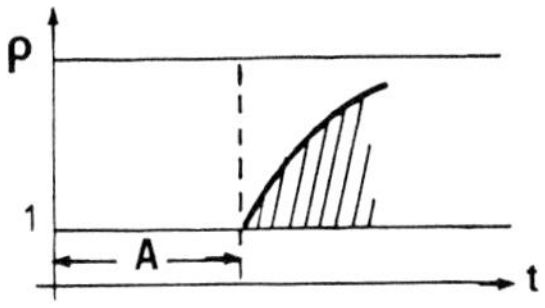

Figure 2 : Fully elastic sphere (A).

In this case, (4) holds with $\varepsilon^{vp} = 0$ (5'). Integration of (4) between r = 1 and r = ρ yields :

$$\sigma_i(t) = A(t)\left(1 - \frac{1}{\rho^3}\right) .$$

This case ends when a viscoplastic zone appears, i.e. $r\, \partial\sigma_r/\partial r = 4\,\omega\, C$.

Such a zone necessarily appears for r = 1, when $\sigma_i = -4\,\omega\, C/3\,(1 - 1/\rho^3)$.

II - FULLY RESIDUAL OR RESIDUAL/ELASTIC SPHERE (D or G)

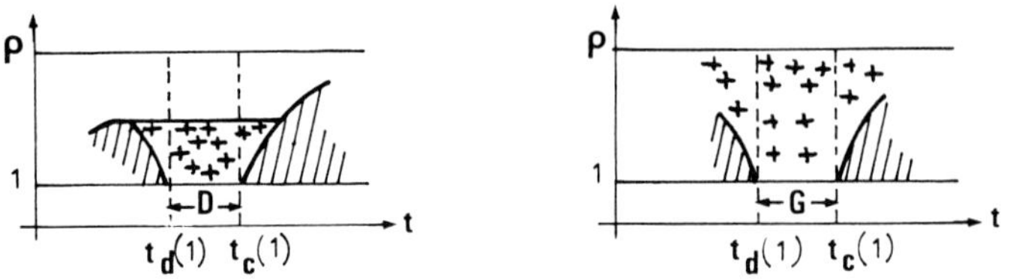

Figure 3 : Fully residual or residual/elastic sphere (D or G).

In those two cases, $\partial\varepsilon^{vp}/\partial t = 0$ (5") in the whole sphere. Then, by derivating (4) with respect to time, then integrating with respect to r between $r = 1$ and $r = \rho$:

$$\dot{\sigma}_i(t) = \dot{A}(t)\left(1 - \frac{1}{\rho^3}\right)$$

This case ends when a viscoplastic zone appears, i.e. :

$$\sigma_i(t_c) + \frac{4C}{3}\left(1 - \frac{1}{\rho^3}\right)\omega(t_c) = \sigma_i(t_d) + \frac{4C}{3}\left(1 - \frac{1}{\rho^3}\right)\omega(t_d) \quad .$$

III - FULLY VISCOPLASTIC SPHERE (E)

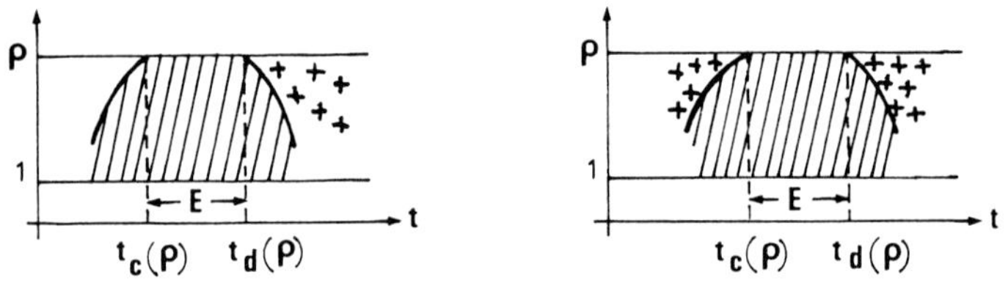

Figure 4 : Fully viscoplastic sphere (E).

In this case, (7) can be used with $x = \rho$, and then $\sigma_x = 0$:

$$\alpha\,\sigma_i + \dot{\sigma}_i = \dot{A}\left(1 - \frac{1}{\rho^3}\right) - \alpha\,\frac{4\,\omega\,C}{3}\,\mathrm{Log}\,\rho^3 \; .$$

Moreover, derivating (6) with respect to time, then multiplying by $\exp(\alpha t)$, and integrating with respect to time between $t_c(\rho)$ and $t_d(\rho)$ yields :

$$\int_{t_c(\rho)}^{t_d(\rho)} \dot{A}(t)\,\exp(\alpha t)\,dt = 0 \quad ,$$

or :

$$0 = \left[\exp(\alpha t) \left\{ \sigma_i(t) + \frac{4\,\omega(t)\,C}{3} \operatorname{Log} \rho^3 \right\} \right]_{t_c(\rho)}^{t_d(\rho)}$$

So the instant t_d when this case ends can easily be calculated.

IV - VISCOPLASTIC/ELASTIC SPHERE (B)

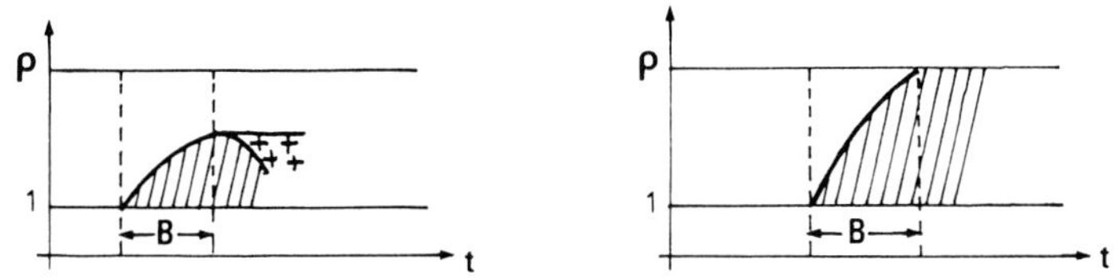

Figure 5 : Viscoplastic/elastic sphere (B).

In the elastic zone $x \leqslant r \leqslant \rho$, (4) and (5') yields :

$$\sigma_x = A(t) \left(\frac{1}{x^3} - \frac{1}{\rho^3} \right)$$

Moreover,

$$\frac{\partial \varepsilon^{vp}}{\partial t} [\, x(t), t \,] = 0, \text{ or } \frac{\partial \sigma_r}{\partial r} [\, x(t), t \,] = \frac{4\,\omega\, C}{x}$$

then $A(t) = -\frac{4\ \omega\ C}{3} x^3$.

On the other hand : $\frac{\partial \sigma_r}{\partial t} [\, x(t), t \,] = \left(\frac{1}{x^3} - \frac{1}{\rho^3} \right) \dot{A}$ then :

$$0 = \alpha\, \sigma_i + \dot{\sigma}_i + \frac{4\ \omega\ C}{3} \left(1 - \frac{1}{\rho^3} \right) \dot{\overline{x^3}} + \alpha\, \frac{4\ \omega\ C}{3} \left(1 + \operatorname{Log} x^3 - \frac{x^3}{\rho^3} \right).$$

This case ends, when $r = \rho$ or when $\dot{x} = 0$. *N.B.* : $\dot{\overline{x^3}}$ means $\frac{d}{dt} x^3$.

V - VISCOPLASTIC/RESIDUAL SPHERE, WITH DECREASING PLASTIC RADIUS (C or F)

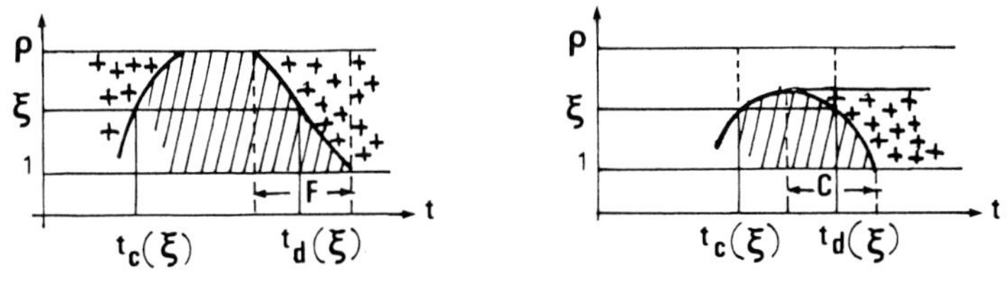

Figure 6 : Viscoplastic/residual sphere, with decreasing plastic radius (C or F).

In the residual zone, $r \leqslant x(t)$, (4) can be derivated with respect to time, then integrated with respect to r between $r = x$ and $r = \rho$:

$$\frac{\partial\sigma_r}{\partial t}[\,x(t),t\,] = \dot{A}(t)\left\{\frac{1}{x^3} - \frac{1}{\rho^3}\right\} \quad .$$

Then (7) can be rewritten in the following way :

$$\alpha\,\sigma_i + \dot{\sigma}_i = \left(1 - \frac{1}{\rho^3}\right)\dot{A}(t) - \alpha\,\frac{4\,\omega\,C}{3}\,\mathrm{Log}\,x^3 + \alpha\,\sigma_x \quad .$$

This latter relation can be derivated with respect to time, as for $\frac{\partial\sigma_r}{\partial t}[\,x(t),t\,]$ is known (see above) and $\frac{\partial\sigma_r}{\partial r}[\,x(t),t\,] = \frac{4\,\omega\,C}{x}$. Then :

$$\alpha\,\dot{\sigma}_i + \ddot{\sigma}_i = \left(1 - \frac{1}{\rho^3}\right)\ddot{A}(t) + \alpha\,\dot{A}(t)\left(\frac{1}{x^3} - \frac{1}{\rho^3}\right) \qquad (8).$$

In this relation remain two unknown quantities, namely A(t) and x(t). In order to eliminate x(t), we integrate (6) as in III :

$$\int_{t_c(\xi)}^{t_d(\xi)} \dot{A}(t)\,\exp(\alpha t)\,dt = 0 \quad .$$

Or, by derivating with respect to ξ^3 :

$$\left[\frac{\dot{A}(t)\ \exp(\alpha t)}{\dot{x}^3(t)}\right]_{t_c(\xi)}^{t_d(\xi)} = 0 \quad .$$

Then (8) can be multiplied by $\exp(\alpha t)$ and integrated with respect to time between $t_c(\xi)$ and $t_d(\xi)$:

$$\left[\exp(\alpha t)\left\{\dot{\sigma}_i(t) - \dot{A}(t)\left(1 - \frac{1}{\rho^3}\right)\right\}\right]_{t_c(\xi)}^{t_d(\xi)} =$$

$$= \alpha \int_{t_c(\xi)}^{t_d(\xi)} \dot{A}(t)\left\{\frac{1}{x^3(t)} - 1\right\}\exp(\alpha t)\,dt \quad .$$

Let x_M be the maximum value of x between $t_c(\xi)$ and $t_d(\xi)$. Then :

$$\int_{t_c(\xi)}^{t_d(\xi)} \dot{A}(t)\left\{\frac{1}{x^3(t)} - 1\right\}\exp(\alpha t)\,dt =$$

$$= \int_{\xi^3}^{x_M^3} -\left(\frac{1}{x^3} - 1\right)\left[\frac{\dot{A}(t)\ \exp(\alpha t)}{\dot{x}^3(t)}\right]_{t_c(x)}^{t_d(x)} dx^3 = 0 \quad .$$

Then the set of equations :

$$\left[\frac{\dot{A}(t)\ \exp(\alpha t)}{\dot{x}^3(t)} \right]_{t_c}^{t_d} = 0 \qquad \left[\exp(\alpha t) \left\{ \dot{\sigma}_i(t) - \left(1 - \frac{1}{\rho^3}\right) \dot{A}(t) \right\} \right]_{t_c}^{t_d} = 0 \quad .$$

allows for calculating $\dot{A}(t_d)$ and $\dot{x}(t_d)$, provided $\dot{A}(t_c)$ and $\dot{x}(t_c)$ have been previously calculated.

VI - VISCOPLASTIC/RESIDUAL SPHERE, WITH INCREASING PLASTIC RADIUS (H)

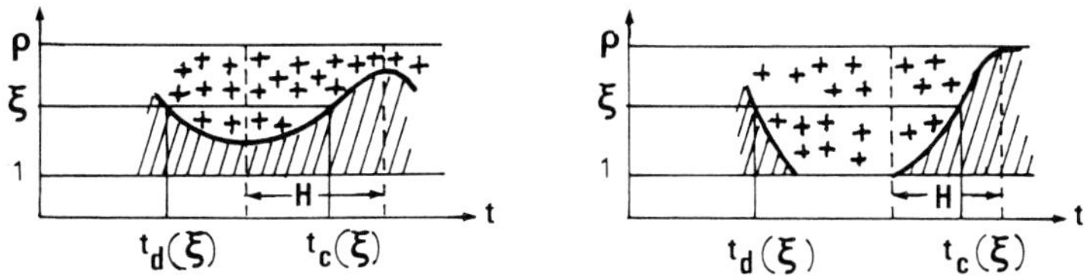

Figure 7 : Viscoplastic/residual sphere, with increasing plastic radius (H)

As for $\varepsilon_{vp}(\xi,t_c(\xi)) = \varepsilon_{vp}(\xi,t_d(\xi))$, (6) yields :

$$A\,[\,t_c(\xi)\,] + \frac{4C}{3}\xi^3\omega[\,t_c(\xi)\,] = A\,[\,t_d(\xi)\,] + \frac{4C}{3}\xi^3\omega[\,t_d(\xi)\,] \qquad ;$$

or, by derivating this relation with respect to ξ :

$$\frac{\dot{A}\,[\,t_c(\xi)\,]}{\dot{x}^3[t_c(\xi)]} - \frac{\dot{A}\,[\,t_d(\xi)\,]}{\dot{x}^3[t_d(\xi)]} = \frac{4C}{3} \left\{ \omega\,[\,t_c(\xi)\,] - \omega\,[\,t_d(\xi)\,] \right\} .$$

Now, (8) can be integrated with respect to time between $t_d(\xi)$ and $t_c(\xi)$:

$$\left[\alpha\ \sigma_i + \dot{\sigma}_i - \alpha\ A \left(1 - \frac{1}{\rho^3}\right) - \dot{A}\left(1 - \frac{1}{\rho^3}\right) \right]_{t_d}^{t_c} = \alpha \int_{t_d}^{t_c} \dot{A}(t) \left(\frac{1}{x^3(t)} - 1 \right) dt \quad .$$

Let x_m be the minimum value of x between $t_d(\xi)$ and $t_c(\xi)$. Then :

$$\int_{t_c(\xi)}^{t_d(\xi)} \dot{A}(t) \left[\frac{1}{x^3(t)} - 1 \right] dt = \int_{\xi^3}^{x_m^3} \left(\frac{1}{x^3} - 1 \right) \left[\frac{\dot{A}(t)}{\dot{x}^3(t)} \right]_{t_c(\xi)}^{t_d(\xi)} dx^3 =$$

$$= \frac{4C}{3} \left[\omega \right]_{t_c(\xi)}^{t_d(\xi)} \left\{ \mathrm{Log}\ x^3 - x^3 \right\}_{\xi^3}^{x_m^3} \quad .$$

Or :

$$\left[\alpha\,\sigma_i + \dot{\sigma}_i - \dot{A}\left(1 - \frac{1}{\rho^3}\right) \right]_{t_d}^{t_c} = \alpha\,\frac{4C}{3}\left[\,\omega\,\right]_{t_c(\xi)}^{t_d(\xi)} \left\{ \text{Log}\; x^3 - \frac{x^3}{\rho^3} + 1 \right\} \quad .$$

Then the set of equations :

$$\left[\frac{\dot{A}(t)}{\dot{x}^3(t)} - \frac{4C}{3}\,\omega(t) \right]_{t_c}^{t_d} = 0 \quad \left[\alpha\,\sigma_i + \dot{\sigma}_i - \dot{A}\left(1 - \frac{1}{\rho^3}\right) + \alpha\,\frac{4C}{3}\,\omega\left(1 + \text{Log}\; x^3 - \frac{x^3}{\rho^3}\right) \right]_{t_c}^{t_d} = 0$$

allows for calculating $\dot{A}(t_c)$ and $\dot{x}(t_c)$, provided $\dot{A}(t_d)$ and $\dot{x}(t_d)$ have been previously calculated.

. *Case of a constant load* :

We consider the particular case of a load of the following form :

$$\sigma_i(t) = \frac{4C}{3}\,F\,Y(t) \quad .$$

Where F is a constant, larger than $(1 - 1/\rho^3)$ and Y(t) is the Heaviside function.
ω has the same sign as F ; without loss of generality, $F > 0$ can be assumed. Then the radius of the viscoplastic zone is given by :

$$F\left\{ \alpha + \frac{d}{dt} \right\} Y(t) = \frac{\rho^3 - 1}{\rho^3}\,\frac{dx^3}{dt} + \alpha\left\{ 1 + \text{Log}\; x^3 - \frac{x^3}{\rho^3} \right\} \quad .$$

Then :

. At the initial instant, x^3 "jumps" to the initial value

$$e^3\,(t = 0) = \frac{\rho^3}{\rho^3 - 1}\,F \quad .$$

. As t tends to infinity, x^3 tends asymptotically to the zero value of :

$$F = 1 + \text{Log}\; x^3 - \frac{x^3}{\rho^3} \quad .$$

If x^3 happens to be equal to ρ^3, unrestricted plastic flow occurs. Then :

. If $F > \rho^3 - 1$, u.p.f. occurs at the initial instant.

. If $F > \text{Log}\,\rho^3$, u.p.f. occurs after a finite interval of time.

. If $F < \text{Log}\,\rho^3$, there is no u.p.f.

The case $\rho = \infty$ will be discussed in the second part of the paper.

SECOND PART

UNDERGROUND CAVITIES IN ROCKSALT

INTRODUCTION

For centuries, rocksalt has been exploited as a raw material, by conventional mining or by leaching. But this material exhibits remarkable properties which are required for large scale underground storage : rocksalt is impervious and large cavities are easy and cheap to be washed out. The existence of salt beds is due to impervious overlying layers which, in return, protect aquifer layers from any pollution by the stored products. These advantages have led to thousands of underground storage cavities in the world excavated in salt beds and salt domes.

Some typical underground cavities have been drawn on figure 8.

. *Loading parameters* :

Many factors are likely to have a major influence on the mechanical behaviour of such cavities, for instance :

- Shape of the cavities.
- Composition of the overlying stratas.
- Vicinity of others cavities, if any.

But we will consider more precisely the influence of more quantitative parameters

- Mean value and variations of the internal pressure.
- Depth of the cavity.
- Mechanical properties of rocksalt.

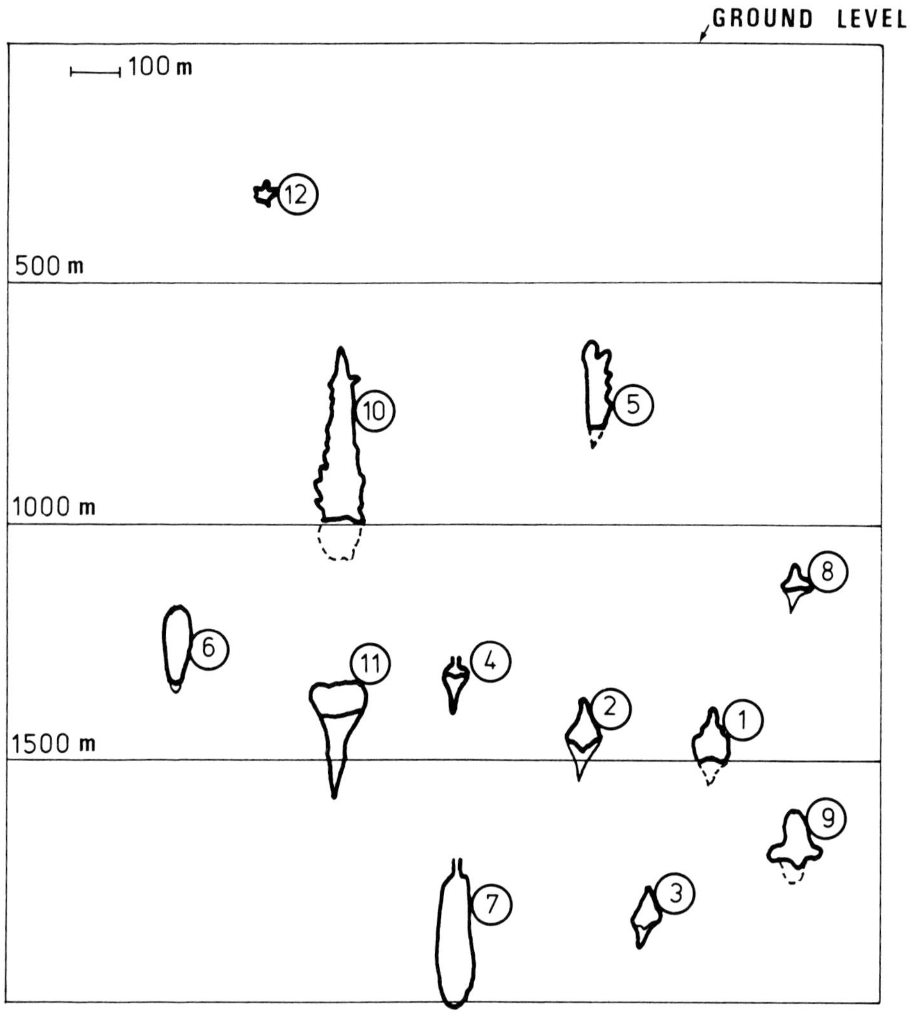

Figure 8 : Salt Caverns :

1. Tersanne (F) - 2. Etrez (F) - 3. Atwick (GB) - 4. Kiel (FRG)
5. Huntorf (FRG) - 6. Epe (FRG) - 7. Eminence (USA -
8. Melville (Can.) - 9. Regina (Can.) - 10. Manosque (F) -
11. Hauterives (F) - 12. Salies de Béarn (F).

. *Internal pressure* :

The internal pressure in an underground storage is strongly dependent on the nature of the stored products :

- for liquid of liquefied products, the central tubing is filled with brine up to the surface, for any movement of products must be balanced by an equivalent movement of brine, so that the cavity and the tubings remain filled up, whatever be the stored quantity of hydrocarbons. The small variations of pressure, due to losses of load during injection or withdrawal, can be neglected, so that the internal pressure is quite close to :
(Pressure in MPa, H in meters)

$$P_i = 0.012 \ H \qquad (1).$$

- For natural gas, the internal pressure can vary in a large extent, only restricted by safety rules. The storage must remain tight and stable ; these two restraints lead to select a relevant maximal pressure (for tightness) and minimal pressure (for stability). In France, for cavities about 1500 meters deep, the selected rules were :

$$8 \text{ MPa} \leqslant P_i \leqslant 0.0166 \ H \qquad (2).$$

The two cases (2) and (3) can be summed up briefly : in a gas (respectively oil) storage the volume (respectively pressure) of the products is kept constant.

. *Depth of the cavity* :

Due to viscoplastic properties of rocksalt, it seems reasonnable to think that in many cases the state of stresses in a salt dome or salt bed is not very different of an isotropic state of stresses, equal to the socalled "geostatic stress" :

$$\sigma_{ij} = - P \ \delta_{ij} \quad , \quad P = 0.023 \ H \qquad (3).$$

. *Mechanical properties of Rock Salt* :

Instead of recent considerable advances in the knowledge of the mechanical behaviour of rock salt, interpretation of laboratory tests appears to be controversial. Many authors have neglected in the past the delayed effects in the behaviour of rocksalt, which has appeared from in situ observations to be of utmost importance ; hasty conclusions sometimes have brought about inadequate interpretations and predictions inconsistent with the actual behaviour of the cavities.

We will propose in the following to use in situ datas in order to back calculate the actual properties of rocksalt.

. *In situ datas* :

Many of the considered cavities have been in operation for more than twenty years and their evolution has been precisely controlled. Some useful conclusions can be pointed out :

i - Liquid or liquefied products storage are affected by a small convergency (less than 5% of the initial volume).

ii- Several gas storage, or, more precisely, storages worked at a very low minimum pressure, have suffered larger damage (figure 9) :

. Eminence Salt Dome, U.S.A. :

The cavity was located between 1725 meters and 1965 meters ; the internal pressure varied between 7 MPa and 28 MPa, with several cycles each year. After two years, the bottom has raised up by 36 meters and the total loss of volume was about 40% (Baar, 1977).

. Kiel, R.F.A. :

This experimental cavity was located between 1300 meters and 1400 meters. After leaching, the brine was pumped out till the tubing emptied, so that the internal pressure vanished to zero. During this test, the roof collapsed ; the loss of volume was 7500 m^3, 45 days after the end of the test (the initial volume was 68 000 m^3, including 28 000 m^3 of sump). Five months after, the cavity exhibited an additional loss of 1900 m^3 (Röhr, 1974).

. Tersanne, France :

Two cavities, situated between 1400 meters and 1500 meters have been operated during 10 years ; the internal pressure varied between 8 MPa and 22 MPa with one or two annual cycles. At the end of this period, the loss of volume appeared to be some 25% or 30% (Boucly, Legreneur, 1980). Two main inferences can be made :

i - The losses of volume, have been larger than expected for these three cavities (or groups of cavities) ;

ii- the deformation have been delayed on several months or years.

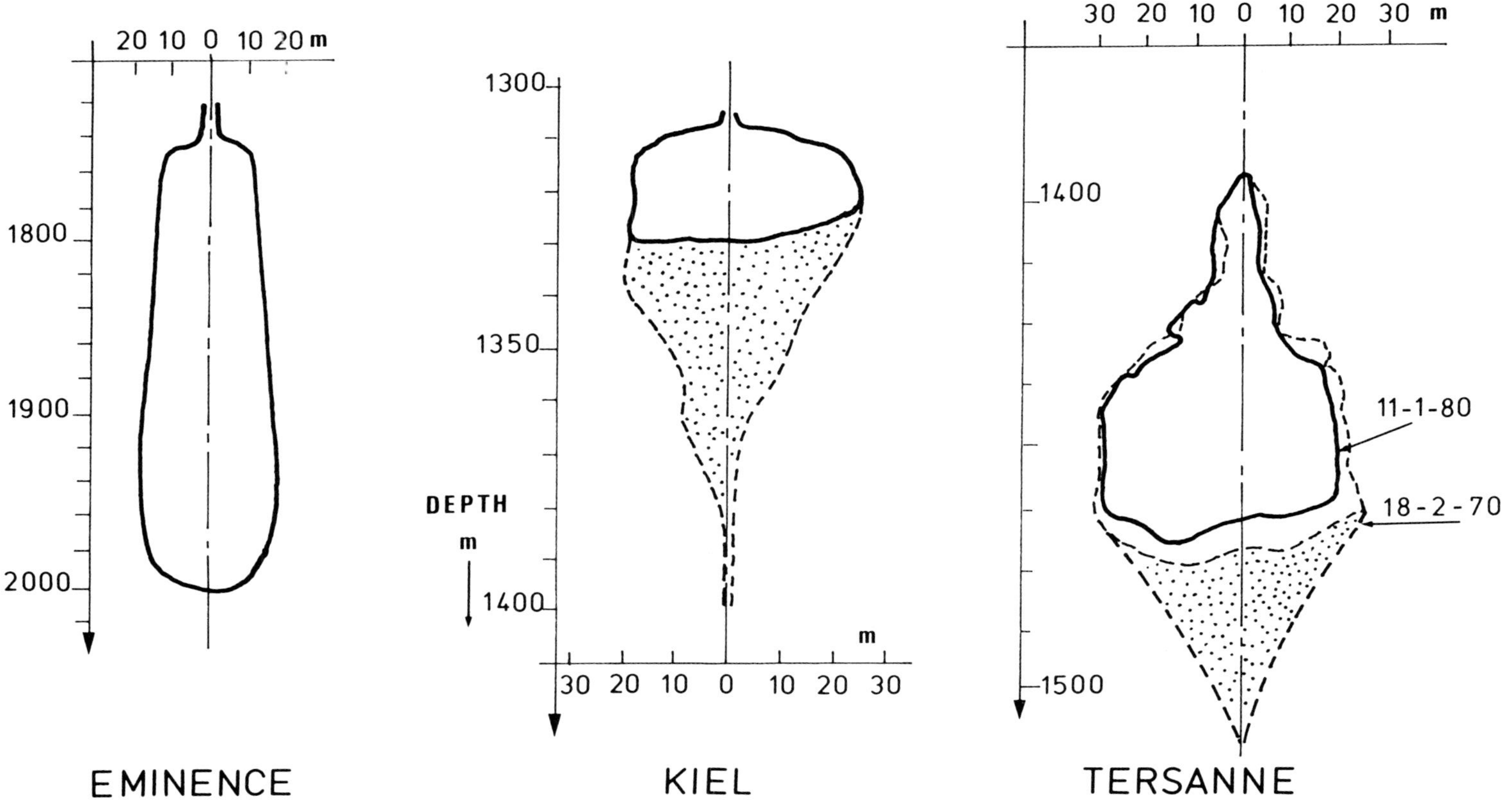

Figure 9 : Eminence, Kiel and Tersanne salt caverns.

A TENTATIVE EXPLANATION

A rough but simple index of the load which the cavities are submitted to, is the difference between the overburden load (formula (1) and the internal pressure (formula (2) or (3)).

The figure 10 shows this index, versus depth, for different cavities. In this diagram, the storages of liquids (oil, LPG, etc...) are standed for by points located along a straight line :

$$P - P_i = (0.023 - 0.012)\ H = 0.011\ H \tag{4}$$

Vertical segments stand for gas storages, because their internal pressure is not constant. Gas storages take up the upper part of the diagram, because they are usually leached in deeper salt beds and their minimal pressure is free and then often very low.

As suggested by figure 10 large losses of volume have been observed when the value of the index is more than 20 MPa.

. *Elastoplasticity* :

The difference $P - P_i$ or applied load intensity appears to have a major influence. But, losses of volume are not proportional to this stress intensity, for minor convergency is observed when $P - P_i$ is less than 20 MPa. These facts suggest a plastic behaviour.

Let us consider the perfect plastic Tresca model, with cohesion C and E Young modulus, ν Poisson's ratio. The loss of volume as a function of $(P - P_i)$ is given by :

$$\frac{1}{3}\frac{\Delta V}{V} = \frac{4C}{E}(1 - 2\nu)\ Q - \frac{3(1-\nu)}{2}\exp(Q - 1)$$
$$Q = \frac{3}{4C}(P - P_i) \tag{5}$$

If likely values of elastic parameters ($\nu = 0.25$, $E = 2.10^4$ MPa) are chosen, this simple model is in good agreement with in situ datas (figure 10) if the cohesion is taken as rather low (less than 3 MPa).

So the framework of plasticity allows for explaining the order of magnitude of the global loss of volume. But in this theoretical framework, the total loss of volume is reached as soon as the cavity is created and filled with the stored products. Such an assumption is not consistent with actual observations. The actual history of the deformation must be explained by resorting to the framework of viscoplasticity.

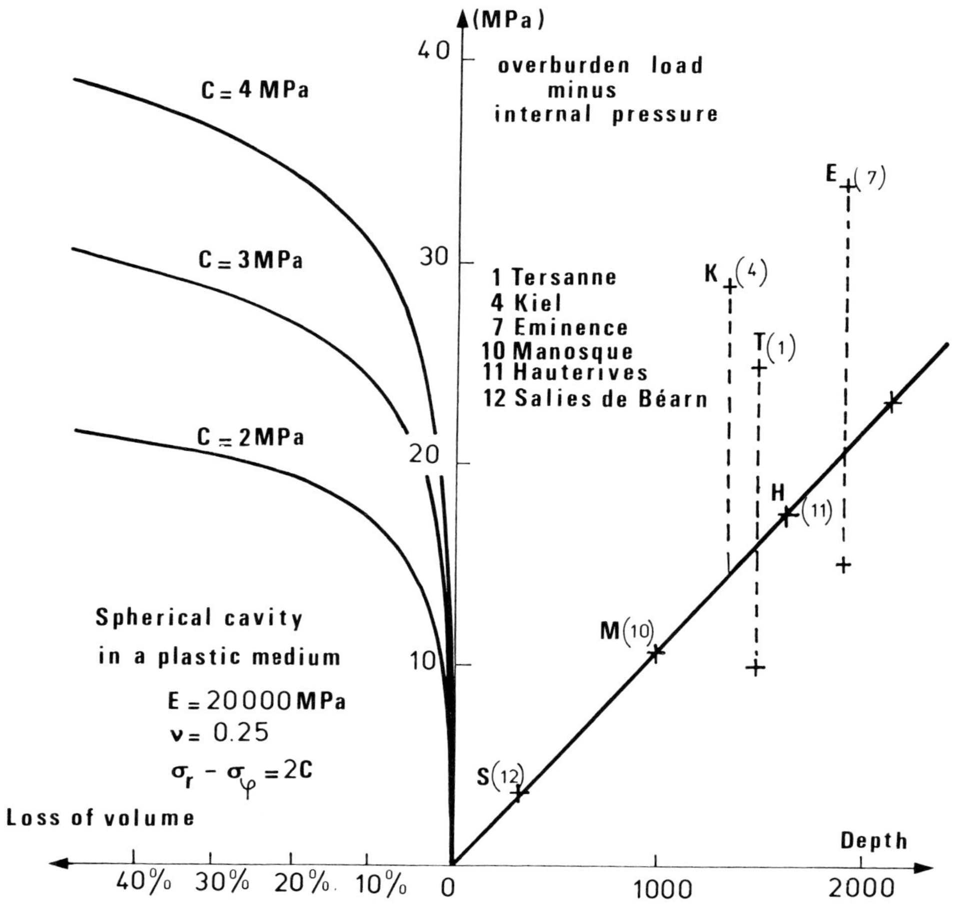

Figure 10 : Mechanical loading versus depths and volume loss.

. *Viscoplasticity* :

Plasticity and viscoplasticity give similar results only if the period between initial and final observations is larger than some viscoplastic time constant, so that most of the convergency has been achieved.

As it has been reported above, in situ experience suggests that this time constant is high ; then it was useful to dispose of a closed form solution for the problem of the evolution of a spherical cavity submitted to a varying internal pressure (Bérest and Nguyen Minh Duc, (1981)).

A more general solution of this problem has been given in the first part of this paper. This solution depends of the values of 3 main parameters (cohesion C, viscosity constant η, and Young modulus E).

Let us select a theoretical example, but rather close to the conditions of the gas storage of Tersanne : the depth is about 1500 meters (P = 34 MPa) and the minimum internal pressure is about 8 MPa. We will try to answer the following question : what triplet of parameters (C, η, E) can explain a 30% loss of volume after ten years under operation ? We must then solve the following equations (t is actual time, and x viscoplastic radius) :

$$\tau = \frac{E\ t}{2(1-\nu)} \quad \begin{cases} \frac{1}{3}\frac{\Delta V}{V} = \frac{4C}{3}\left\{ (1-2\nu)\ Q - \frac{3}{2}\ (1-\nu)\ x^3 \right\} \\ \frac{dQ}{d\tau} + Q - 1 = \frac{dx^3}{d\tau} + \mathrm{Log}\ x^3 \end{cases} \qquad (6).$$

The internal pressure is constant in our example ; then let s be defined by $Q = (3/4)\ (P - P_i)/C = (1+s)\ .\ Y(t)$, where Y(t) is the Heaviside function ; so the solution takes the form :

$$\tau = \int_s^{x^3} \frac{d\ \zeta}{s - \mathrm{Log}\ \zeta} \qquad (7).$$

Then, given t = 10 years and $\Delta V/V = 0.3$, a differential relation can be obtained for C and when E is fixed. The representative surface in (E, C, η) space of possible values for the parameters is shown on figure 11.

Now, several authors (Clerc-Renaud, Dubois (1978), Boucly (1980)) (personnal communication), infer values for Young's modulus in the range E = 10 000 to E = 30 000 MPa.

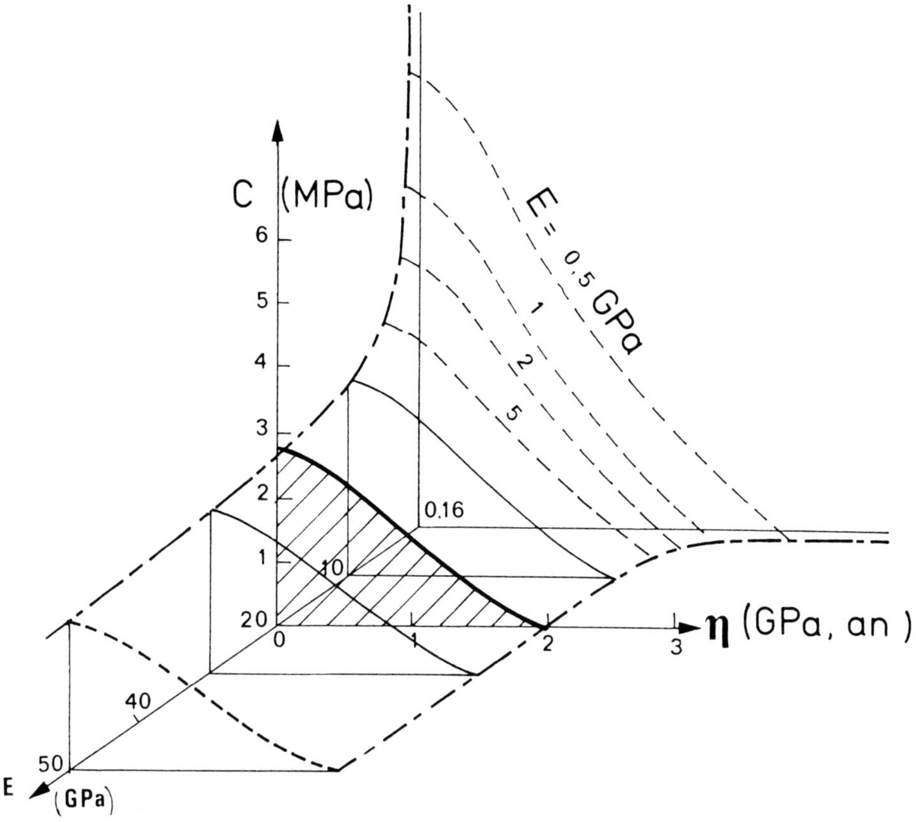

Figure 11 : Values of C, E, η accounting for a 30% loss of volume after 10 years.

Figure 11 shows that inside these limits E is of limited importance. So, the error will be negligible is we choose the mean value E = 20 000 MPa.

Then, the outlined curve C versus η on figure 11 shows that :

- either the cohesion C is high : we are close to perfect plasticity ; the additional loss of volume, after 10 years, can be predicted to be small.

- Or, the cohesion C is low, and the viscoplasticity constant is high but remains finite : rocksalt behaves like a Newton's fluid ; convergency will go on during centuries and up to total closure.

REFERENCES :

Aufaure, M., 1975, "Etude élastoviscoplastique d'un réservoir sphérique épais soumis à des charges lentement croissantes", Journal de Mécanique, Gauthier-Villars Ed., vol. 14, n° 2, pp. 221-235.

Baar, C.A., 1977, "Applied Salt Rock Mechanics - I", Elsevier Scientific Publication Company.

Bérest, P., Nguyen Minh, D., 1983, "Response of a spherical cavity in an elastic viscoplastic medium under a variable internal pressure" Int. J. Solids and Structures, Pergamon Press, vol. 19, pp. 1035-1048.

Bérest, P., Nguyen Minh, D., 1984, "Deep Underground Storage Cavities in Rock Salt : Interpretation of In Situ Data from French and Foreign Sites", The Mechanical Behavior of Salt, Trans. Tech. Publications, H.R. Hardy and M. Langer Ed., pp. 555-572.

Boucly, Ph., Legreneur, J., 1980, "Hydrocarbon storage in cavities leached out of salt formations", Subsurface Space, Rockstore 80, M. Bergman ed. Pergamon Press, pp. 251-258.

Clerc-Renaud, A., Dubois, D., 1978, "Long term operation of underground storage in salt", V Symposium on Salt, Hambourg.

Madejski, J., 1960, "Theory of non stationary plasticity explained on the example of thick-walled reservoir loaded with internal pressure", Arch. Mech. Stos., n° 5/6, vol. 12, pp. 775-788.

Ottosen, J., 1985, "Behaviour of viscoelastic-viscoplastic spheres and cylinders - partly plastic vessel walls", Int. J. Solids and Structures, Pergamon Press, vol. 21, n° 6, pp. 573-595.

Röhr, H.U., 1974, "Mechanical behaviour of a gas storage cavern in evaporitic rocks", IV Symposium on Salt.

Tijani, S.M., 1978, "Résolution numérique des problèmes d'élastoviscoplasticité. Application aux cavités de stockage en couches salines profondes", Thèse de Docteur-Ingénieur, Université Paris VI.

Wierzbicki, T., 1963, "A thick-walled elastoviscoplastic spherical container under stress and displacement boundary value condition", Arch. Mech. Stos., n° 2, vol. 15, pp. 297-308.

CHARACTERISTIC STATE AND INFRARED VIBROTHERMOGRAPHY OF SAND

Minh Phong Luong

Laboratoire de Mécanique des Solides,
CNRS UA 317 Ecole Polytechnique
91128 Palaiseau Cedex France

Summary

Rheological properties of granular soils subjected to vibratory and transient loading can be interpreted at the granular level where the solid particles interact leading to a global irreversible aggregation (contractancy) or disaggregation (dilatancy).

A cohesionless granular soil is considered as a grain assembly. Observed macroscopic deformations are derived essentially from their structural modifications, i.e. rearrangement of the constitutive grains inducing irreversible contractive or dilative volume changes :

- Compaction mechanism corresponding to the mutual tightening of solid particles which induces a contractive behaviour,

- Distortion mechanism due to grain slidings leading initially to a contractive behaviour followed by a dilative behaviour when the deviatoric stress level $\eta = q/p$ exceeds the grain interlocking threshold called characteristic state (stationary volume change or zero dilatancy rate),

- Attrition mechanism subsequent to asperity breakage and grain crushing modifying the relative density under high stresses. The resultant effect is a contractive behaviour.

The characteristic threshold is revealed by the appearance of a dilatancy loop when the load cycle crosses the deviatoric stress level η_c . Such observations enable the determination of the entanglement capacity of granular soil.

Below the characteristic threshold, the intergranular contacts are stable. The limited slips tend to a maximal aggregation. In this subcharacteristic domain or contractancy zone, a hysteresis loop occurs when reloading. The mechanical behaviour depends upon the load history.

Above the characteristic threshold, the grain contacts become unstable, leading to significant slidings due to interlocking breakdown. A reload shows a dilatancy loop with memory loss of load history and a softening phenomenon occurs.

The infrared vibrothermography used in our laboratory LMS demonstrates the thermal dissipation of sheared granular soil characterizing the sliding mechanism of grains when the granular interlocking structure breaks down on exceeding the characteristic threshold.

This non-destructive and non-contact testing technique allows records and observations in real time of heat patterns produced by the dissipation of energy due to friction between grains.

The infrared vibrothermographic test couples mechanical and thermal energy. Additionally it offers the potential of directly monitoring the stress state of particle rearrangements or characteristic threshold and of predicting the degradation or damage of granular materials by active heating.

I - INTRODUCTION

Research in granular material behaviour has been considerably developed using constitutive laws in the case of dynamic, vibratory, cyclic and transient loading. The applicability of the stress-strain analysis requires realistic understanding and ready determination of the significant factors of the mechanical performance of granular material response.

This paper describes an experimental approach in order to :

1.- Grasp the basic aspects of stress-strain response of granular materials under various loadings ;

2.- Consolidate the experimental data for a comprehensive definition of a stress domain in which the resultant effect of repeated loading is contractancy (contractive behaviour) or dilatancy (dilative behaviour) ;

3.- Describe the deformation process which influences the overall macroscopic behaviour of granular materials ;

4.- Visualize the distortion mechanism occurring in the grain structure by infrared thermographic analysis giving a physical meaning to parameters introduced in the constitutive law ;

5.- Suggest the development of relevant parameters for use in analytical and numerical models ;

6.- Guide and interpret the main features of the cyclic behaviour of granular materials.

2. DEFORMATION MECHANISMS

In order to analyze and predict the macroscopic behaviour of granular material, it is necessary to understand how the individual microscopic constituent elements interact at the grain level.

A cohesionless granular material can be considered as a grain assembly where the discrete and solid granules are in contact and free to move with respect to their neighbours. It is often assumed that the constituent granules are in direct, elastic contact with one another. The inherent non linearity of Hertz relationships between two elastic bodies indicates great difficulties in the application of contact theory to the study of granular media.

Nevertheless, observed macroscopic deformations of the material are derived essentially from their structural modifications, i.e. rearrangements of the constitutive grains inducing irreversible contractive or dilative volume changes :

1.- Compaction mechanism corresponding to the mutual tightening of solid particles inducing a contractive behaviour ;
2.- Distortion mechanism due to irreversible grain slidings leading initially to a contractive behaviour, then interlocking disrupture - where the individual particles are plucked from their interlocking seats and made to slide over the adjacent particles with large distortion of the grain arrangement - inducing significant dilative volume changes ;
3.- Attrition mechanism subsequent to asperity breakage and grain crushing which modifies the relative density under high stresses. The resulting effect is a contractive behaviour.

Granular material deformation under load is due in part to elastic deformation of the solid particles. This elastic deformation often constitutes only a small fraction of the total deformation and is often obscured by deformation resulting from slippage, rearrangement and crushing of particles.

Classical elastic plastic theory assumes that the elastic and plastic components of strain can be isolated by loading and subsequent unloading. The recoverable strain is elastic. The total strain ε, is the sum of the elastic strain ε^e and the plastic strain ε^p. However, in granular materials it is not possible to separate the elastic strains simply by unloading.

Even when recovery of strain in granular materials is a result of stored elastic energy, the strains recovered are not always purely elastic. Slippage at particle contacts may accompany strain recovery. Sometimes elastic and plastic deformations occur in parallel and cannot be isolated from each other experimentally.

3. MECHANICAL CHARACTERIZATION

Extensive laboratory tests are presented to assess various rheological responses of cohesionless granular materials subjected to vibratory, cyclic and transient loading under drained or undrained conditions.

When tests are performed at varying confining pressures or changing pore water pressures, the penetration of the protective membrane into the solid skeleton may cause significant errors in the measurements of volume changes and pore pressure developments. To overcome this difficulty, it was necessary to set up the arrangement shown in Fig. 1, the purpose of which is to check the calibration and the accuracy of the volumetric strains or pore pressures obtained usually with conventional test procedure. Thanks to the double membrane device, the thin annular part allows a perfect control of effective confining pressure and avoids the penetration of the inner protective membrane by maintaining the pore pressure u' of the outer part always equal to the pore pressure u of the inner part. For this test, membrane penetrations are located only in the annular part of the sample.

The loading parameters of axisymmetric triaxial tests are :

- mean stress $p = (\sigma_a + 2\,\sigma_r)/3$
- deviatoric stress $q = \sigma_a - \sigma_r$
- deviatoric level $\eta = q/p$.

The corresponding deformation parameters are defined by :

. volumetric strain $\varepsilon_v = \varepsilon_a + 2\,\varepsilon_r$

. deviatoric strain $\varepsilon_q = 2(\varepsilon_a - \varepsilon_r)/3$

. dilatancy rate $\delta = \dot{\varepsilon}^p_v/\dot{\varepsilon}^p_q$.

Tests are conducted either with a constant stress rate $\dot{\sigma}$ or a constant strain rate $\dot{\varepsilon}$ on saturated materials under drained or undrained conditions. ε, ε^e and ε^p denote respectively total, elastic and plastic strain. For example, contraction or dilatation ε_v may be the resulting value of elastic volume increase or decrease ε^e and plastic volume change ε^p_v due to dilatancy and/or contractancy.

3.1.- CHARACTERISTIC THRESHOLD

Extensive laboratory tests on several sands show that the lowest point on the volume change axial strain curves (Fig. 2), - that is the point of minimum volume of the sample -, corresponds to a constant stress ratio η_c [1] .

The stress peak or maximum of shear resistance occurring at maximum dilatancy rate has been analyzed and interpreted by the stress dilatancy theory [2] .

The asymptotic part of the stress strain curves determining the ultimate strength has suggested the well-known critical state concept [3].

For our concern, the transient and cyclic loading cases require the analysis of the prepeak part where the stress ratio η_c at zero dilatancy rate (δ = 0) defines evidently the characteristic state of the granular material associated with an angle of aggregate friction φ_c [1] .

Under undrained conditions on a saturated sand a pore pressure increase characterizes the first stage of any undrained triaxial test irrespective of the initial sand density. As the deviatoric stress $q = \sigma_a - \sigma_r$ increases, the pore pressure generation rate $\dot{u}$ decreases, passing through zero to become negative, followed by large irreversible axial strains. This pore pressure generation behaviour parallels phases of contractancy and dilatancy in a drained test [4] .

Thus the deviatoric stress level corresponding to either the inversion of pore pressure generation rate in an undrained test, or the zero of the volume change rate in a drained one, determines the characteristic threshold CT unambiguously : the granular material is in a "characteristic state" having the following properties :

i. the volume change rate is zero ;

ii. the stress level reached by the material is an intrinsic parameter which defines a characteristic friction angle $\varphi_c = \sin^{-1}[3\eta_c/(\eta_c + 6)]$ determining the interlocking capacity of the grain assembly.

The value of φ_c is independant of the initial sand density. At any point on the characteristic line CL (Fig. 4) herely determined, the rate of the irreversible volume change is stricly zero.

The position of the effective loading point relative to the CL line determines its mechanical behaviour. In particular, the material undergoes large deformations after crossing the characteristic threshold CT into the surcharacteristic domain in which the characteristic value η_c is exceeded.

The characteristic line CL divides the allowable stress-space into two regions :(1) subcharacteristic region corresponding to an interlocking of grain structure or contractancy, (2) surcharacteristic region where disaggregation of granular material or dilatancy occurs. Thus every closed stress path in the subcharacteristic domain exhibits a contracting soil behaviour illustrated by an irreversible compaction (or an irreversible increase of pore water pressure under undrained condition) whereas a closed load cycle in the surcharacteristic domain leads to an irreversible volume expansion (or an irreversible decrease of pore water pressure).

A very accurate experimental procedure of determination of the characteristic threshold η_c follows which is readily available under either drained or undrained conditions : η_c is revealed by the appearance of a dilatancy loop with volume change or pore water pressure during a load cycle crossing the characteristic line CL (Fig. 5).

3.2.- CHARACTERISTIC PROPERTIES OF GRANULAR SOILS

Extensive laboratory tests using the axisymmetric triaxial apparatus on various sands : Fontainebleau sand, Loire sand, carbonate Channel sand [4], carbonate marine sediments [5] and Hostun sand [6] substantiate the following rheological properties :

. Under Drained Conditions

1.- Adaptation or elastic response may be considered as obtained after a finite number of cyclic isotropic loadings :

2.- Accommodation with stable hysteresis loop appears under radial loading or conventional loading at a stress level $\eta = q/p$ smaller than the characteristic threshold η_c. Stress-volume change curves of sandy soils exhibit a clockwise hysteresis loop after unloading and reloading. This hysteresis susceptibility becomes negligible when the number of cycles increases.

3.- For η greater than η_c, the hysteresis loop moves toward large deformations and cyclic loadings cause ratcheting behaviour with increasing cumulative strains. The soil volume dilates and reflects the phenomenon of dilatancy of the grain structure. After unloading, a dilatancy loop is seen in an anticlockwise direction on any diagram where volume change is plotted. The dilatancy loop is a very practical and useful criterion for the detection of the characteristic threshold.

4.- Densification of dense sands may be obtained easily by cyclic loading at large amplitude exceeding both triaxial compression and extension characteristic thresholds. The high amplitude loading benefits in partial loss of strain-hardening during the dilating phase in the surcharacteristic domain leading to a breakdown of the granular interlocking assembly. On each reload, the tightening mechanism induces new irreversible volumetric strains and recurs each time with a renewed denser material.

. Under Undrained Conditions

5.- Sand liquefaction occurs only when load is cycled alternately on both sides of zero deviatoric stress and has reached the characteristic levels.

6.- Cyclic non-alternated deviatoric stress tests show a progressive tendency of the effective stress state to move toward the characteristic level and to stabilize there, i.e. cyclic softening is occurring.

7.- Cyclic hardening of sandy soils may be observed when undrained loads are cycled in the surcharacteristic region bounded by the failure line FL and the characteristic line CL. It leads to a stabilization of the granular material on the characteristic thresthold as shown in Fig. 6. Irreversible strains accumulated during undrained loadings depend on the stress amplitude of cycles.

4. THERMAL DISSIPATION

Infrared thermography has been successfully employed as an experimental method for detection of plastic deformation during crack propagation under monotonic loading of a steel plate or as a laboratory technique for investigating damage, fatigue and creep mechanisms [7].

This experimental tool is used here to illustrate the grain sliding mechanism of a granular material subjected to a shear loading which exceeds the characteristic threshold. The heat dissipation evidenced here is associated with a plastic work of distortion.

4.1.- HEAT PRODUCING MECHANISM

A consideration of the forces and deformations at each contact surface [8] may serve as one starting point in interpreting the thermomechanical coupling of sand behaviour under vibratory shearing.

For the simplest case of two like spheres, each of radius R compressed statically by a force N which is directed along their line of centers, normal to their initial common tangent plane, the contact theory due to Hertz predict a plane, circular contact of radius $a = [3(1 - \nu^2)NR/4E]^{1/3}$ where ν, and E denote respectively Poisson's ratio and Young's modulus of the sphere.

The normal pressure on the contact area is given by :

$$\sigma = 3N(a^2 - \rho^2)^{1/2}/2\pi a^3$$

where ρ represents the radial distance from the center of the contact circle.

An additional force $\mathcal{T}$ is assumed to act in the plane of contact and its magnitude rises monotonically from zero to a given value. Because of symmetry the distribution of normal pressure remains unchanged. If there is no slip or relative displacement of contiguous points on a portion of the contact surface, the displacement of the contact surface in its plane is constant. The Mindlin's solution of the appropriate boundary-value problem shows that the tangential traction is parallel to the displacement (and to the applied force $\mathcal{T}$) axially symmetric in magnitude, and increases without limit on the bounding curve of the contact area. Slip is assumed to be initiated at the edge of the contact and to progress radially inward, covering an annular area. In accordance with Coulomb's law of sliding friction $\tau = f\sigma$ where f is a constant coefficient of static friction. The radius c of the adhered portion or the inner radius of the annulus of slip is given by $c = a(1 - \tau/fN)^{1/3}$ (Fig. 1).

When the tangential force $\mathcal{T}$ decreases from a peak value $\mathcal{T}^*$, $0 < \mathcal{T}^* < fN$, slip, once again, occurs, but its direction will be opposite to that of the initial slip. An annulus of counter-slip is formed and spreads radially inward as the tangential force is gradually decreased. Its inner radius is $b = a[1 - (\mathcal{T}^* - \mathcal{T})/2fN]^{1/3}$. The inelastic character of the unloading process appears evident since the annulus of counter-slip does not vanish when the tangential force is completely removed.

Under oscillating tangential force, the load-displacement curve forms a closed loop traversed during subsequent oscillation of $\mathcal{T}$ between the limits $\pm\mathcal{T}^*$ providing that N is maintained constant. The area enclosed in the loop represents the frictional energy dissipated in each cycle of loading. For small tangential forces, it has been suggested by Johnson [9] that the tangential displacement necessary to relieve the singularity in traction takes the form of an elastic deformation of the asperities. An increase in applied tangential force causes the asperities at the edge of the contact surface to deform plastically through relatively large strains, a process which leads to a marked increase in energy dissipation and to severe damage to the surfaces.

Thus at small amplitudes of the tangential force, energy is dissipated as a result of plastic deformation of a small portion of the contact surface, whereas, at large amplitudes, the Coulomb-sliding effect predominates.

In the conventional triaxial test, if the load is cycled within the subcharacteristic domain below the characteristic threshold η_c the intergranular contacts are stable. Small slips lead to a maximum entanglement due to the relative approach of constituent granules. The dissipated work given by the hysteresis loop (a) in the (q,ε_a) diagram is relatively small. The corresponding heat production is relatively low and negligible. On the contrary when the shear load is cycled above the characteristic trheshold, the intergranular contacts become unstable, leading to significant slidings due to interlocking breakdown. A large frictional energy (B) is dissipated as shown in the Figure 8 and is transformed almost entirely into heat owing to the thermomechanical conversion.

4.2.- VIBROTHERMOGRAPHY

Infrared thermography utilizes a photon-effect detector in a sophisticated electronics system in order to detect radiated energy and to convert it into a detailed real-time thermal picture on a video system. Temperature differences in heat patterns as fine as 0,1°C are discernible instantly and represented by several distinct hues.

This technique is sensitive, non-destructive and non-contact, thus ideally suited for records and observations in real time of heat patterns produced by the heat transformation of energy due to friction between grains of sheared sand. No interaction at all (with the specimen) is required to monitor the thermal gradient.

The quantity of energy W emitted by infrared radiation is a function of the temperature and the emissivity of the specimen. The higher the temperature, the more important is the emitted energy. Differences of radiated energy correspond to differences of temperature, since $W = h\ T^4\ \omega$ where h denotes a constant, T the absolute temperature and ω the emissivity.

Soils present a low thermomechanical conversion under monotonic loading. However plastic deformation - whereby sliding between grains occurs creating permanent changes globally or locally - is one of the most efficient heat production mechanisms. Most of the energy which is required to cause such plastic deformation is dissipated as heat. Such heat development is more easily observed when it is produced in a fixed location by reversed or alternating slidings due to vibratory reversed applied loads. These considerations define the use of vibrothermography as a non-destructive method for observing granular material damage.

4.3.- EXPERIMENTAL SET-UP AND RESULTS

The thermomechanical behaviour of a Stampian sand (Fontainebleau sand) is studied when subjected to two types of vibratory loading.

a. Conventional triaxial loading : indirect shearing

A cylindrical dry sand sample (dry unit weight γ_d = 15.7 kN/m3 ; void ratio e = 0.72 ; relative density I_D = 0.62) confined under a constant isotropic pressure of 100 kPa is subjected to a vibratory force generated by a steel mass located at the top of the specimen excited by an electro-dynamic vibrator.
When the frequency reaches 87 Hz with a controlled displacement of 1 mm at the base, the specimen (70 mm diameter - 150 mm high) is subjected to stationary stress waves and presents a striction zone where the deviatoric stress level η exceeds the characteristic threshold η_c of interlocking breakdown of the granular structure (Fig. 9a).

b. Cylindrical loading : direct shearing

A tubular sand sample at the same initial density confined under a pressure of 50 kPa is directly sheared by a concentric steel cylinder exciter in an axial vibratory motion by the electrodynamic generator.
In this case of loading, the principal stress axes rotate during loading.
At the frequency of 80 Hz and with a controlled displacement of 1 mm, the characteristic threshold is exceeded and hot colours due to heat production by friction appear as shown in the Figure 9b. The temperature increase is about 6° Celsius for a test duration of 20 secondes.

3.4.- THERMOMECHANICAL COUPLING

Thermal effects occurring in these tests may be interpreted on the basis of the theory of internal variables. Assuming the case of small disturbances, the two laws of thermodynamics and Fourier's law of heat conduction can lead to the following equation of thermomechanical evolution written with tensorial quantities k,α and σ [10] :

$$C\dot{T} = \text{div}(k \text{ grad } T) - t\,\alpha\,\dot{\sigma} + D + \phi$$

C denotes the specific heat per unit volume

T the absolute temperature of reference

k the thermal conductivity tensor

α the coefficient of thermal expansion

σ the stress tensor

D the intrinsic energy dissipation

ϕ the density of heat sources.

In the second member of the transient heat flow equation, the first term governs the thermal diffusion. The second term represents the thermo-elastic effect which may be significant in the cases of isentropic loading. The third term concerns the thermal dissipation due to viscosity or irreversibility phenomena. The last term denotes the presence of thermal sources.

Only friction energy dissipation has been shown in the presented tests. It may be interesting to study further how the other parameters k and α can be affected in the course of loading.

5. CONCLUDING REMARKS

The characteristic state concept, defined easily using conventional tests, allows the prediction of drained or undrained properties of granular materials.

a. The essential parameter for studying the rheological behaviour of granular materials subjected to cyclic and transient loading is the development of the volumetric strain during loading stages.

b. The friction angle φ_c is an intrinsic factor characterizing the interlocking capacity of grain structure under drained conditions and the average mobilized friction angle under undrained loading.

c. The characteristic threshold is defined by a stress threshold η_c at which level a dilatancy loop appears after unloading.

d. This criterion becomes all important when considering questions of cyclic loading, facilitating the definition of a region of contractive behaviour for the granular material. Beyond that region of contractancy, in the surcharacteristic domain and up to the failure limit the granular material behaviour during load cycles is dilative.

e. Infrared vibrothermography demonstrates the thermal dissipation of sheared granular material characterizing the sliding mechanism of grains when the granular interlocking structure breaks down on exceeding the characteristic threshold.
This non-destructive testing technique allows records and observations in real time of heat patterns produced by the dissipation of energy due to friction between grains.
The infrared vibrothermographic test couples mechanical and thermal energies. Additionally it offers the potential of directly monitoring the stress state of particle rearrangements or characteristic threshold and of predicting the degradation or damage of granular materials by active heating.

REFERENCES

[1] KIRKPATRICK, W.M., (1961), "*Discussion on Soil Properties and their Measurement*", Proc. 5th Conf. Soil Mech. and Found. Eng., III, pp. 131-133, Paris.

[2] ROWE, P.W., (1971), "*Theoretical Meaning and Observed Values of Deformation for Soils. Stress Strain Behaviour of Soils*", Cambridge, March 1971, pp. 143-194, ed. R.G.H. Parry.

[3] SCHOFIELD, A.N. & WROTH, C.P., (1968), "*Critical State Soil Mechanics*", Mc Graw Hill, London, G.B.

[4] LUONG M.P., (1980), "*Stress-Strain Aspects of Cohesionless Soils under Cyclic and Transient Loading*", Proc. Int. Symp. on Soil under Cyclic and Transient Loading, Swansea, Jan. 1980, U.K., PP. 315-324.

[5] NAUROY, J.F. & LE TIRANT, P., (1981), "*Comportement des Sédiments Carbonatés sous l'Action de Chargements Cycliques*", Proc. Xth ICSMFE, June 1981, Stockhölm, Sweden.

[6] THANOPOULOS, I., (1981), "*Contribution à l'Etude du Comportement Cyclique des Milieux Pulvérulents*", Thèse de Docteur-Ingénieur, I.M.G. Grenoble.

[7] REIFSNIDER, K.L., HENNEKE, E.G. & SINTCHCOM, W.W., (1980), "*The mechanics of Vibrothermography*" in Mechanics of Non destructive Testing, ed. by W.W. Stinchcom, pp. 249-276.

[8] MINDLIN, R.D. & DERISIEWICZ, H., (1953), "*Elastic Spheres in Contact under Varying Oblique Forces*", J. Appl. Mech. 20, pp. 327-344.

[9] JOHNSON, K.L., (1955), "*Surface Interaction between Elastically Loaded Bodies under Tangential Forces*", Proc. Roy. Soc., London, ser. 1, 230, pp. 531-548.

[10] NGUYEN, Q.S., (1984), "*Thermodynamique des Milieux Continus*", Cours DEA-ENPC (1984).

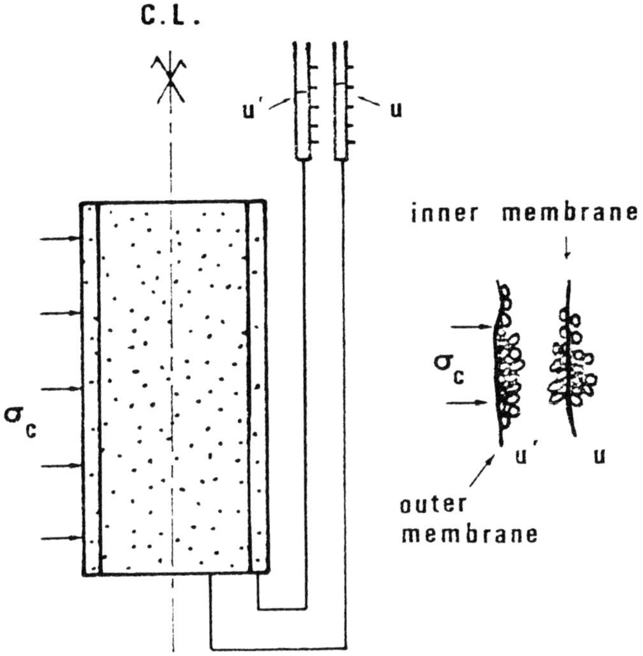

Fig.1 - Double membrane device for tests under varying confinement condition

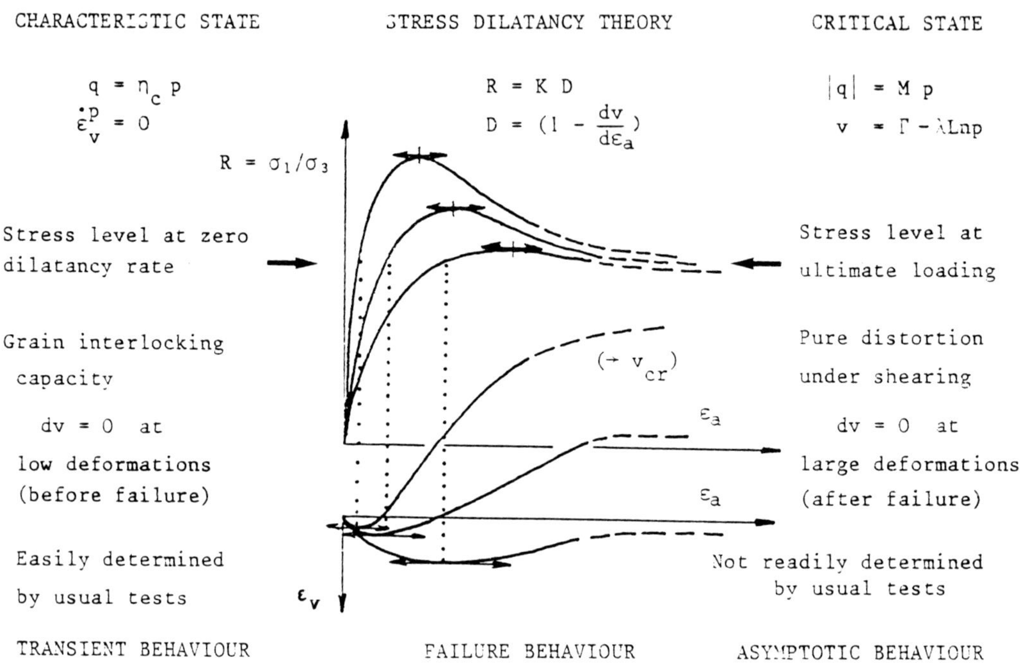

Fig. 2. Location of basic concepts.

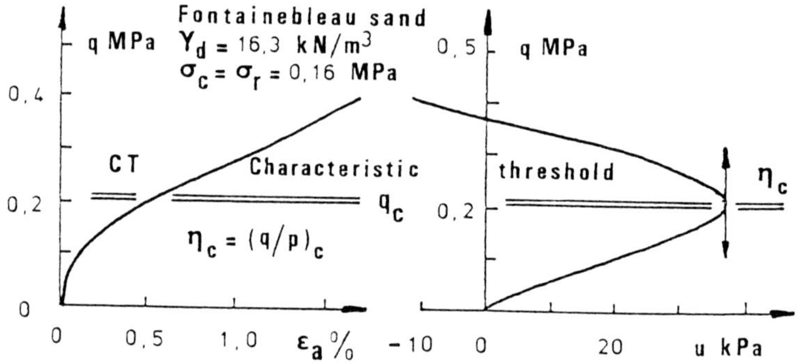

Fig. 3. Definition of the Characteristic Threshold under undrained condition.

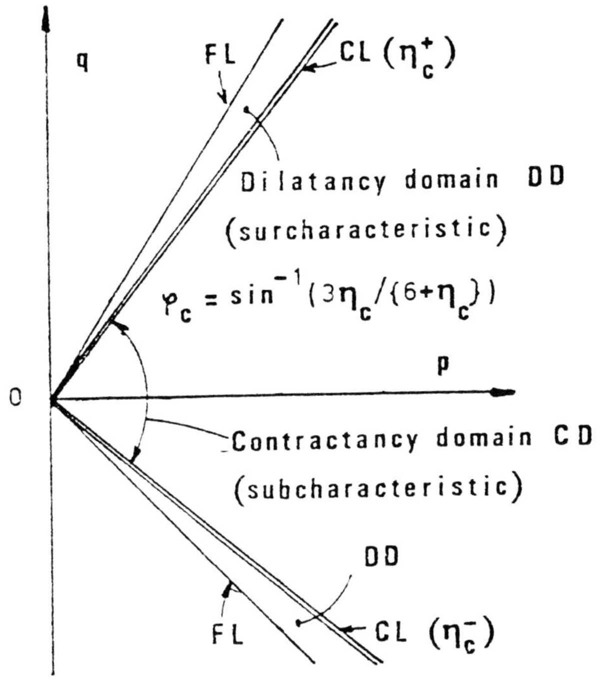

Fig. 4. Characteristic Criterion.

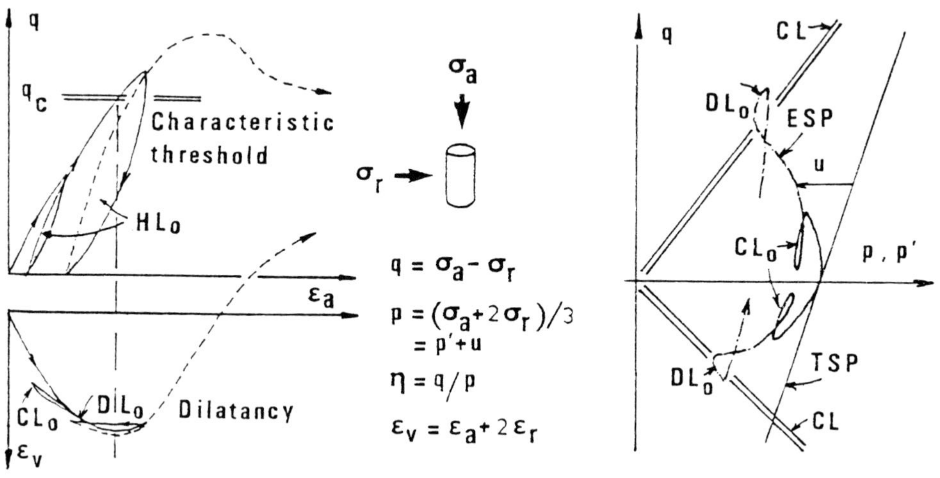

HL_o Hysteresis loop CL_o Contractancy loop DL_o Dilatancy

(5a) Drained test (5b) Undrained test

Fig. 5. Dilatancy loop observed after an effective load cycle exceeding the characteristic threshold.

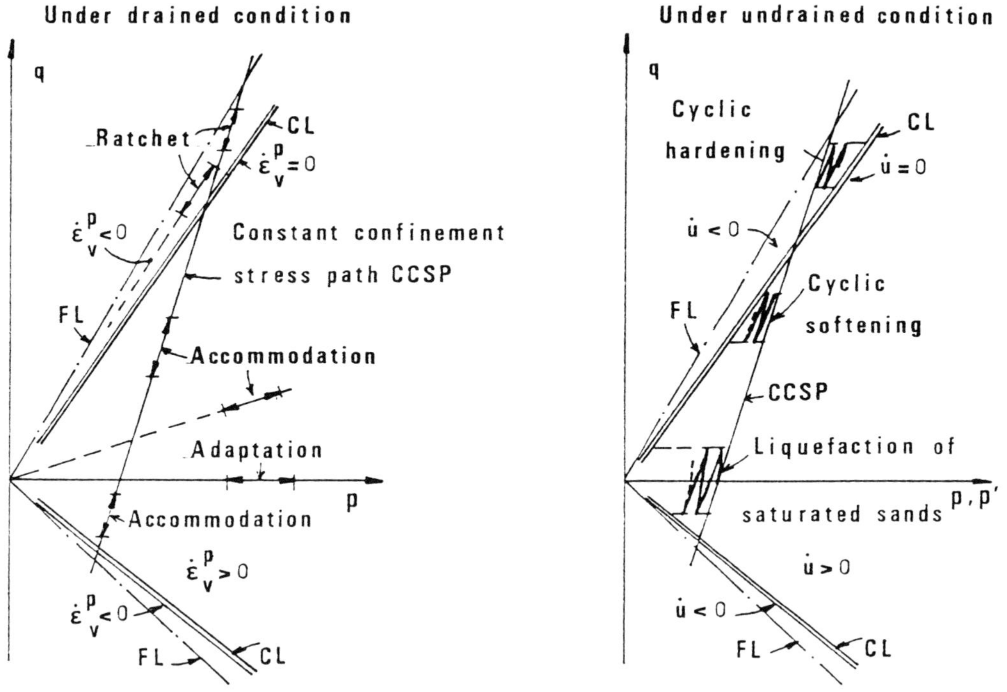

Fig. 6. Diverse cyclic behaviours of cohesionless soils readily obtained from the conventional axisymmetric triaxial apparatus.

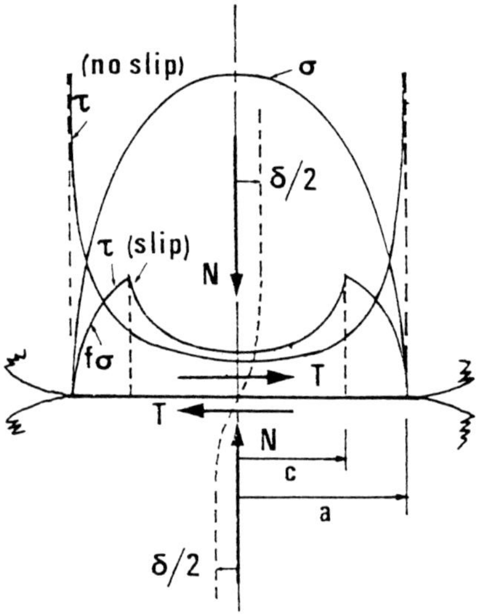

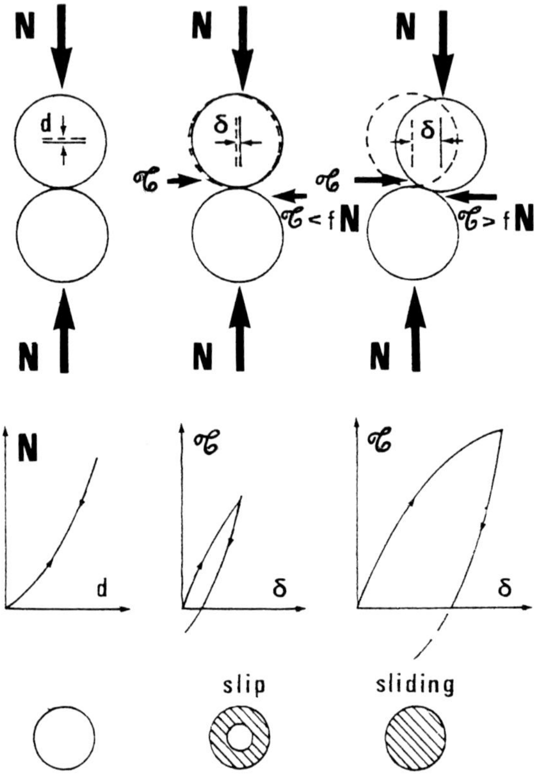

Fig. 7.

Stress and strain on contact surface of two like spheres subjected to normal force N followed by a tangential force $\mathcal{T}$ (after Mindlin 1953).

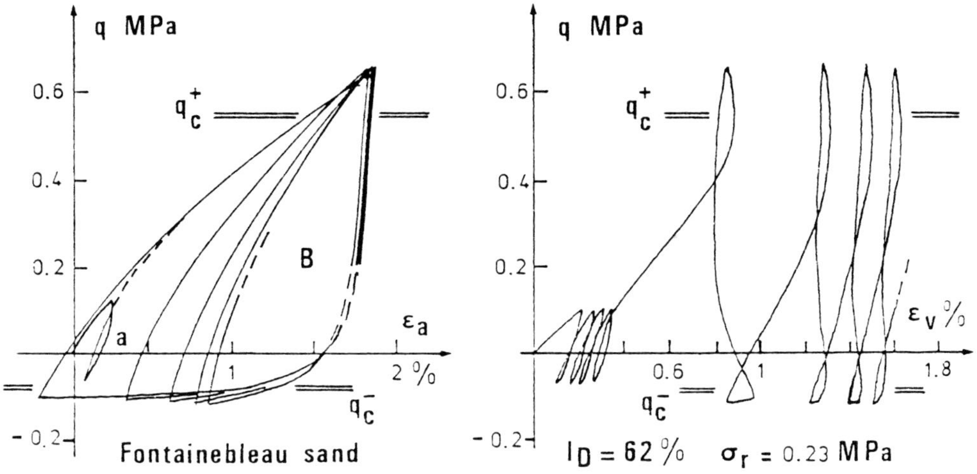

Fig. 8. Triaxial test under small amplitude (a) and large amplitude (B) of deviatoric stress.

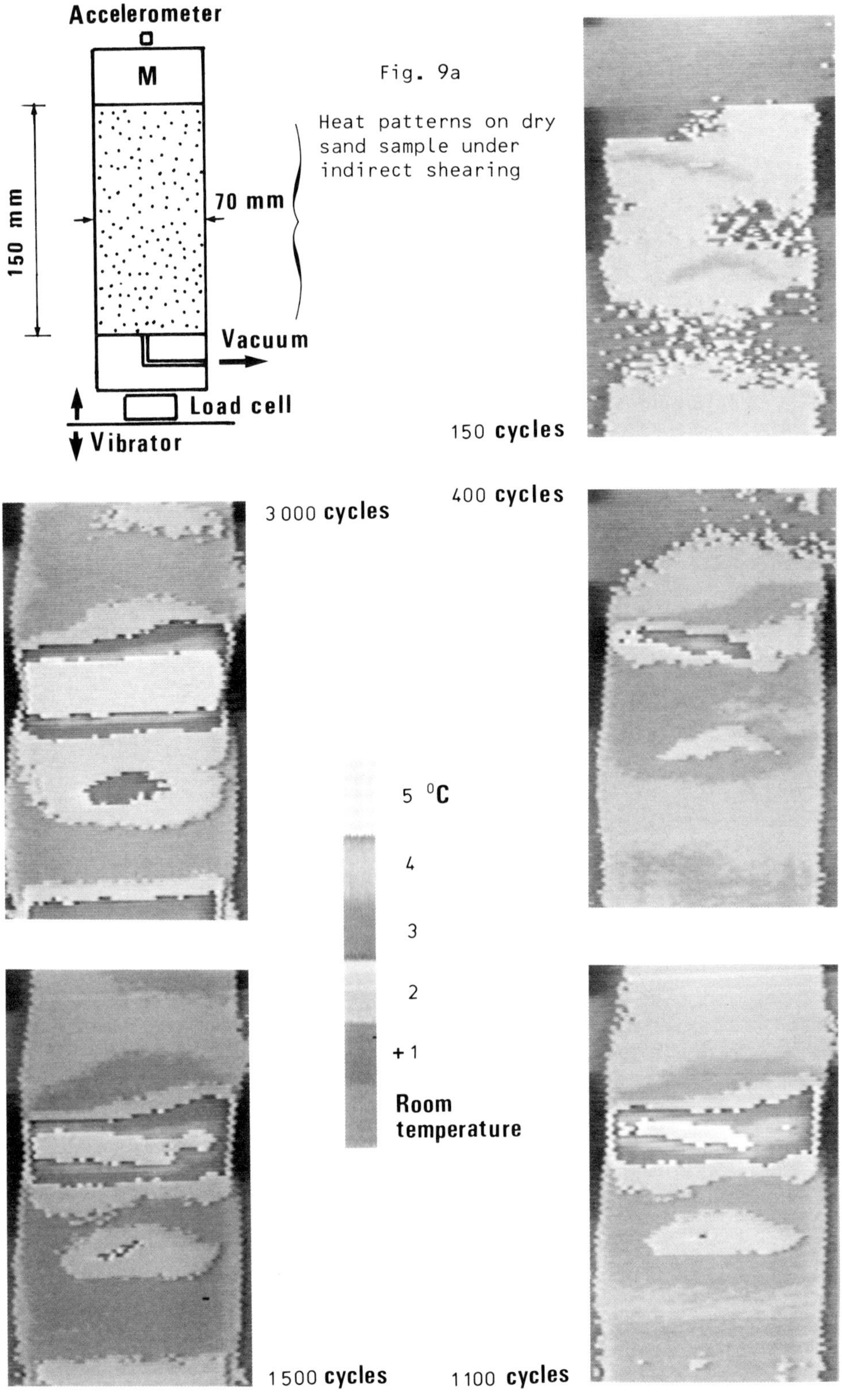

Fig. 9a

Heat patterns on dry sand sample under indirect shearing

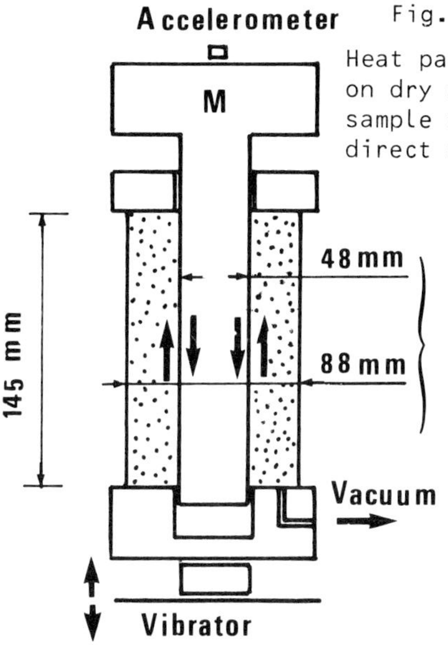

Fig. 9b Heat patterns on dry sand sample under direct shearing

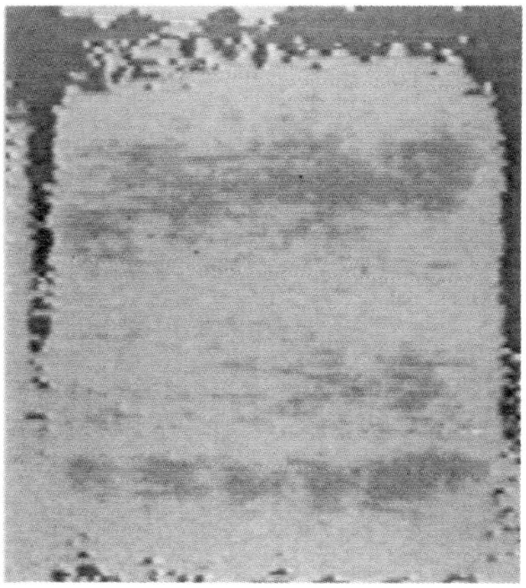

50 cycles

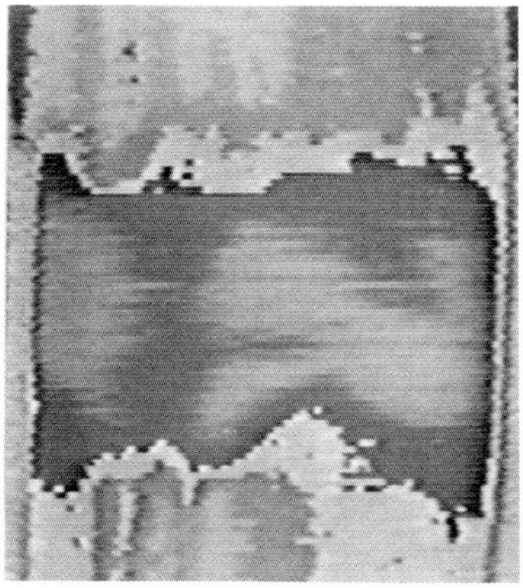

2 000 cycles

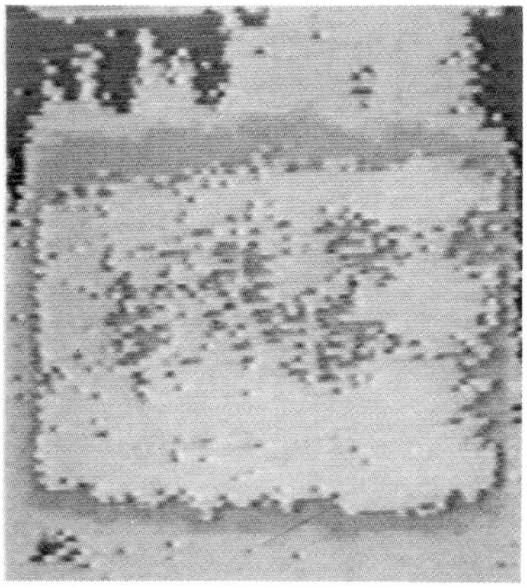

250 cycles

1 500 cycles

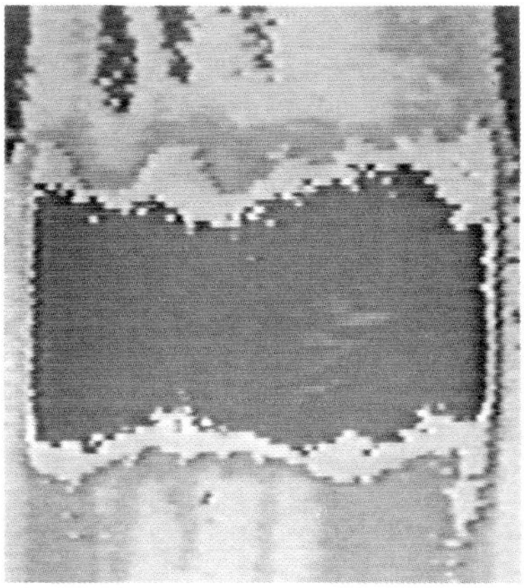

1 000 cycles

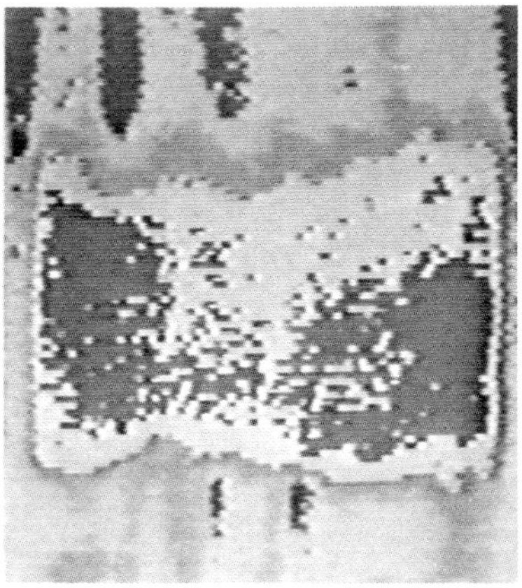

NON LINEAR BEHAVIOUR OF ANISOTROPIC ROCKS

R. Ribacchi
University of Rome

1. INTRODUCTION

Large engineering structures, in particular caverns for power plants, have been built in recent years in the metamorphic formations of the Alps (Martinetti, 1977; Dolcetta, 1982) and others are now in the design stage (Piedilago cavern).

During the investigations for the geomechanical characterization of these sites the constituent anisotropic rocks were found to have a complex behaviour, and sometimes surprising discrepancies between the results of in situ and laboratory tests were observed.

In this paper, after a review of simplified theoretical models, some interesting experimental results are presented. It is to be taken into account that such experimental studies have developed over a number of years and that the most of them had a technical rather than a research-oriented approach. This explains why the data are often not exhaustive and are insufficient for interpreting some ambiguous situations. In some cases, with the knowledge available today, a well-designed plan of tests could clarify many problems and produce more reliable models of the behaviour of these anisotropic rocks.

2. ENGINEERING MODELS OF ANISOTROPIC ROCKS

In an anisotropic rock the engineering strain and stress vectors are connected by a symmetric compliance matrix **M**

$$\boldsymbol{\varepsilon} = \mathbf{M}\boldsymbol{\sigma}$$

where

$$\boldsymbol{\varepsilon} = \begin{matrix} \varepsilon_1 \\ \varepsilon_2 \\ \varepsilon_3 \\ \gamma_{23} \\ \gamma_{31} \\ \gamma_{32} \end{matrix} \qquad \boldsymbol{\sigma} = \begin{matrix} \sigma_1 \\ \sigma_2 \\ \sigma_3 \\ \tau_{23} \\ \tau_{31} \\ \tau_{32} \end{matrix} \qquad (1)$$

In Rock Mechanics problems only simplified models of anisotropy, that is the orthotropy and the polar anisotropy (or transversal isotropy) are usually utilized. The latter corresponds to the macroscopic textural symmetry of many stratified and metamorphic rocks.

Assuming as a reference system 1,2,3 the principal orthotropy directions, the compliance matrix **M** of an orthotropic rock can be written in terms of the engineering elastic parameters E (Young moduli), G (shear moduli) and ν (Poisson's coefficients).

$$\mathbf{M} = \begin{bmatrix} 1/E_1 & -\nu_{21}/E_2 & -\nu_{31}/E_3 & 0 & 0 & 0 \\ & 1/E_2 & -\nu_{32}/E_3 & 0 & 0 & 0 \\ & & 1/E_3 & 0 & 0 & 0 \\ & & & 1/G_{23} & 0 & 0 \\ & & & & 1/G_{13} & 0 \\ & & & & & 1/G_{12} \end{bmatrix} \qquad (2)$$

For a transversally isotropic rock, if axis 3 is the axis of circular symmetriy, axes 1 and 2 are equivalent, and the rock is characterized by 5 independent elastic parameters (2 Young moduli, E_1 and E_3, 1 shear modulus G_{13}, 2 Poisson's coefficients ν_{31} and ν_{21}).

In the following, 1, 2, 3 will be a reference system corresponding to the principal directions of elasticity. In the case of polar anisotropy, 3 is the polar axis, that is in practice the direction corresponding to the minor principal modulus. The directions lying respectively in the plane of transversal isotropy (1 or 2) and normal to it (3) will also be indicated by // and ⊥. For transversal isotropy the angle between the polar axis and a generic direction will be indicated by θ (Fig.1).

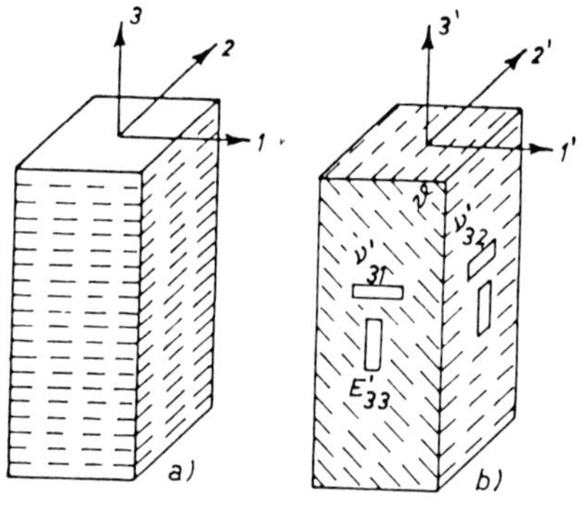

Fig. 1 - Orientation of the reference axes in a transversely isotropic rock

A generic reference system will be indicated by 1', 2', 3' and **M'** will be the corresponding compliance matrix. The transformation formulas containing the direction cosines of 1', 2', 3' with respect to 1, 2, 3 are given by Lekhnintski, (1963).

Even in the simplest anisotropic models (transversal isotropy), the determination of three main elastic parameters (Poisson's coefficients often have a small influence and are estimated a priori) poses a difficult problem. In some cases, often without any experimental evidence, the assumption is made that the shear modulus G_{13} is not an independent parameter, but is related to the other constant by the following relationship (Saint Venant relationship)

$$1/\hat{G}_{13} = 1/E_1 + 1/E_3 + 2\nu_{31}/E_3. \tag{3}$$

For an orthotropic rock the three principal shear moduli can be related to the other elastic properties by relations similar to rel. (3); the independent elastic constants would be reduced to 6.

The deviation of the shear modulus of a transversally isotropic rock from the value predicted by the Saint Venant relationship can be expressed by the ratio :

$$R_G = \hat{G}_{13}/G_{13} = \frac{M_{44}}{M_{11} + M_{33} - 2M_{13}} \tag{4}$$

where $\hat{G}_{13}$ is the value satisfying the Saint Venant relationship (rel. 3). Therefore, the main characteristics of the anisotropy can be expressed by the anisotropy ratios $R_E = E_1/E_3$ and R_G.

When $R_G = 1$, the shear modulus and the lateral compliance $M'_{13} = -\nu'_{31}/E'_3$ are invariant for every couple of axes lying in a plane containing the polar axis; the normal compliance varies as a function of the orientation in the same way as the normal component of a tensor, that is

$$M'_{33} = \frac{1}{E'} = \frac{\cos^2\theta}{E_3} + \frac{\sin^2\theta}{E_1} . \tag{5}$$

In the general case, instead, the normal compliance is given by:

$$\frac{1}{E'} = \frac{\cos^2\theta}{E_3} + \frac{\sin^2\theta}{E_1} + \frac{1}{G_{13}} - \frac{1}{\hat{G}_{13}} \frac{\sin^2\theta}{4}. \tag{6}$$

If G_{13} becomes small enough $(G_{13} < E_3/2(1 + \nu_{31})$, the minimum value of $E'(\theta)$ occurs not on the polar axis $(\theta = 0)$ but for some intermediate angle θ.

In a similar way, the lateral compliance is given by:

$$M'_{13} = \frac{\nu'_{31}}{E_3} = \frac{\nu_{31}}{E_3} + (\frac{1}{G_{13}} - \frac{1}{\hat{G}_{13}})\frac{\sin^2\theta}{4} . \tag{7}$$

The anisotropy of a rock being the result of the preferred orientation of its constituent minerals is the "intrinsic" or "solid matrix" anisotropy, and its compliance will be indicated by $\mathbf{M}^m$. Various averaging techniques, which cannot be discussed here, have been utilized to estimate $\mathbf{M}^m$. In the following, when the orientation of the minerals is random (statistically isotropic) the elastic parameters K_m and G_m (bulk modulus and shear modulus) will be simply evaluated by averaging the so-called Voigt and Reuss estimates of K and G.

In many types of schistous metamorphic rocks the minerals are not uniformly distributed in the mass, but the rock is composed of levels mainly formed by mica plates, and of levels of quartz and/or feldspaths. Whilst the micas always show a preferred planar orientation, the other minerals are often not markedly oriented.

To obtain an indication about the tipical intrinsic anisotropy of such schistous rocks, the values of R_E and R_G of a rock consisting of alternating layers of perfectly oriented mica plates and of randomly oriented quartz crystals, were determined by means of the averaging technique proposed by Salomon (1968). The results are shown in Fig. 2. The mica content of typical gneisses is 10-20%, while for schists it may be as much as 50% or more.

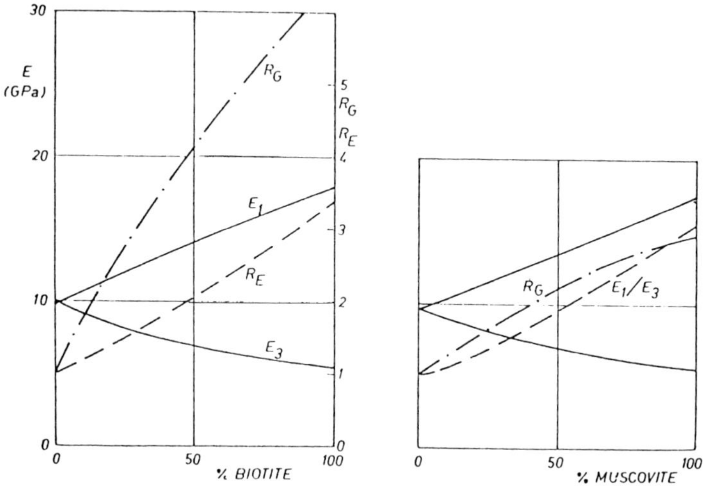

Fig. 2 - Theoretical values of elastic parameters for a biotitic and a muscovitic schist

This kind of calculation appears to support a comment of a classical test on Rock Mechanics (Jaeger and Cook, 1969), that, for rocks, the anisotropy ratio R_E "is rarely as high as 2".

However, experimental data on metamorphic rocks furnish quite a different picture. For instance, Fig. 3 summarizes the anisotropy

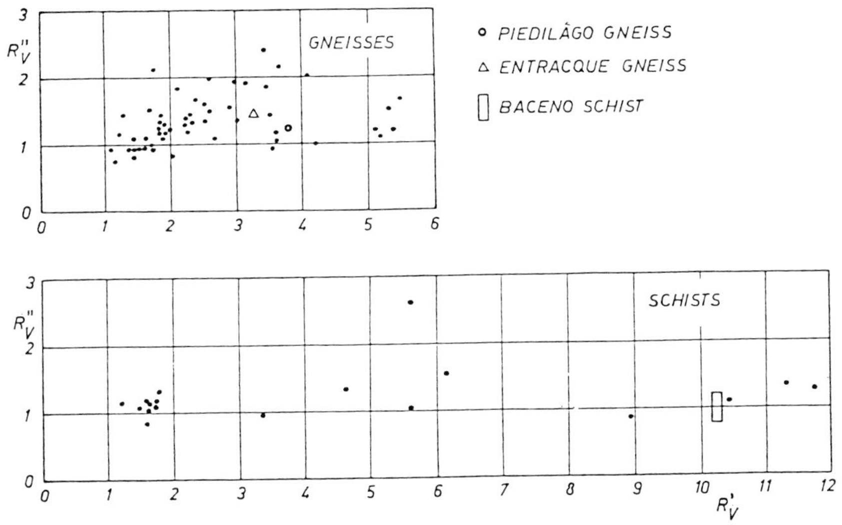

Fig. 3 - Anisotropy ratios for Alpine gneisses and schists determined on the basis of the squared ratios of principal velocities. Axis 3 is perpendicular to the main planar structure, axis 1 is aligned with the lineation

ratios obtained by Johnson and Wenk (1974) for Alpine gneisses and schists. They are based upon seismic velocity measurements:

$$R'_V = (V_1/V_3)^2$$

$$R''_V = (V_1/V_2)^2$$

where axis 3 is perpendicular to the main planar structure of the rock, and axis 1 is directed along a possible lineation. The values of R_V probably underestimate the corresponding static anisotropy ratios R_E.

The figure shows that very high anisotropy ratios are possible even in gneisses for which the intrinsic anisotropy is certainly very low. In some cases the data are compatible with a model of transversal isotropy ($R''_V = 1$), whereas some other rocks require at least an orthotropic model.

These obsevations suggest that the main factor influencing the anisotropic properties of rocks is the presence of oriented sets of microfractures (or cracks); their effect will be discussed in the following paragraphs.

3. ANALYTICAL MODELS OF CRACK-INDUCED ANISOTROPY

3.1. Open cracks

The influence of open cracks has been more frequently investigated, probably because of their importance for the geophysical problems related to seismic wave propagation.

The simplest theoretical model is based on the assumption that there is no interaction between the various cracks in the rock. This D.C. (Dilute Cracks) Model was utilized for instance by Walsh (1965 and 1969). For large crack concentrations the so-called S.C. (Self Consistent) Model is applied (O'Connel and Budianskii, 1974); the effect of the interaction between cracks is approximately accounted for by assuming that each crack behaves as if it were embedded in a homogeneous matrix having the average elastic properties of the cracked body.

The global compliance $\mathbf{M}$ of the cracked rock is obtained by adding to the compliance $\mathbf{M}^m$ of the matrix the contribution $\mathbf{M}^{oc}$ of the open cracks

$$\mathbf{M} = \mathbf{M}^m + \mathbf{M}^{oc} \tag{8}$$

Let us assume that the anisotropy of the solid matrix is negligible, and that the rock contains a set of non-interacting penny-shaped cracks of variable radius, r; the crack density parameter, e, of the set is expressed by:

$$e = \frac{1}{V}\sum r^3 = N\langle r^3\rangle \tag{9}$$

where N is the number of open cracks in unit volume and $\langle r^3\rangle$ the average value of r^3.

In the D.C. Model the only non-zero terms of the $\mathbf{M}^{oc}$ matrix are:

$$M_{33}^{oc} = \frac{16}{3}\,\frac{1 - \nu^2{}_m}{E_m}\,e_{oc} \tag{10}$$

$$M_{44}^{oc} = M_{55}^{oc} = \frac{16}{3}\,\frac{1 - \nu^2{}_m}{E_m(1 - \nu_m/2)}\,e_{oc} \tag{11}$$

where E_m and ν_m are respectively the intrinsic modulus and Poisson's coefficient. The characteristics of the anisotropy are evidenced in Fig. 4 and 5 (left side).

Rel. (10) and (11) show that the anisotropy induced by a system of parallel cracks is a particular type of transversal isotropy, because it is characterized by 3 elastic parameters only, instead of the 5 of the general transversal isotropy. The value of R_G is only slightly greater than 1.

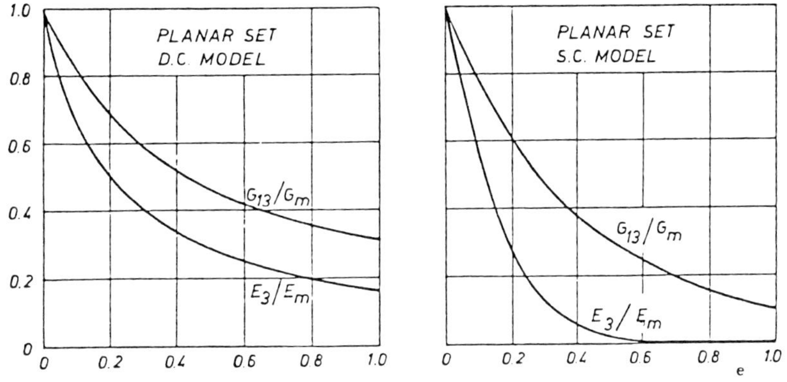

Fig. 4 - Young and shear moduli predicted by the D.C. and S.C. models for a rock containing a planar sey of cracks with a crack density e

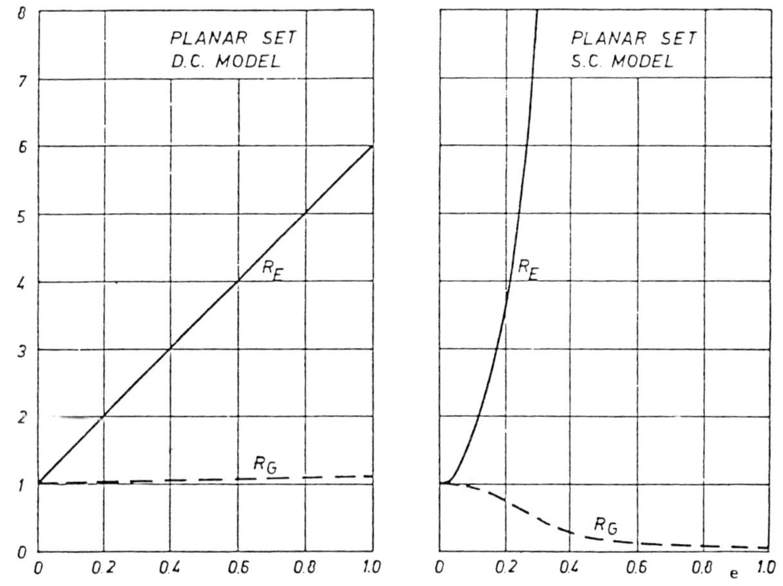

Fig. 5 - R_E and R_G ratios predicted by the D.C. and S.C. models for a rock containing a planar set of cracks

An important simplification is obtained if the term $\nu_o/2$ is neglected with respect to unity; in this case rel. (11) becomes:

$$M_{44}^{oc} = M_{55}^{oc} = M_{33}^{oc} \tag{12}$$

Now, the shear modulus exactly satisfies the Saint Venant condition ($R_G = 1$) and, whatever the orientation of the reference axes, the lateral strain compliance terms M'_{12}, M'_{13} and M'_{23} are the same and are equal to the uncracked matrix.

The above relationships can be utilized with a good approximation even for cracks whose shape, in the plane of the crack, is not circular, provided that an equivalent radius (which is the function of the area and the perimeter of the crack) is adopted (O'Connel and Budianskii, 1974).

Even for the two-dimensional cracks which were considered by Walsh (1965), only a modification of the numerical coefficients in rel. (10) and (11) is required:

$$M_{33}^{oc} = 4\pi \frac{1 - \nu^2 m}{E} e_{oc} \tag{13}$$

$$M_{44}^{oc} = M_{55}^{oc} = M_{33}^{oc}.$$

In this last case the Saint Venant relationship is exactly verified.

When various crack sets are present, the global compliance matrix M^{oc} can be immediately determined (in the D.C. formulation) as a sum of the $\mathbf{M}^{oc}$ of the single sets, after they have referred to a common reference system.

The elastic symmetry, in general, corresponds to the geometric symmetry of the fabric deriving from the various sets of cracks. For instance, a common situation in schistous rocks is the presence of two microfissure sets having different characteristics, and at oblique angles from each other possibly superimposed onto randomly oriented cracks. Both the fabric and the elastic symmetry of this rock would be of monoclinic type.

However, if the approximate relation (12) is accepted, it can be easily shown that the elastic symmetry will be at least of the orthotropic type. Besides, it will be characterized by $R_G = 1$, and by a constant lateral strain coefficient:

$$\frac{\nu_{ij}}{E_i} = \frac{\nu_m}{E_m} \tag{14}$$

and therefore by 4 elastic parameters only (Berry *et al.*, 1974; Crea *et al.*, 1976).

The above simple considerations have been expressed in a more formal way by Kachanov (1980), Oda (1982), Oda *et al.* (1984). The latter introduces a "fabric tensor", F, defined by

$$F_{ij} = \frac{1}{V}\sum 2\pi r^3 n_i n_j \tag{15}$$

where n is the unit vector normal to the plane of each crack. Accepting the approximation (12), the principal directions of this tensor correspond to the principal direction of the orthotropic compliance matrix. In this reference system, the compliance matrix is given by:

$$\mathbf{M}^{oc} = \frac{8(1-\nu^2{}_m)}{3\pi E_m}\begin{bmatrix} F_{11} & 0 & 0 & 0 & 0 & 0 \\ & F_{22} & 0 & 0 & 0 & 0 \\ & & F_{33} & 0 & 0 & 0 \\ & & & F_{22}+F_{33} & 0 & 0 \\ & & & & F_{33}+F_{11} & 0 \\ & & & & & F_{11}+F_{22} \end{bmatrix} \tag{16}$$

The crack density tensor $\boldsymbol{\alpha}$ introduced by Kachanov is closely related to the fabric tensor $\mathbf{F}$. For instance, for penny-shaped cracks:

$$\alpha_{ij} = \frac{1}{V}\pi\sqrt{\pi}\sum r^3 n_i n_j. \tag{17}$$

A particularly important crack distribution corresponds to a statistically isotropic crack orientation ("random cracks"). By means

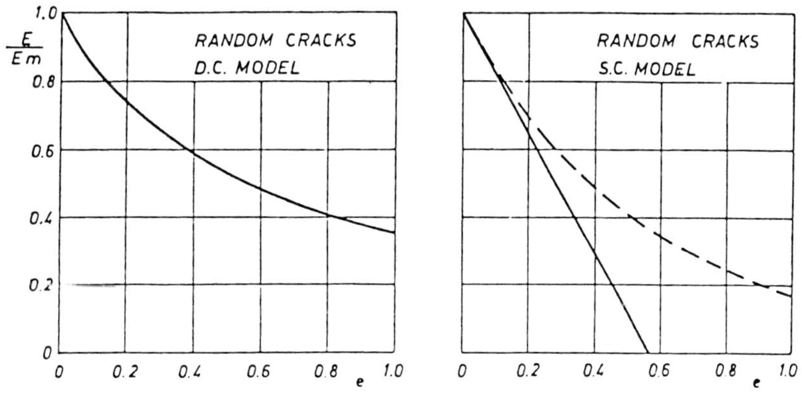

Fig. 6 - Young modulus of a randomly cracked body predicted by D.C. and S.C. models as a function of the crack density parameter e. Dashed line indicates the S.C. model proposed by BRUNER

of averaging techniques we obtain the relationships (Fig. 6, left):

$$\frac{1}{E} = \frac{1}{E_m} + \frac{16}{9}\,\frac{1-\nu_m^2}{E_m}\,\frac{1-0.3\nu_m}{1-0.5\nu_m}\,e_{oc} \tag{18}$$

$$\frac{1}{G} = \frac{1}{G_m} + \frac{16}{9}\frac{1-\nu}{G_m}\frac{1-0.2\nu_m}{1-0.5\nu_m} e_{oc} \ . \tag{19}$$

If the assumption of no interaction between cracks cannot be maintained, the situation becomes much more complex.

Explicit self-consistent formulations of the compliance were obtained by O'Connel and Budianskii (1974) in the case of random cracks (Fig. 6, right) and by Hoening (1979) in the case of a set of parallel cracks (Fig. 4, right).

The following indications can be drawn:

- the effect of cracks on the modulus reduction is by far stronger in the S.C. models than in the D.C. models, at least in the field of crack density which is of practical interest for most types of rock;
- for a single planar set of cracks, the induced anisotropy is characterized by values of R_G markedly lower than 1, a trend which is totally different from that indicated by the D.C. formulations.

However, it was found by Kachanov (1980) that even when S.C. models are adopted, the elastic compliance induced by any anisotropic crack distribution has (in a first approximation) a symmetry which is orthotropic, with principal directions corresponding to that of the **F** of $\boldsymbol{\alpha}$ tensors.

The big differences between the predictions of the D.C. and S. C. models cast some doubts on the validity of the quantitative application of either theories. The application of the S.C. models for the evaluation of the elastic properties of composites was discussed by Hashin (1970), who showed that their predictions are not necessarily more accurate than those furnished by other averaging procedures.

The D.C. model undoubtedly provides a lower limit to the compliance of the cracked rock, but it cannot be said that the S.C. predictions always come closer to the true results. For instance, the S.C. prediction that the moduli become zero at some finite crack concentration, appears physically unreasonable. The modification proposed by Bruner (1976), in which an incremental procedure is adopted for the introduction of the cracks, seemingly gives a more satisfactory relationship between moduli and crack concentration (Fig.6, left, dashed curve).

3.2. Closed Cracks

The presence of a system closed cracks induces an anisotropic behaviour which, however, cannot be represented by an elastic compliance, and which is strongly dependent on the stress path adopted for loading the rock.

Only in the (non realistic) case of zero friction between the crack surfaces is the behaviour elastic; the non-zero terms of the matrix $\mathbf{M}^{cc}$ for a material containing a single set of parallel cracks are only

$$M_{44}^{cc} = M_{55}^{cc} = M_{44}^{oc} . \tag{19}$$

In these conditions the anisotropy would be characterized by a value of $R_E = 1$ and a value of R_G markedly greater than 1.

When friction between the crack surfaces is taken into account, only a fraction of the shear stress is active for the deformation of the crack; this fraction is given by:

$$\tau_{act} = \tau - \sigma_n \, tg\phi \tag{20}$$

where σ_n is the compressive stress normal to the plane of the crack and ϕ is the friction angle.

In the D.C. models it is therefore possible to calculate the deformability for a given stress path and a given distribution of the crack orientation (Brady, 1969). For instance when a cylindrical sample of rock containing a planar set of closed cracks is loaded in the so-called triaxial conditions, at a constant confining pressure σ_c and with an increasing axial stress σ_a, the contribution of the cracks to the axial compliance and to the lateral compliance are given by:

$$M_{33}^{'cc} = \frac{16}{3} \frac{1 - \nu_m^2}{E_m(1 - 0.5\nu_m)} K(\theta) e_{cc} \tag{21}$$

$$M_{13}^{'cc} = -M_3^{'cc} . \tag{22}$$

If

$$(\sigma_a - \sigma_c)(\sin\theta \cos\theta - tg\phi \cos^2\alpha) - \sigma_c \, tg\phi > 0 \tag{23}$$

$K(\theta)$ is given by

$$K(\theta) = \sin\theta \cos\theta \, (\sin\theta \cos\theta - tg\phi \cos^2\theta) . \tag{24}$$

Otherwise $K(\theta) = 0$.

The stress-strain curve will show an abrupt change of slope because the closed cracks will be mobilized only above a certain value of σ_a. Only in uniaxial loading, the stress-strain curve is rectilinear. Walsh (1965) evaluated the apparent compliance in uniaxial stress conditions of a material containing a random distribution of closed cracks; for the same value of the crack density, the closed cracks are less effective than the open ones in reducing the moduli of the rock. A complete analysis for plain strain conditions and for a stress path σ_c/σ_a = const, was presented by Horii and Nemat-Nasser (1982 and 1983).

The aforementioned results are all based on the hypothesis that no residual normal stress exists between the surfaces of a crack when the rock is macroscopically unstressed. However, the dishomogeneity of the elastic characteristics of the grains originates variable residual stresses at the microscopic level, which in some cases could be measured by X-ray diffraction techniques (Friedman, 1972); compressive stress should therefore be present between the faces of closed cracks, which were produced probably by the high shear stress at the boundary between different mineral grains.

If this residual stress is indicated by P_c, rel. (21) and (24) still hold, but rel. (23) is modified in the following way:

$$(\sigma_a - \sigma_c)(\sin\theta \cos\theta - \mathrm{tg}\phi \cos^2\theta) - (p_c + \sigma_c)\,\mathrm{tg}\phi > 0. \tag{25}$$

Now, even in uniaxial loading conditions, a non-linearity of the stress-strain curve will be observed, with a decrease of the modulus at increasing stress.

3.3. Effects of Crack Closure

Penny-shaped cracks and 2D elliptical cracks close when the critical normal stress reaches respectively the critical values given by (Berg, 1965):

$$p_{oc} = \frac{\pi E_m}{4(1 - \nu_m^2)}\,\alpha \tag{26}$$

$$p_{oc} = \frac{E}{2(1 - \nu_m^2)}\,\alpha \tag{27}$$

where α is the aspect ratio, that is the ratio between the maximum thickness and the diameter (or length) of the crack.

When the normal stress is gradually increased but stays below this critical value, the aspect ratio decreases but the diameter of the crack is not modified and therefore, the compliance M^{oc} remains constant. At the critical stress the crack completely closes and its effects disappear.

This simple situation would permit to determine the spectra of the aspect ratio of the cracks, by subjecting the sample to an isotropic loading (for which the closed cracks are inactive). Such evaluations were carried out for instance by Morlier (1971), Simmons *et al.* (1975), Feves and Simmons (1976). The results are interesting but they should be considered merely as being indicative.

The main drawback in these analyses is the effect of the shape of the cracks (Mavko and Nur, 1978). It is true that at a given stress the effect of thin cracks depends only upon their length and not upon their aspect ratio; however, non-elliptical cracks, with tapered ends, gradually close at increasing pressure, varying their length; therefore, even a rock containing cracks of the same aspect ratio would show a gradual increase of the stiffness at increasing (isotropic) load, thus simulating the presence of a continuous spectrum of crack aspect ratios, and in particular the cracks having very low values of α.

In general, therefore, an isotropic loading test in a randomly cracked rock will simply permit to evaluate the integral distribution of the crack density parameters e as a function of the closure pressure. The distribution, representing the open crack density surviving at the pressure p, will be indicated by $e^+_{oc}(p)$; obviously $e^+_{oc}(0) = e_{oc}$. For an anisotropic crack distribution, it will be possible to define, in a similar way, the integral distribution $F^+_{ij}(p)$ of the fabric tensor.

For stress paths different from the isotropic one the situation is more complex. At a given stress level, the global compliance is the sum total of the following compliances:

- the compliance of the matrix;
- the compliance deriving from the surviving open cracks;
- the compliance deriving from the sliding of some of the cracks initially open which were closed by the stress;
- the compliance deriving from the sliding of initially closed cracks when the shear stress induced by the load overcomes the friction between the crack surfaces.

In a D.C. formulation it is easy to evaluate the contribution of the deformability of each crack (or set of cracks) to the global deformability. The final results will obviously not correspond to that valid for an elastic body.

For instance, let us suppose that a rock contains a planar set of cracks, some of which are open, with a given crack density distribution, $e^+_{oc}(p)$. In the closed cracks of the set, different residual normal stresses between the faces will be present; the set will therefore be characterized by an integral distribution $e^+_{cc}(p)$, which represents the crack density relative only to the cracks with a residual normal stress greater than p.

For a uniaxial compressive loading path, the axial and lateral compliance are given by:

$$M'_{33} = \frac{1}{E_m} + \frac{16}{3}\frac{1-\nu_m^2}{E_m} e^+_{oc}(p_A)\cos^2\theta + \frac{16}{3}\frac{1-\nu_m^2}{E_m}\{e_{oc} - e^+_{oc}(p_A) + e_{cc} - e^+_{cc}(p_B)\}K(\theta) \qquad (28)$$

$$M'_{13} = -\frac{\nu_m}{E_m} - \frac{16}{3}\frac{1-\nu_m^2}{E_m}\{e_{oc} - e^{+}_{oc}(p_A) + e_{cc} - e^{+}_{cc}(p_B)\}K(\theta) \quad (29)$$

where

$$K(\theta) = \sin\theta\cos\theta\,(\sin\theta\cos\theta - tg\phi\cos^2\theta) \qquad \theta > \phi$$

$$K(\theta) = 0 \qquad \theta < \phi \quad (30)$$

$$p_A = \sigma_a \cos^2\theta$$

$$p_B = \sigma_a \frac{\sin\theta\cos\theta - tg\phi\cos^2\theta}{tg\phi} \; . \quad (31)$$

3.4. "Undrained Loading" in Saturated Rocks

Up to now the deformability of the rock has been evaluated only in the case that the cracks are dry or "drained", that is when the load is so slow that no water overpressures are induced within the cracks.

In undrained isobaric conditions, that is when no flow of water is allowed out of bulk regions of the rock, the normal compliance for a planar set of cracks depends not only on the crack density, but also on the aspect ratio of the cracks through the stiffness parameter ω.

$$\omega = \frac{1}{\alpha}\frac{K_w}{K_m} \simeq \frac{0.03}{\alpha} \quad (32)$$

where K_m and K_w are the bulk moduli respectively of the solid matrix and of the pore water (O'Connel and Budianskii, 1976 and 1977).

For a typical value $\omega = 10$, corresponding to a crack aspect ratio of an order of magnitude 10^{-3}, the results of the D.C. and S.C. models are shown in Fig. 7. The shear modulus G_{13} is not at all modi-

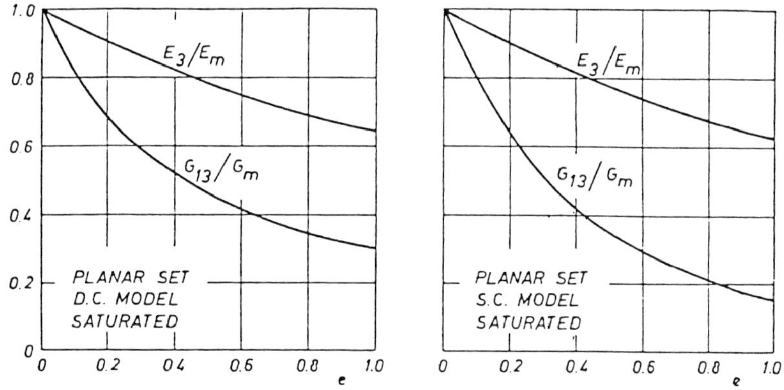

Fig. 7 - Young and shear moduli predicted by the D.C. and S.C. models for a saturated rock containing a planar set of cracks with a stiffness ratio ω = 10

fied (in the D.C. model) or is only slightly modified (in the S.C. model). As a consequence, the anisotropy ratio R_E of the rock is much reduced in undrained loading conditions, whereas R_G becomes markedly greater than 1 (Fig.8).

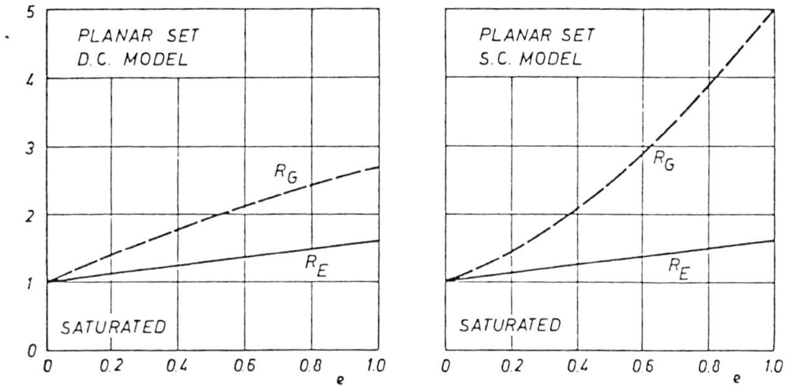

Fig. 8 - R_E and R_G ratios in undrained conditions predicted by the D.C. and S.C. models for a saturated rock containing a planar set of cracks

The bulk modulus of a randomly cracked rock rises markedly in undrained loading conditions, whereas, the shear modulus is not modified; as a consequence the Young modulus is only slightly affected by the loading conditions (Fig.9), whereas Poisson's coefficient beco-

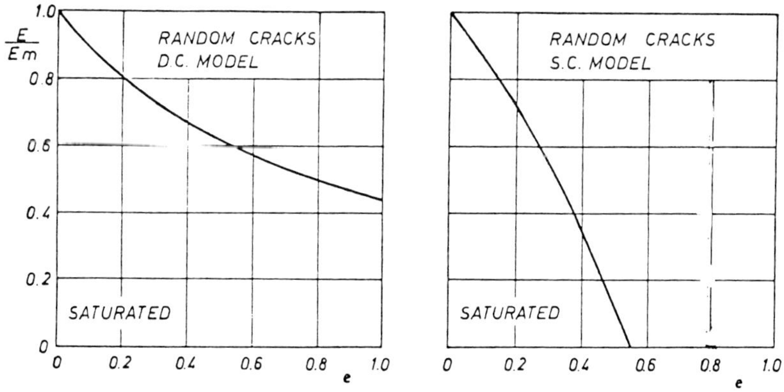

Fig. 9 - Young modulus of a randomly cracked saturated body in undrained conditions predicted by D.C. and S.C. models

mes greater than that of the solid matrix and tends towards 0.5 at increasing crack density.

3.5. Seismic Velocities

The measurement of the seismic velocity in laboratory samples or in situ is commonly used for the exploration of the characteristics of the rocks. For instance, many correlations have been proposed for estimating the "static" deformability of the rock mass on the basis of the "dynamic" elastic parameters, which can be by far more easily determined. However, the results have often been inconclusive and the dispersion of the correlations very high. A summary of the in situ results obtained in many Italian sites is presented in Fig.10.

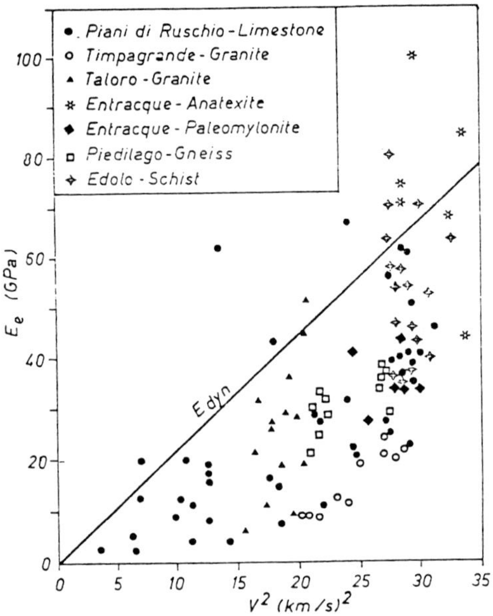

Fig. 10 - Correlation between seismic velocity in situ and Young modulus of the rock mass determined in various Italian sites (LEMBO-FAZIO and RIBACCHI, 1984)

The seismic velocities are essentially related to the compliances corresponding to the solid matrix and to the open cracks. In fact, the stress level corresponding to the wave propagation is low with respect to the in situ stress or to the residual stress between the crack surfaces and, therefore, most of the closed microfissures remain inactive.

The variation of the seismic velocity as a function of the pulse orientation must present a symmetry corresponding to that of the elastic compliance deriving from the open cracks (if the instrinsic anisotropy is negligible); therefore, in dry rocks we should observe an orthotropic symmetry in the more general situation, or a polar anisotropy when a single set of cracks is superimposed onto a random distribution. The determination of the seismic velocity is therefore a simple and powerful technique to determine the symmetry and the principal directions of the fabric tensor of the open cracks (Anderson *et al.*, 1974).

A strong difference in seismic velocities (P waves) is to be expected between dry and saturated conditions; in the latter case the velocities are to be calculated on the basis of the undrained compliance, owing to the rapid loading conditions. In general, the compliance for undrained loading conditions will not show orthotropic symmetry except when:

- the geometrical symmetry of the crack systems is itself orthotropic;
- the stiffness ratio w is very high ($\omega > 100$), that is, the crack aspect ratio is very low.

When in situ measurements are carried out above the water table, the velocity will be strongly influenced by varying saturation conditions. This factor can partly explain why the relationships between the "static" and the "dynamic" moduli are always dispersed and why the difference between the static and the dynamic moduli increases as the fissuring increases.

For a rock containing a planar set of cracks, the principal values of the P-waves are shown in Fig.11 for dry and saturated conditions.

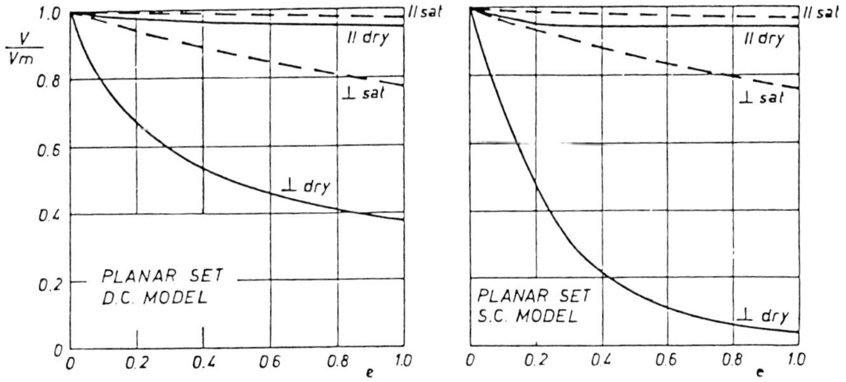

Fig. 11 - Seismic velocity (P-waves) in dry and saturated conditions for a material containing a planar set of crack. The results of the D.C. and S.C. models are compared

We notice that the velocity along the plane of the cracks, V_1, is only slightly influenced by the saturation and is almost equal to that of the intact matrix. The difference between dry and saturated conditions is instead very strong for the minor principal velocity, V_3.

The variation of the seismic velocity with the pulse orientation is shown in Fig.12. We notice that the trend of the curves

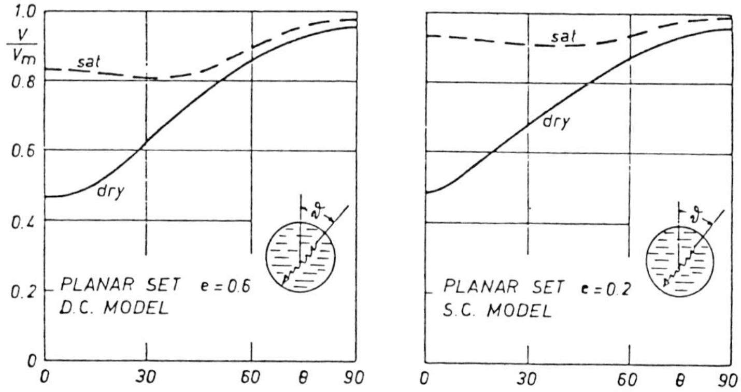

Fig. 12 - Variation of seismic velocity as a function of the orientation of pulse for a material containing a planar set of cracks. Crack density parameters were chosen in order to obtain approximately the same value of V_3 in dry conditions for the D.C. and S.C. models

corresponding to the S.C. and D.C. models is somewhat different. For dry cracks Nur (1971) proved that with the D.C. model, the variation of V^2 with θ should be similar to that valid for the normal component of a tensor; this is confirmed by the shape of the curve in Fig.12.

The seismic velocities (P waves) for randomly cracked rock are shown in Fig.13. It is apparent that in this case, as in the preceding

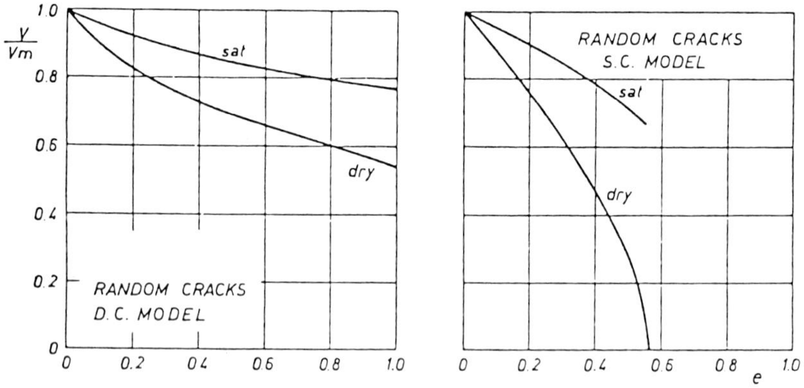

Fig. 13 - Seismic velocity (P-waves) in dry and saturated conditions for a material containing random cracks. The results of D.C. and S.C. models are compared

one, the saturated seismic velocity is not in itself a sensitive indicator of the microfissuring conditions. The velocity variation ΔV

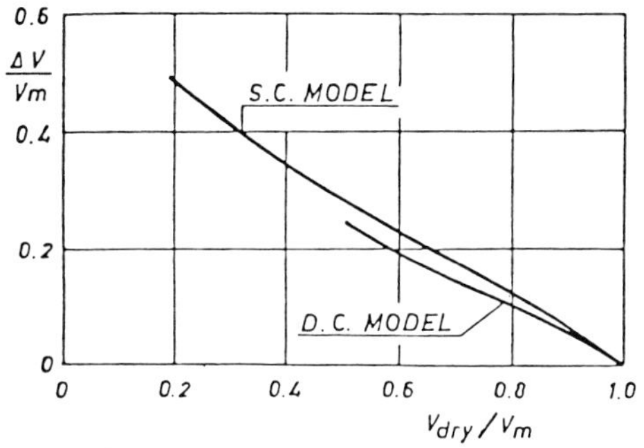

Fig. 14 - Variation of seismic velocity between dry and saturated conditions predicted by D.C. and S.C. models for a material with random cracks

between dry and saturated conditions is almost linearly related to the value of dry velocity and is almost independent of the model adopted for the analysis (Fig.14).

4. EXPERIMENTAL RESULTS

4.1. Tested Rocks

The tested rocks were gneisses and schists belonging to the metamorphic formations of the Central and Western Alps.

1. *Piedilago gneiss.* It shows a roughly planar oriented texture deriving from undulating layers of mica (biotite) enveloping porphyroblasts of feldspath. Porosity is about 0.9%. Mineral composition is: Biotite (B) 20%; Orthoclase (O) 10%; Oligoclase (Ol) 10%; Quartz (Q) 25%. The samples were taken from an exploratory drift for a power plant cavern, at a depth of about 400 m.
2. *"Serizzo" gneiss.* Appearance and texture are very similar to that of the Piedilago gneiss. Mineral composition is: B = 16%; O = 26%; Ol = 38%; Quartz = 18%. The rock was taken from a quarry for ornamental stones, near the site of rock 1.
3. *Beola gneiss.* It is a fine-grained rock, with a markedly planar oriented texture in spite of the low mica content. Mineral composition is: B = 12%; O = 34%; Ol = 28%; Q = 26%.
4. *Entracque gneiss* ("anatexite"). It is a fine-grained rock with a poorly oriented texture. Mineral composition is: B = 8%; O = 30%; Ol = 29%; Q = 28%.
5. *Edolo schist.* Markedly fissile rock with an oriented but

undulating texture. Mica content is high (50%). Other minerals are quartz (25%) and feldspar (25%).

6. *San Fiorano schist.* Quite similar in composition and texture to the Edolo schist.

7. *Baceno schist.* It is a fine-grained markedly fissile rock with an oriented planar texture. Mica content is about 50%. Other minerals are quartz (35%), oligoclase (15%).

4.2. Uniaxial Compression Tests on Gneisses

A summary of the results of uniaxial compression tests on laboratory samples of various Alpine gneisses are presented in Fig.15;

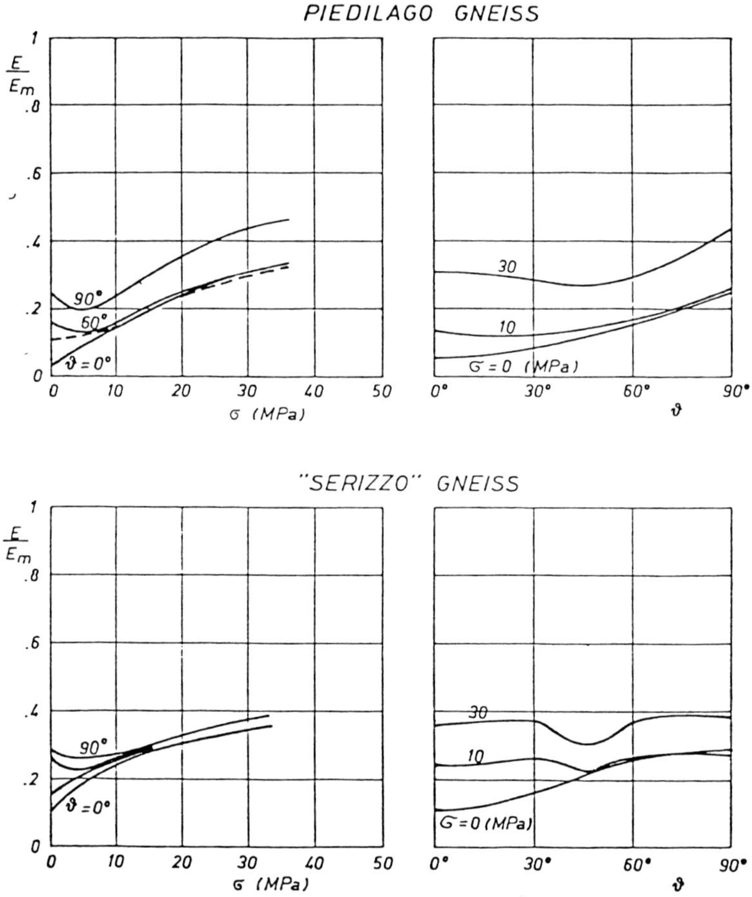

Fig. 15 - Results of uniaxial compression tests on various Alpine gneisses. The moduli are scaled with respect to the intrinsic values, calculated on the assumption that the mineral orientation is random

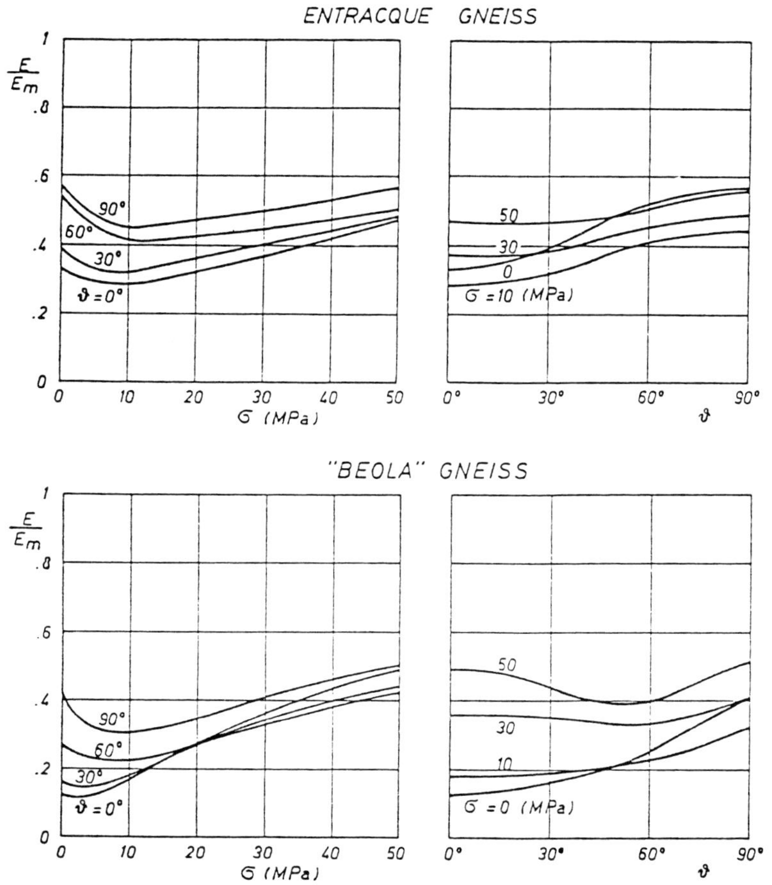

Fig. 15 - continued

all the curves have been scaled with respect to the estimated elastic modulus E_m of the intact matrix, which, for the various rocks is comprised between 70 and 80 GPa.

Although the macroscopic texture is planar, the elastic symmetry was generally found to be orthotropic; the samples analysed in Fig.15 were taken with their axes lying in one of the planes passing through the polar symmetry axis of the macroscopic fabric.

Notwithstanding the peculiarities of the various rocks, certain general trends are apparent.

- The anisotropy ratio R_E is often very high (up to 6) at low load levels, but decreases when the stress increases, at a rate varying with the types of rock; at 20 MPa R_E is already lower than 1.4 (Fig.16).
- At low stress levels, the moduli gradually increase with the angle

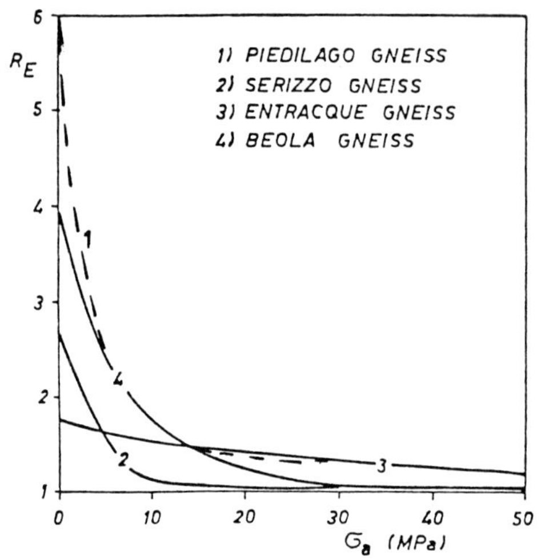

Fig. 16 - Anisotropy ratios R_E in uniaxial compression tests for the rocks of fig. 15

θ; at high stress levels, instead, the lowest modulus corresponds to an orientation of the sample with an angle θ of about 45-50°.

- Above a certain value of the stress level, whatever the orientation of the sample, the modulus increases, at a decreasing rate, with the applied stress; for a stress of 50 MPa, the moduli are equal to 40-50% those of the solid matrix.
- At low stress levels, the moduli often decrease at increasing stress; the fall is very sharp for samples loaded along the schistosity plane (θ = 90°) or at high θ angles. The minimum value of the modulus occurs at a uniaxial stress level of 5-10 MPa. For samples loaded along the polar axis or at low θ angles, this behaviour is absent or less important.

It is apparent that the behaviour of these rocks is mainly controlled by the characteristics of the sets of cracks.

For a better visualization of the various factors, theoretical models of bodies containing both open and/or closed cracks with variable density were analyzed, utilizing - for sake of simplicity - - the D.C. model.

With the available data, the problem is obviously not fully determined and the theoretical simulations can only have a qualitative correspondence with the real behaviour of the rock; more complete information could be obtained by utilizing various stress-paths (in particular the isotropic one) for loading the samples.

The simplest model (Fig.17A), shows an anisotropic behaviour deriving from a planar set of cracks only; to take into account the dispersion of the orientation of the mica plates (to which most of the cracks are probably connected) a uniform distribution of the orienta-

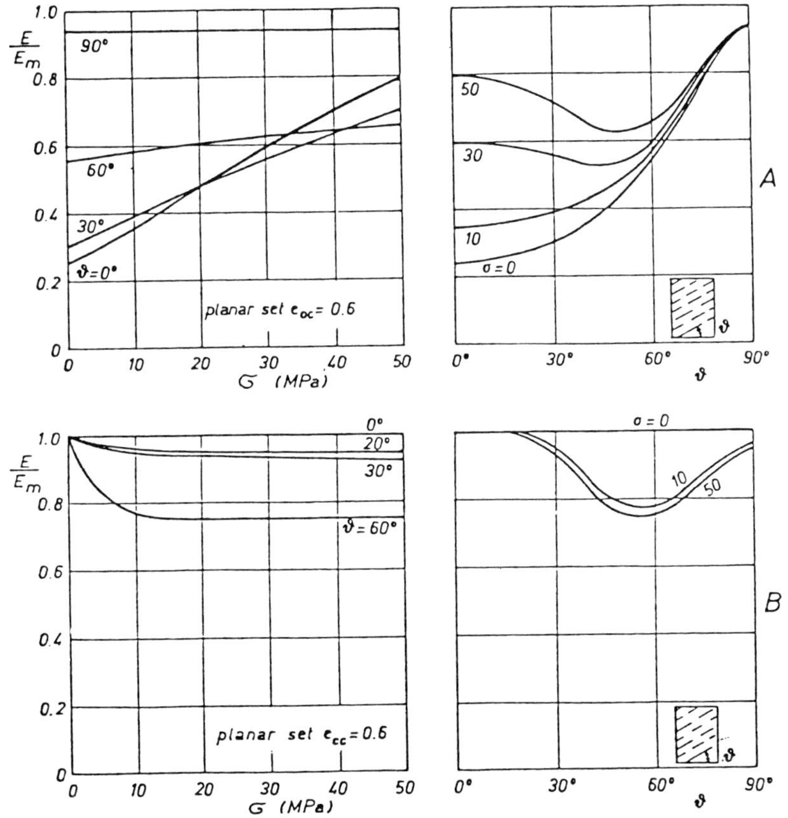

Fig. 17 - Simulations of anisotropic behaviour in uniaxial compression tests induced by oriented tests of microfissures aither open (with crack density e_{oc}) and/or closed (with a crack density e_{cc})

tion within an angle ±15° around the average value was assumed. The variability of the aspect ratio of the cracks was represented by a negative exponential distribution

$$e^{+}_{oc}(p) = e_{oc} \cdot \exp(-p/p_A) \tag{33}$$

in which p_A was assumed to be equal to 20 MPa.

The fall of the modulus in the initial phase of the loading observed in the curves of Fig.15, indicates that closed microcracks with a residual stress between the crack surfaces must be present. A distribution of the crack density parameter similar to that of rel.(33) can be assumed

$$e_{cc}^{+}(p) = e_{cc} \cdot \exp(-p/p_B) \ . \tag{34}$$

The position of the minimum of the E - σ_a curves indicates that p_B must be markedly lower than p_A; a value of 2 MPa appears to be a reasonable choice. The effects of the presence of only closed cracks or of both open and closed cracks are shown in Figs. 17B and 17C; the

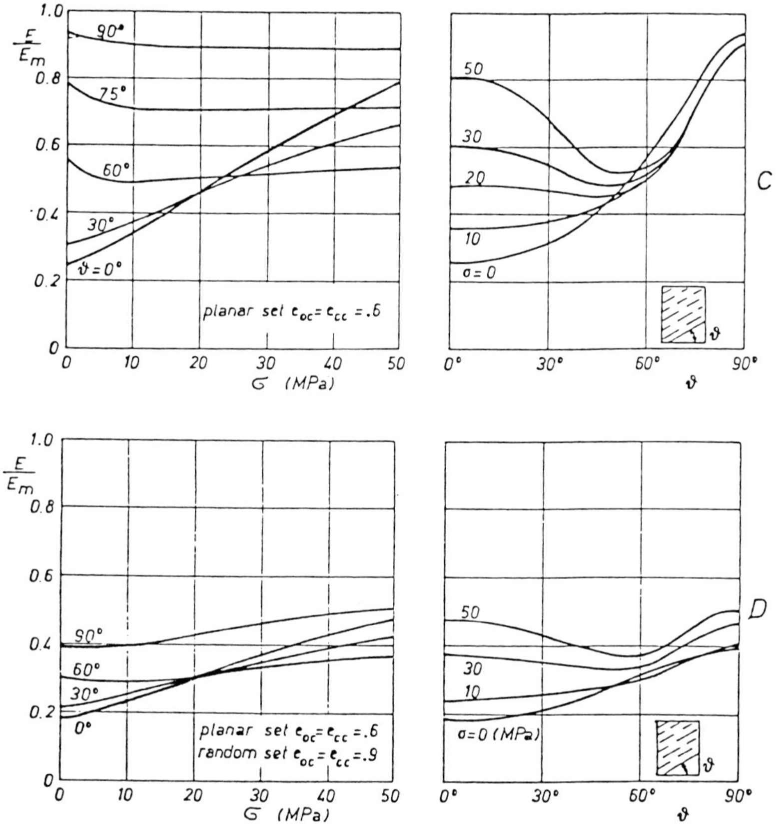

Fig. 17 - continued

latter already shows many characteristics of the experimental curves. However, the tested rocks present low values of the major principal moduli with respect to the intrinsic modulus, and therefore, also a random set of open and closed cracks must be present. It was assumed that their aspects ratio was represented by the same distributions (33) and (34) adopted for the planar set. The results of this simulation is shown in Fig. 17D; the curves are rather similar to the experimental curves obtained for the "Beola" gneiss.

An interesting indication of these simulations is that the marked decrease of R_E for increasing stresses observed in the samples does not necessarily imply that the cracks of the planar set have an aspect ratio lower than that of the random cracks.

Another observation is that the trend of the experimental curves in the samples compressed normally to the planar structure, requires the presence of a large number of cracks with a very low closure pressure (2-5 MPa). Application of rel.(26) or (27) would indicate unreasonably low aspect ratios for some cracks, but it is probable that the effect is due to the presence of tapered-end cracks, as discussed in 3.3.

For the "Serizzo" gneiss, a more complete investigation was aimed at the evaluation of the complete compliance matrix (always in conditions of uniaxial loading); the tests were carried out in a block taken from the quarry.

The initial reference system X Y Z connected to the block had the Z axis perpendicular to the planar structure of the rock and the X and Y axes in an arbitrary position.

From the blocks, 27 cylindrical samples oriented in different direction, were extracted and tested; in each sample the longitudinal strain and 2 transverse strains (according to the scheme in Fig.1) were determined. Each of these strains is linearly related to the 21 terms of the compliance matrix, which could therefore be estimated (together with their variance and covariance) by means of a regression analysis. For instance at a stress level of 10 MPa the compliance terms and their standard deviations are (10^{-3} GPa^{-1})

$$
\begin{matrix} M'_{ij} = \\ (s.d.) \end{matrix}
\left[\begin{array}{cccccc}
45.7 & -5.5 & -6.6 & 0.0 & -6.0 & -0.6 \\
(2.2) & (1.2) & (1.1) & (2.9) & (3.8) & (3.2) \\
 & 50.6 & -6.6 & -3.4 & -1.3 & -3.1 \\
 & (1.7) & (1.1) & (2.7) & (3.6) & (3.2) \\
 & & 57.9 & -6.6 & -9.1 & -0.9 \\
 & & (1.5) & (2.5) & (3.1) & (2.5) \\
 & & & 137.2 & 5.1 & 4.9 \\
 & & & (4.7) & (7.8) & (1.0) \\
 & & & & 121.8 & 0.9 \\
 & & & & (5.8) & (10.7) \\
 & & & & & 107.7 \\
 & & & & & (5.4)
\end{array}\right]
$$

By means of statistical tests it is possible to analyse the type of anisotropy of the rock. For instance, the hypothesis that the Z axis is a principal axis of anisotropy can be tested with the F test, by comparing the residual mean square of the complete regression (21 coefficient) with that of a 13-coefficient regression, in which 8 regressors (M'_{14}, M'_{15}, M'_{24}, M'_{25}, M'_{34}, M'_{35}, M'_{46}, M'_{56}) are eliminated.

To determine whether the behaviour of the rock can be assimilated to that of an orthotropic material, it is firstly necessary to identify the principal strain directions $\bar{1}\ \bar{2}\ \bar{3}$, that is the principal direction of the linear compressibility tensor which is defined by the relation

$$S_{ij} = \begin{bmatrix} S'_i & S'_6/2 & S'_5/2 \\ & S'_2 & S'_4/2 \\ \text{symm} & & S'_3 \end{bmatrix} \qquad (35)$$

where

$$S'_i = \sum_{j=1}^{3} M'_{ji}$$

The compliance matrix M of the rock is then referred to the coordinate system $\bar{1}\ \bar{2}\ \bar{3}$. Orthotropic conditions are satisfied if the elimination of 13 regressors, corresponding to the zero compliance terms in rel.(2), does not significantly increase the residual variance of the observations. In a similar way, the validity of Saint Venant conditions for the rock (R_G = 1) is tested.

Fig.18 shows the best estimates of the principal strain direc-

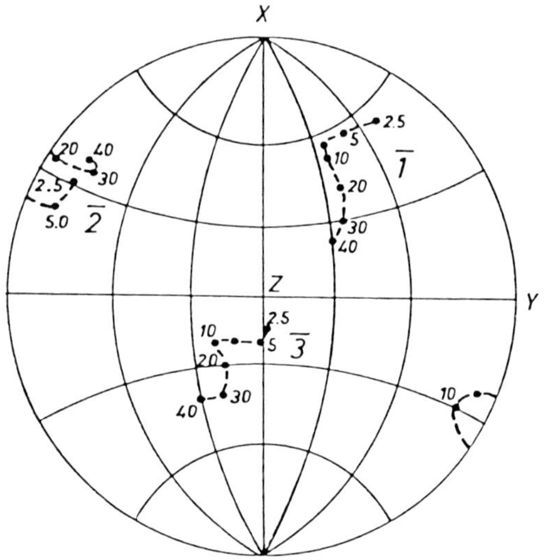

Fig. 18 - Principal "strain directions" for a block of "Serizzo" gneiss as a function of the stress level in uniaxial compression tests

tions as a function of the stress level. Statistical tests indicate that up to a stress level of 20 MPa the anisotropy of the rock may be considered of the orthotropic type and that the Saint

Venant condition is valid. However, the rock cannot be considered as transversally isotropic and, besides, axis Z corresponds to a principal axis of the compliance matrix only at very low stress levels (less than 2.5 MPa).

The values of the moduli relative to the principal strain directions $\bar{1}$ $\bar{2}$ $\bar{3}$ are given in Fig.19. The trend of the curves and the

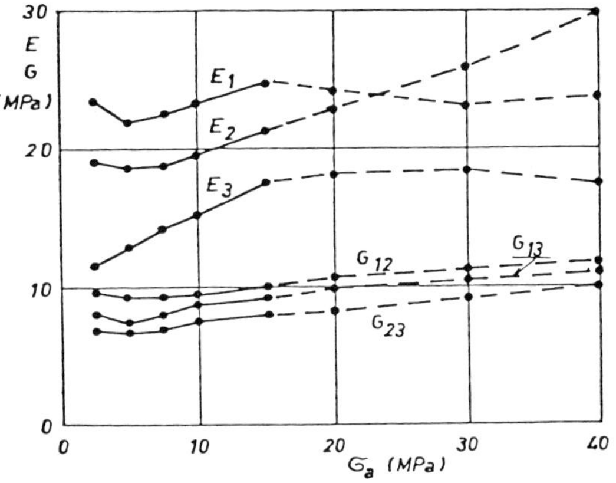

Fig. 19 - Values of the Young and shear moduli for the principal strain directions obtained by the statistical analysis of the uniaxial compression tests

observation that, at increasing stresses, the orientation of axes 1 and 3 varies from values roughly in agreement with the macroscopic structure of the rock towards values at an angle of about 45°, probably indicates that the compliance of the rock can no longer be represented well by means of elastic models when the stress in greater than 20 MPa; if the experimental data are forced into an elastic model, abnormal results are obtained.

4.3. Seismic Velocity

The variation of the seismic velocity as a function of the pulse orientation was measured in spherical samples, both in dry and in saturated conditions (Borelli, 1983). This technique was proposed by Pros and Babuska (1968), and it has been utilized since then by many researchers (Bur *et al.*, 1973; Thill *et al.*, 1973; Friedman and Bur, 1974).

Dry conditions were obtained by storing the samples at a temperature of 80°C for a few days. To obtain saturation conditions the sample was first subjected to vacuum and then deaerated water was introduced in the vessel; the measurements were carried out while the sample was kept underwater.

The results obtained for the Piedilago gneiss are presented further down.

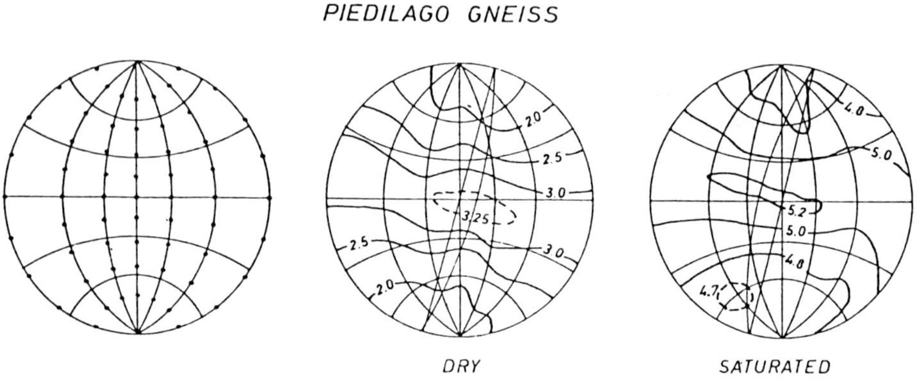

Fig. 20 - Isolines of seismic velocity (km/s) in dry and saturated spheres of the Piedilago gneiss. The orientation of the measurements is shown on the left

In each sample the velocity (P waves) was measured in 55 directions (Fig.20). The shape of the isolines correspond to an almost orthotropic symmetry: the principal velocities in dry conditions were equal to 1.75, 3.15 and 3.45 km/s, and in saturated conditions, to 4.70, 5.15 and 5.25 km/s. On the basis of the mineralogical composition, the seismic velocity of the solid matrix (neglecting its intrinsic anisotropy) was estimated equal to 5.9 km/s.

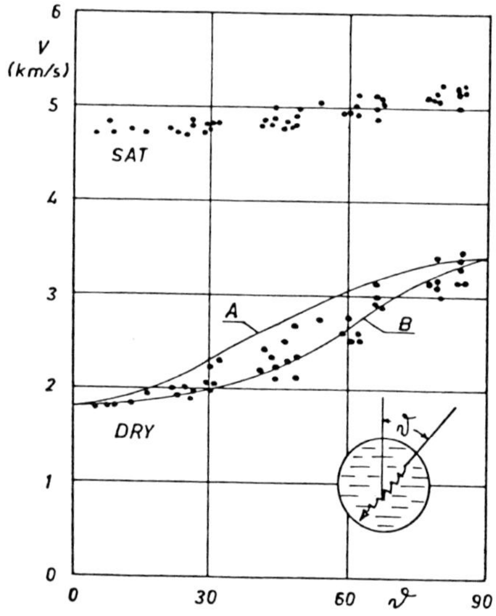

Fig. 21 - Seismic velocity for the Piedilago gneiss as a function of the pulse orientation will respect to the (approximate) polar axis. Lines A and B are the trends given by the relations

A) $V^2 = V_3^2 \cos^2 \theta + V_1^2 \sin^2 \theta$

B) $1/V^2 = \cos^2\theta/V_3^2 + \sin^2\theta/V_1^2$

The deviations from axysymmetric conditions are relatively weak for this rock and can be neglected in a first approximation analysis. Therefore, in Fig.21 the seismic velocity is represented as a function of the angle θ. The shape of the curves is different from that predicted by the theoretical D.C. model (line A) and even more from that of the S.C. model; it appears to be better represented by a relationship of the type (line B)

$$\frac{1}{V^2} = \frac{\cos^2\theta}{V_3^2} + \frac{\sin^2\theta}{V_1^2} \qquad (36)$$

which is similar to the relation valid for the Young modulus of a material satisfying the Saint Venant relation.

If the variation of the seismic velocity between dry and saturated conditions for samples of all orientations are plotted as a function of the corresponding dry velocity, a linear trend becomes apparent (Fig.22). This behaviour is not in agreement with the theore-

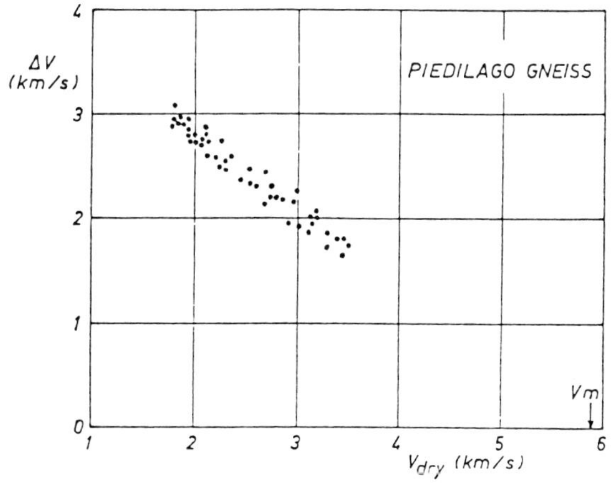

Fig. 22 - Variation of seismic velocity between dry and saturated conditions as a function of the pulse orientation (Piedilago gneiss)

tical prediction for rocks containing a planar set of cracks, and probably derives from the combination of various factors (influence of random cracks, slight dishomogeneities of the rock).

On the basis of dry velocity values, an approximate S.C. analysis, in which the planar set of cracks is assumed to be embedded in a matrix whose increased compliance derives from the presence of the random cracks, leads to estimate a crack density parameter respectively equal to 0.3 for the random set and to 0.2 for the planar set.

If instead a D.C. model is adopted, estimates are equal to 0.85 for the random set and to 1.25 for the planar set.

The indications given by the saturated seismic velocities are less reliable because even a slight intrinsic anisotropy or incomplete saturation can lead to serious errors in estimating the microfissuring conditions.

If the "dynamic" principal moduli derived from the seismic velocities, which are equal respectively to 8.5 GPa and to about 25 GPa, are compared with the "static" moduli found in uniaxial or triaxial compression tests (Fig.23), the latter prove to be markedly

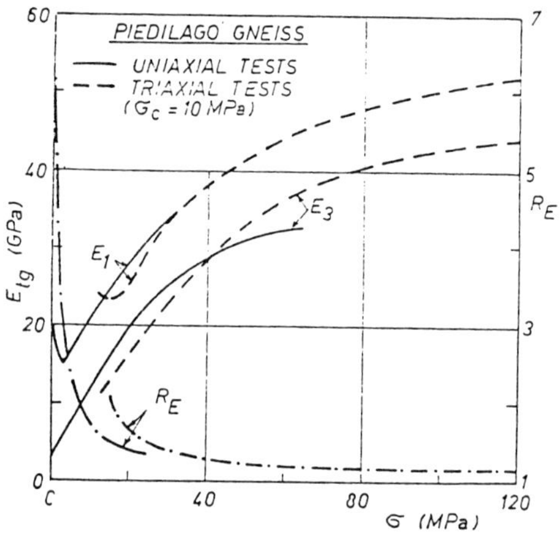

Fig. 23 - Static principal moduli of the Piedilago gneiss (as a function of the stress level. The curves for uniaxial conditions correspond to those shown in scaled form in fig. 15

lower; the discrepancy is surprisingly high for the minor principal modulus E_3 and no convincing explanation can be advanced.

4.4. In Situ Plate-loading Tests

It is interesting to compare the deformation behaviour of anisotropic rocks at the scale of laboratory samples with that evidenced by in situ plate-loading tests (Martinetti and Ribacchi, 1983); Lembo-Fazio and Ribacchi, 1984). For in situ tests, careful techniques were adopted which included:

- the flattening and smoothing of the loaded area (having a diameter equal to 0.5 m) by means of a flat faced rotary bit;
- the application of the load by means of a flat-jack to ensure a uniform pressure distribution;
- the measurement of the displacement at various points (up to 3 m depth) in a central borehole below the loaded area.

The load is applied in three stages (usually 4, 8 and 12 MPa). For each stage the load is brought to its maximum level at a rate of about 1 MPa/min; it is maintained constant for 10 or 20 minutes in order to evidence any possible rheologic behaviour, then it is reduced to (almost) zero. This cycle is repeated twice. When the rock is anisotropic, tests are performed by loading the rock firstly in a direc-

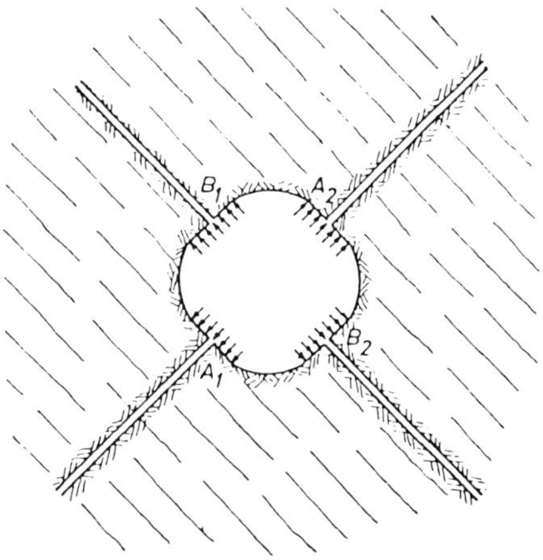

Fig. 24 - Plate-loading tests inside a drift in an anisotropic rock

tion perpendicular to the planar structure (A in Fig.24) and subsequently in a parallel direction (B in Fig.24).

A typical result of a test is shown in Fig.25.

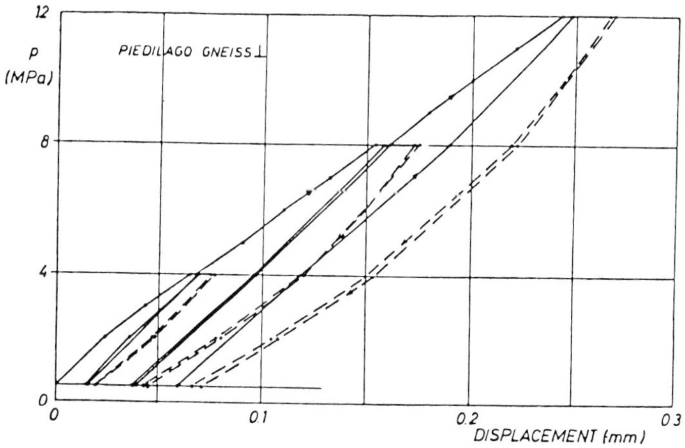

Fig. 25 - Example of the results obtained in a plate-loading test in the Piedilago gneiss (load applied ⊥ to the planar structure, displacements measured at the surface below the centre of the plate)

In Rock Mechanics it is customary practice to evaluate a "deformation modulus" E_d on the basis of the total displacements from the beginning of the test, and an "elastic" modulus E_e on the basis of the immediately recoverable displacements measured in an unloading phase. The moduli are calculated on the basis of the relationships which are valid for a homogeneous and isotropic medium (Steinbrenner approximation).

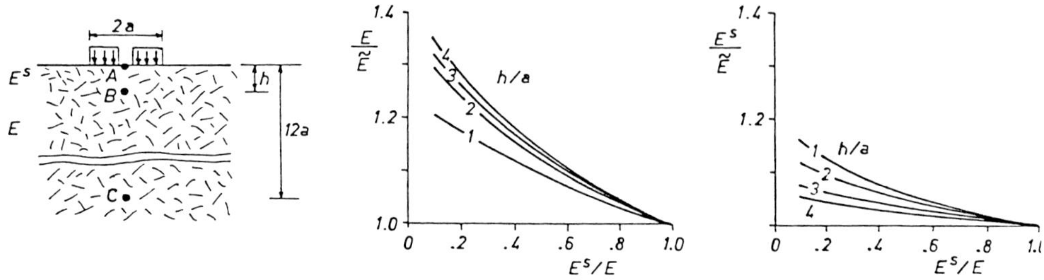

Fig. 26 - Corrective factors for a better estimate of the true moduli of the superficial layer E^s and of the in depth rock E, on the basis of the apparent moduli $\tilde{E}^s$ and $\tilde{E}$ calculated with the Steinbrenner approximation from the relative displacements respectively of the points A and B and of the points B and C

The presence of a low modulus superficial layer below the plate has been evidenced in all the in situ investigations carried out in Italy. With the layout adopted in our tests, the modulus E^s of the superficial layer and the modulus of the underlying rock mass E can be estimated separately.

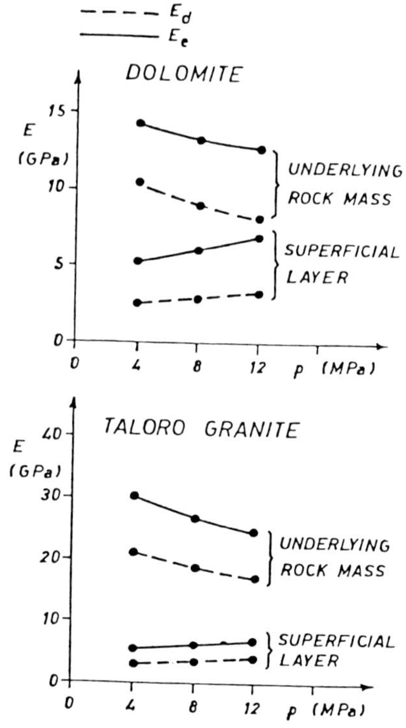

Fig. 27 - Moduli of the superficial layer and of the underlying rock mass obtained in two types of strongly fractured rock masses.

Numerical and experimental investigations were carried out to analyse the errors which can be introduced by simplified interpretation procedures.

The influence of the Steinbrenner approximation on the moduli evaluation in a dishomogeneous two-layer medium is shown in Fig.26; the apparent moduli $\tilde{E}$ underestimate the true moduli both for the superficial layer and in depth, but the maximum error in most cases is lower than 20-25%.

The tests always show a non-linear behaviour of the rock mass (Fig.27); the moduli of the in-depth rock always decrease at increasing loads, whereas in the superficial layer they sometimes increase (as in Fig.27) or sometimes decrease. The errors introduced by this factor are difficult to quantify; it is to be taken into account that the state of stress underneath the plate depends not only on the load applied by the test but also on the initial state of stress around the exploration drift.

When tests are carried out in transversely isotropic rocks, as shown in Fig.24, a simplified interpretation is based on the assumption that the apparent moduli $\tilde{E}$, obtained in configuration A, correspond to the minor principal modulus E_3 and that those obtained in configuration B correspond to the major modulus E_1. For a more accurate evaluation, the corrective coefficients of Fig.28 can be utilized

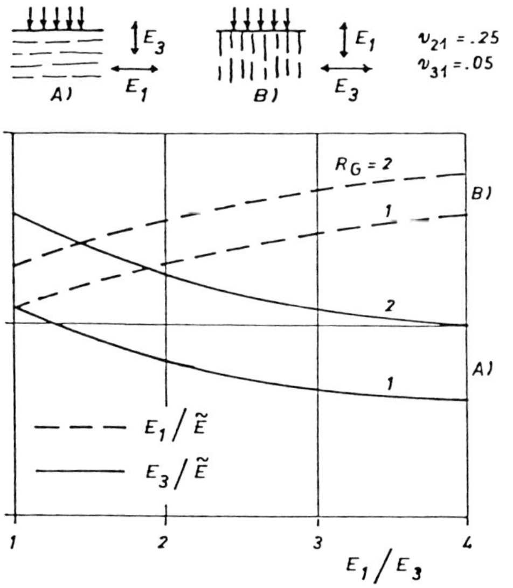

Fig. 28 - Relationship between the true moduli E_1 and E_3 and the apparent moduli $\tilde{E}_1$ and $\tilde{E}_3$ for plate-loading tests in trasversely isotropic rocks

(Lembo Fazio and Ribacchi, 1984).

However, the two types of test shown in Fig.24 are not sufficient for determining all the relevant elastic parameters of the rock, that is E_1, E_3 and G_{13} (or R_G) and therefore an a priori value of R_G is usually assumed. When $R_G = 1$ the true moduli E_1 and E_3 are respectively greater and smaller than the apparent values $\tilde{E}_1$ and $\tilde{E}_3$, and therefore, the true anisotropy ratio R_E is greater than the apparent value.

The results of the in situ plate-loading tests performed in the Piedilago gneiss (6 tests in configuration A and 6 tests in configuration B) will be discussed in detail, because for this rock more complete data are available. The site of the test was a drift (2.5 m wide) at a depth of 380 m below ground surface. Average fracture spa-

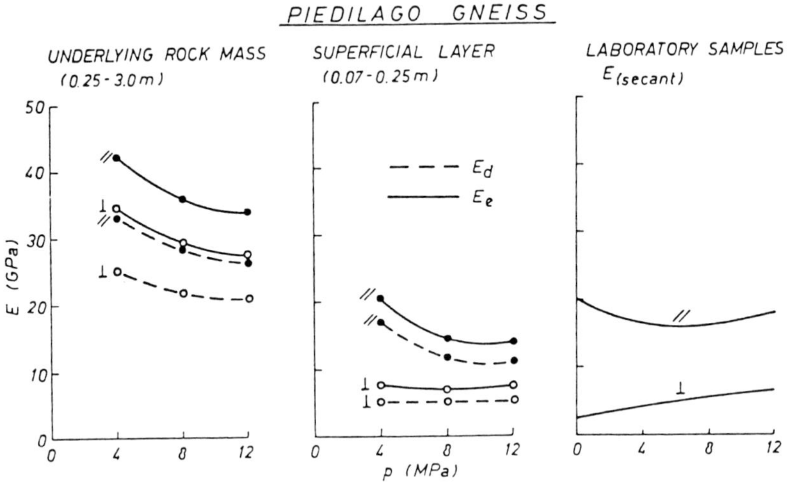

Fig. 29 - Young moduli (secant values) for the Piedilago gneiss obtained in situ and for laboratory samples

cing of the rock mass was about 0.45 m. The drift was situated below the water table, but the pore pressure modification induced by the excavation of the drifts and the saturation conditions of the rock mass below the loading plates are not known.

For each test the seismic velocity was determined by means of a sonic log in the axial borehole. Average values V_1 = 5.25 km/s and V_3 = 4.65 km/s were obtained which are practically coincident with the values found in the laboratory tests on **saturated** samples.

The results of the tests are summarized in Figs. 29 and 30. For each load level the moduli are evaluated on the basis of the average displacements obtained in the 6 tests; the conventional relations deriving from the Steinbrenner approximation were utilized, without any attempt at introducing the corrective factors of Figs. 26 and 28.

In the same figures the behaviour of laboratory samples is represented by means of the **secant** moduli in uniaxial compression, at a value of the stress corresponding to the load applied by the plate. This choice is to a large extent arbitrary, owing to the variable and uncertain state of stress below the loading plate.

Form the analysis of Figs. 29 and 30, the following indications can be drawn (they would not be substantially modified if the correction factors were introduced for the calculation of the moduli)

- the anisotropy ratio of the rock mass (R_E = 1.2) is much lower than that of the laboratory samples; it is practically independent of the load applied on the plate and is in good agreement with the value derived from the mineralogical composition of the rock;
- the moduli of the rock mass are much higher than those of the laboratory samples; they are about 2 times greater for the major modulus and 10-13 times higher for the minor modulus; it can be said, therefore, that the marked anisotropy of laboratory samples is mainly due to the strong reduction of the minor modulus with respect to the in situ conditions;
- both the in situ major and minor moduli decrease at increasing

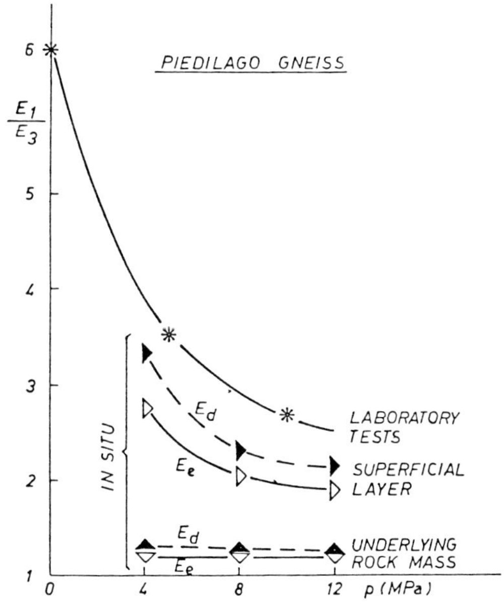

Fig. 30 - Anisotropy ratios fot the Piedilago gneiss obtained in situ and in laboratory samples

plate pressure in agreement with the general behaviour observed in all the types of rock masses (see Fig.27);

- the behaviour of the superficial layer is intermediate between that of the rock mass at depth and that of the laboratory samples.

The results obtained in other types of markedly schistous rocks are shown in Fig.31; the in situ behaviour of the rock mass is similar

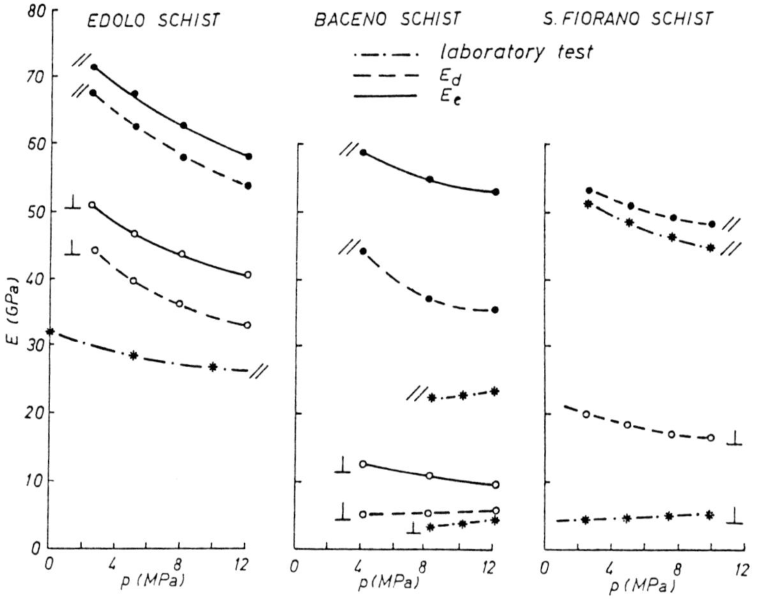

Fig. 31 - Young moduli for various schistous rocks of the Alps obtained in situ and for laboratory samples. Only the values of the rock mass below the superficial layer are indicated for the in situ tests

in many respects to that shown by the Piedilago gneiss. The anisotropy ratio varies greatly from one rock type to another; in spite of this, the mica content is approximately the same (about 50%).

In the plate-loading test carried out in a direction perpendicular to the planar fabric, a rheologic behaviour was often observed. A typical example taken from a test in the Piedilago gneiss is represented in Fig.32; it may be noticed that after the first cycle, the deformation of the rock is almost fully reversible.

The delayed displacements observed, both in the superficial layer and at depth, after the load had been maintained constant for 10 minutes (a convenient arbitrary value), were scaled down with respect to the immediately recovered displacements upon unloading. The average values are represented in Fig.33 as a function of the load level. It is apparent that the delayed deformations are especially important for the superficial layer when the load is applied normally to the planar

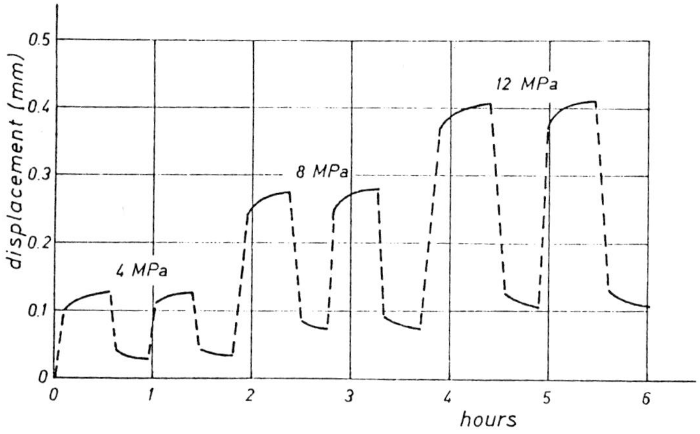

Fig. 32 - Rheologic behaviour shown in situ by the Piedilago gneiss when loaded ⊥ to the planar structure. The surface displacements at the center of the loaded area are indicated

structure. In the underlying rock mass the (short term) rheologic behaviour is less important and is not much influenced by the load direction.

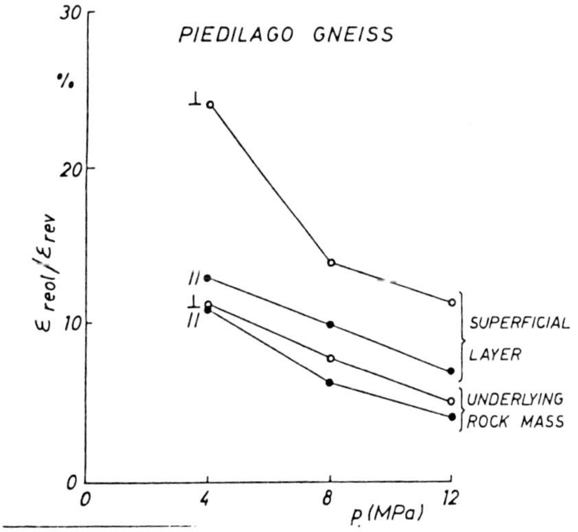

Fig. 33 Delayed deformations (after 10 minutes) scaled down with respect to the immediate recovered deformations, for the superficial layer and the underlying rock mass.

To explain the behaviour of the rock masses and especially the discrepancies between the in situ and laboratory values, one (or more) of the following factors can be invoked:

i) an influence of the in situ state of stress on the compliance of the rock mass;

ii) an influence of the saturation conditions in the rock mass in situ (the laboratory samples were always tested in "dry" conditions);

iii) an irreversible modification of the microcrack fabric deriving from the disturbance of the coring operation;

iv) a similar fabric modification, but occurring gradually during storage after the coring of the samples ("delayed rebound").

As to the first point, a rough evaluation of the pre-existing state of stress on the walls of the drifts can be made by assuming an initial isotropic stress (corresponding to the overburden load) and an isotropic behaviour of the rock mass.

Fig.34 shows that in the zone comprised between a distance of

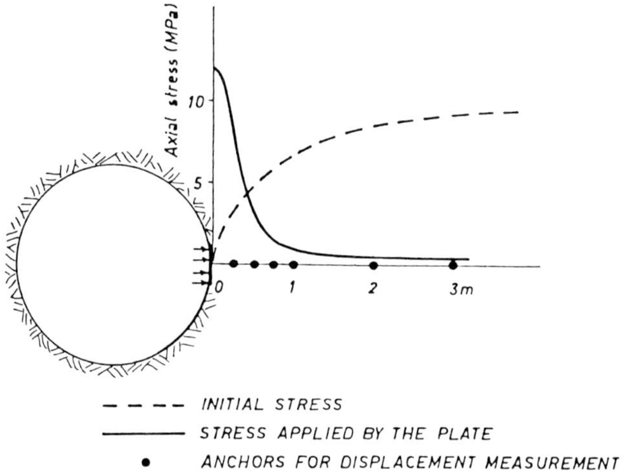

Fig.34 Comparison between the axial preexisting stress on the walls of the drifts and the stress applied by the plate at the maximum load.

0.25 and 1.0 m below the plate (which gives the greatest contribution to the estimation of the moduli), the axial stress applied by tests is comparable and even lower than the pre-existing stress. Therefore, some of the microcracks which are oriented normally to the plate axis would be closed in the in situ conditions and do not contribute to the compliance of the rock.

However, an examination of Fig.23 shows that such a factor, while undoubtedly effective, cannot by itself explain the high E_3

value obtained in situ.

As to the influence of the saturation conditions, it cannot be excluded that water overpressure in the microcraks (conditions of partly "undrained" loading) are induced during the plate-loading tests.

In a rock containing microcracks with a random orientation, the results of plate-loading tests would be similar in drained or undrained conditions, because the Young modulus is practically the same in the two conditions (Figs. 6 and 8). The situation is quite different for a rock containing a planar set of cracks, as is shown by the following example based on the D.C. model.

Let us suppose that the rock contains a planar set of open cracks with a density e = 0.6; Fig.4 shows that in drained conditions $R_E = 4$ and $R_G = 1$; according to Fig.28, a plate-loading test would indicate apparent moduli $\tilde{E}_3 = 1.25\ E_3$ and $\tilde{E}_1 = 0.77\ E_1$, with an apparent $\tilde{R}_E = 2.50$. In undrained conditions the minor modulus becomes 3 times greater than the value in drained conditions (see Fig.7), and R_G rises by about 2. In this condition plate-loading tests would indicate apparent moduli $E_3^u = 2.4\ E_3$ and $E_1^u = 0.83\ E_1$ with an apparent anisotropy ratio $\tilde{R}_E = 1.38$.

The effect is therefore quite important and would be even greater in the S.C. models. It is to be noted that an apparent rheologic behaviour could also be caused by the dissipation of water overpressure.

However, with available data it is not possible to decide whether the premeability of the rock materials was low enough to induce partially undrained conditions during the test, at least in the rock mass below the superficial layer.

Regarding the possibility of a fabric modification, it is well-known that high stress concetrations occur at the bottom of a core stub, when coring is effected in a rock mass subjected to high initial stress (Durelli *et al.*, 1968). In particular, Stacey (1982) evidenced that tensile strains in a direction parallel to the core axis are induced, which sometimes cause the core to break down into thin disks.

It is quite possible therefore that, in an anisotropic rock, these stress concentrations cause the formation or extension of cracks aligned along the fissility plane of the rock, thereby increasing its anisotropy.

Also mechanical stresses and the vibrations induced by the coring bit can produce some effects on the crack geometry, as was suggested by Tullis (1977), in order to explain the strains measured by various researchers during the overcoring of isolated (macroscopically stress-free) blocks of rock.

The above-mentioned phenomena occur just after the coring, but also a slow modification of the microfissuring conditions of a rock after its extraction from its original environment has been suggested by many researchers, although direct evidence is not always clear-cut.

Because of the non-homogeneity and anisotropy of the constituent minerals, high residual stresses at the macroscopic level can be present in a stress-free sample (Nur and Simons, 1970). Values of differential stress up to 30-40 MPa have been measured in quartzites and granites by means of X-ray diffraction techniques (Friedman, 1972).

A widening and extension of microfissures are likely to accompany the gradual release of some residual stresses; a slight thermal or humidity variation of the samples during storage may facilitate this process. For instance, Buen (1979) observed that the seismic velocity of gneiss samples, and sometimes also their strength, decreased with the time of storage. The velocity variation was about 10-20% and it continued at a decreasing rate throughout the observation period (3 months).

A good indication of fabric variations would be the observation of increasing strain in unloaded samples of rock (delayed rebound). Such observations are difficult because of the influence of instrumentation drift, and are easily perturbed by variations of temperature and humidity. However, a recent accurate investigation (Teufel, 1982) confirmed the occurrence of delayed rebound strains, which in some cases have reached a magnitude of about 50% the instant strains.

In the case of the Piedilago site, however it is difficult to explain how an extension and widening of the cracks, important enough to cause such a large modification of the compliance characteristics, is not felt at all by the seismic velocity, even if allowance is made for the scarce sensitivity of this parameter to cracking intensity in saturated conditions; a possible explanation is that the in situ conditions were not fully saturated.

5. CONCLUSIONS

A marked difference in compliance between in situ rock and laboratory samples has been evidenced by the tests, but the effective mechanism controlling such behaviour could not be precisely identified.

On the other hand, similar difficulties and uncertainties regarding the real in situ microfissuring conditions and the effects of sampling procedures were met by other researchers, as is shown by a review of many experimental data presented by Tullis (1977) and by the discussion in a paper by Wang and Simmons (1978).

Practical implications vary greatly depending on which of the mechanisms discussed in the preceding paragraph applies.

In fact, if the microfissuring variations are mainly caused by the coring disturbance or by delayed rebound in unstressed cores, the characteristics of the rock materials are not important at all for the behaviour of large engineering structures, and the choice of design parameters should be based on in situ measurements only; non-linear effects will be less important.

On the contrary, if the observed behaviour is due to pore pressure effects in cracks, the undrained moduli obtained by the in situ tests would be much greater than those relevant for the deformation of the rock mass during the excavation of the caverns. In fact, the behaviour of the rock mass corresponds to that of a double porosity medium, with a network of spaced fractures isolating blocks of intact rock having very low permeability. Piezometric measurements in one of the caverns (Edolo plant) showed that "drained" water pressure conditions in the fracture network are quickly reached after the excavations. Pore pressure variations within the blocks are not known; however, given that the natural drainage lengths depend on the size of the blocks and not on the size of the excavations, and taking into account the slow rate of excavation, fully drained conditions probably occur also within the blocks.

It cannot be excluded however, that the not negligible delayed deformations observed in some caverns may be partly due to the transition between partially undrained and fully drained conditions.

The complex mechanical behaviour of the anisotropic rocks have an important influence also in the interpretation of initial stress determinations in rock masses which are based on the measurement of the strain due to a complete stress release, obtained with "overcoring".

In Italy, the state of stress at various sites of the Alps was measured by means of the "doorstopper" method (Leeman, 1965): strain gauges are glued to the flattened end of a borehole and the stress is released by deepending the hole with a coring crown.

The interpretation is based on linear elastic models and requires the knowledge both of the elastic constants of the rock (at the scale of about 10 mm) and of the stress concentration coefficients, relating the initial stress to the induced stress at the bottom of the hole.

Until recently, the values of the stress concentration coefficients were known only for isotropic media, and approximate methods of interpretation were utilized (Ribacchi, 1977; Martinetti and Ribacchi, 1980). Recently, with the development of efficient three-dimensional finite element techniques, theoretical elastic solutions can be found also for anisotropic materials (Borsetto *et al.*, 1984; Rahn, 1984).

Unfortunately this is not sufficient to solve all the difficulties, because it is not easy to decide which is the best technique capable of taking into account the non-linear stress-strain behaviour found in anisotropic rocks, and even worse, possible crack modifications in the fabric caused by overcoring. It seemes reasonable to utilize the moduli found for the cores in the laboratory, and in particular the "secant" moduli (starting from unstressed conditions); the stress path and the final stress level in the tests should be chosen taking into account the values of the in situ stress, but they are somewhat arbitrary.

This procedure will certainly give incorrect results if a delayed rebound occurs in the storage time between the coring and the testing of the sample; in this situation the compliance should be determined as soon as possible after the coring, or even better at the site itself.

It is true that similar difficulties can be found also when dealing with brittle isotropic rock types, but they appear to be much more severe in anisotropic rocks because of the greater density of the microfissure sets and of the markedly non-linear behaviour they induce in the rock.

REFERENCES

Anderson, D.L.; Minster, B.; Cole, D. (1974) The effect of oriented cracks on seismic velocities. *J. Geophys. Res.* **79**, 4011.

Berg, C.A. (1965) Deformation of fine cracks under high pressure and shear. *J. Geophys. Res.* **70**, 3447-3457.

Berry, P.; Crea, G.; Martino, D.; Ribacchi, R. (1974) The influence of fabric on the deformability of anisotropic rocks. *32d Congr. Int. Soc. Rock Mech.* **II A**, 105-110, Denver.

Borelli, G.B. (1983) Misura delle velocita P ed S in laboratorio e in situ. *Atti del 2 Convegno del Gruppo Nazionale di Geofisica della Terra Solida*, 485-495, Roma.

Brady, B.T. (1969) The non linear mechanical behaviour of brittle rock, Part I. Stress-strain behaviour during region I and II. *Int. J. Rock Mech. Min. Sci.* **6**, 211-225.

Bruner, W.M. (1976) Comments on seismic velocities in dry and saturated cracked solids, *J. Geophys. Res.* **14**, 2573-2576.

Buen, B. (1979) Changes in residual stress due to rheology. *4th Congr. Int. Soc. Rock Mech.* 1, 81-83, Montreux.

Bur, T.R.; Thill, R.E.; Hjelmstadt, K.E. (1969) An ultrasonic method for determining the elastic symmetry of materials. *U.S. Bu. Mines*, R.I. 7333.

Crea, G.; Martino, D.; Ribacchi, R. (1976) Determinazione delle caratteristiche di una roccia anisotropa, *Riv. It. Geotecnica* **10**, 259-268.

Dolcetta, M. (1982) ENEL's recent experiences in the construction of large underground power houses, shafts and pressure tunnels. *Proc. Int. Symp. on Rock Mechanics related to Caverns and Pressure Shafts*, Aachen.

Durelli, A.J.; Obert, L.; Parls, V.J. (1968) Stress required to initiate core discing. *Trans SME* **241**, 269-276.

Feves, M.; Simmons, G. (1976) Effects of stress on cracks in Westerly granite. *Bull. Seism. Soc. America* **66**, 1755-1765.

Friedman, M. (1972) Residual elastic strain in rocks. *Tectonophysics* 15, 293-330.

Friedman, M; Bur, T.R. (1974) Investigations of the relations among strain, fabric, fracture and ultrasonic attenuation and velocity in rocks. *Int. J. Rock Mech. Min. Sci. & Geomech. Abstr.* 11, 221-234.

Hashin, Z. (1970) Theory of composite materials, in "*Mechanics of composite materials*" ed. by F.W.Wendt, H. Liebowitz, N. Perrone, 201-242.

Hoenig, A. (1979) Elastic moduli of a non-randomly cracked body. *Int. J. Solid & Structures* **15**, 2, 137-154.

Horii, H.; Nemat-Nasser, S. (1982) Deformation and fracture of rocks containing microcracks, in "*Problems in Mechanics of Materials and Structures*" edited by F. Maceri, Roma.

Horii, H.; Nemat-Nasser, S. (1983) Overall moduli of solids with microcracks: Load induced anisotropy. *J. Mech. Phys. Solids* **31**, 2, 155-171.

Jaeger, J.C.; Cook, N.G.V. (1969) *Fundamentals of Rock Mechanics.* Methuen, London.

Johnson, L.R.; Wenk, H.R. (1974) Anisotropy of physical properties in metamorphic rocks. *Tectonophysics* **23**, 79-98.

Kachanov, M. (1980) Continuum model of medium with cracks. *J. Engrg. Mech. Div. , Proc. ASCE* **106**, SM5, 1039-1051.

Leeman, E.R. (1964) The measurement of stress in rock. *J South African Inst. Min. Metall.* **65**, 45-114 and 254-285.

Lekhtniskii, S.G. (1963) *Theory of elasticity of an anisotropic elastic body*, Holden-Day Series in Mathematical Physics, S. Francisco.

Lembo Fazio, A.; Ribacchi, R. (1984) Progressi nella realizzazione e nella interpretazione delle prove di carico su piastra negli ammassi rocciosi. *Riv. It. Geotecnica* **18**, 21-31.

Martinetti, S. (1977) Experience in field measurements for underground power stations in Italy. *Proc. Int. Symp. Field Measurements in Rock Mechanics*, **2**, 509-534, Zurich.

Martinetti, S.; Ribacchi, R. (1980) In situ stress measurement in Italy. Rock Mech., Suppl. **9**, 31-47.

Martinetti, S.; Ribacchi, R. (1983) Plate loading tests in anisotropic rock masses. *Int. Symp. "In situ testing"*, **2**, 89-94, Paris.

Mavko, G.H.; Nura, A. (1978) The effect of non elliptical cracks on the compressibility of rocks. *J. Geophys. Res.* **83**, 4459-4468.

Morlier, P. (1971) Description de l'etat de fissuration d'une roche a partir d'essais non-destructifs simples. *Rock Mech.* **3**, 125-138.

Nur, A.; Simmons, G. (1970) The origin of small cracks in ignenous rocks. *Int. J. Rock Mech. Min. Sci.* **7**, 307-314.

Nur, A. (1971) Effects of stress on velocity anisotropy in rock with cracks. *J. Geophys. Res.* **76**, 2022-2034.

O'Connel, R.J.; Budianskii, B. (1974) Seismic velocities in dry and saturated cracked solids. *J. Geophys. Res.* **79**, 5412-5426.

O'Connel, R.J.; Budianskii, B. (1977) Viscoelastic properties of fluid--saturated craked solids. *J. Geophys. Res.* **82**, 5719-5735.

Oda, M. (1982) Fabric tensor for discontinuous geological materials. *Soils Found.* **22**, 4, 96-108.

Oda, M.; Suzuki, K.; Maxshibu, T. (1984) Elastic compliance for rock--like materials with random cracks. *Soils Found.* **24**, 3, 27-40.

Pros, Z.; Z. Babuska, V. (1967) An apparatus for investigating the elastic anisotropy of spherical rock samples. *Studio geoph. et geod.* **12A**, 192-198.

Rahn, W. (1983) Untersuchung moglicher Ausvetungsfehler bei der Interpretation von In-situ Spannungsmessungen in anisotropen Gesteneinen. *Proc. 5th Congr. Int. Soc. Rock Mech.*, F 63-68, Melbourne.

Ribacchi, R. (1977) Rock Stress Measurements in Anisotropic Rock Masses. *Proc. Int. Symp. "Field Measurements in Rock Mechanics"* **2**, 183-196, Zurich.

Salamon, M.D.G. (1968) Elastic moduli of a stratified rock mass. *Int. J. Rock Mech. Min. Sci.* **5**, 519-527.

Simmons, G.; Siegfried, R.W.; Feves, M. (1974) Differential strain analysis: a new method for examining cracks in rocks. *J. Geophys. Res.* **79**, 4383-4385.

Simmons, G.; Todd, T.; Baldridge, W.S. (1975) Toward a quantitative relationship between elastic properties and cracks in low porosity rocks. *American Jour. of Sci.* **275**, 318-345.

Stacey, T.R. (1982) Contribution to the mechanism of core discing. *J. South Afr. Inst. Min. Metall.* **83**, 269-274.

Talebi, S. (1983) Une methode precise pour la mise en evidence et l'etude de l'anisotropie dans les roches. *Rev. Inst. Francais Petrole* **38**, 4, 439-453.

Teufel, L.W. (1982) Prediction of hydraulic fracture azimuth from anelastic strain recovery measurements of oriented cores. *Proc. 23rd U.S Symp. Rock Mech.*, 238-245.

Thill, R.E.; Bur, T.R.; Steckley, R.C. (1973) Velocity anisotropy in dry and saturated rock spheres and its relation to rock fabric. *Int. J. Rock Mech. Min. Sci. & Geomech. Abstr.* **10**, 535-557.

Tullis, T.E. (1977) Reflections on measurement of residual stress in rock. *Pageoph.* **115**, 57-68.

Walsh, J.B. (1965 a) The effect of cracks on the compressibility of rock. *J. Geophys. Res.* **70**, 381-389.

Walsh, J.B. (1965 b) The effect of cracks on the uniaxial elastic compression of rocks. *J. Geophys. Res.* **70**, 399-411.

Walsh, J.B. (1965 c) The effect of cracks in rocks on Poisson's ratio.*J. Geophys. Res.* **70**, 5249-5257.

Walsh, J.B. (1969) New analysis of attenuation in partially malted rock. *J. Geophys. Res.* **74**, 4333-4337.

ROCK-SUPPORT INTERACTION IN LINED TUNNELS

N. Cristescu
University of Bucharest, Department of Mathematics, Bucharest, Romania
D. Fotă and **E. Medveș**
Geomechanics Laboratory, Bucharest, Romania

1. INTRODUCTION AND HISTORICAL OUTLINE

After the excavation of a deep tunnel, a support (or lining) is generally applied to limit the closure of the section of the tunnel in a squeezing rock and to ensure the security of the opening. If the rock is not competent, in the absence of the lining, very often failure may occur as a final result of excessive deformation (wall convergence). Sometimes, even in the presence of a lining structural failure may occur as a result of excessive pressure exerted on the lining by the rock. For an appropriate design of the tunnel support one has to choose the best excavation layout and sequence. Since the convergence of the tunnel takes place slowly in time and failure may occur either quite soon after excavation or sometimes after a long period of time possibly destroying also the support, any analysis of the rock-support interaction is to be based on the *rheological properties* of the rock and on the mechanical properties of the support. On the other hand, it is well known that the sequence and excavation procedure and application of the support influence the tunnel performances and its ultimate shape; therefore the excavation layout and support application sequence are essential factors to be considered in the tunnel support analysis.

Most authors who have considered the tunnel support analysis have used *linear viscoelastic* constitutive equations in order to describe the behavour of the rock. For instance Baoshen (1979) has used a linear standard model and a linear Maxwell model associated with the assumption of volume incompressibility; Popcvic et al. (1979) have used a uniaxial linear standard model. Panet (1979) has revised various rheological models used by several authors; Berest and Nguyen Minh (1983), Nguyen Minh et al. (1983) (1984) have used linear visco-plastic models to describe the rock-support interaction.

A simplified approach (from physical point of view) of rock--support interaction is due to Ardashev et al. (1985); the finite element method is used in conjunction with a nonlinear stress-strain relation. The finite element method is used also by Moore and Booker (1982), Rowe et al. (1982), Yufin et al. (1985), Sharma et al.(1985), Rodriguez-Roa (1985), Kimura et al. (1985), Swoboda (1985) and many others in order to study the rock-support interaction. The same method is used in conjunction with an elastic/viscoplastic constitutive equation by Sun and Lee (1985) for rocks having orthogonal families of joints. A quite complete time-independent theory of rock--support interaction analysis can be found in Hoek and Brown (1980). The tunnel stability was studied by Mühlhaus (1985) assuming the rock to be elastic-plastic (satisfying a Mohr-Coulomb yield condition) and considering uniform internal pressure due to a lining or to rock bolts. Model tests of tunnel excavation in which lining and rock bolt were simulated by thin papers, were carried in dry sand by Adachi et al. (1982).

In another approach the concept of *ground reaction curve*, i.e. the relationship between the ultimate ground displacement and the tunnel support pressure, was introduced without making specific the mechanical properties of the support. This was done either by introducing the concept of equivalent stiffness for an elastic rock (Kaiser (1981)) or a plastic cohesive-frictional dilatant model (Detournay and Fairhurst (1982), Vardoulakis and Detournay (1982), Detournay and Vardoulakis (1985)), or a Mohr-Colomb yield criterion (Cividini et al. (1985)), this time this curve is called "characteristic curve". In some of the examples given by these authors the assumption concerning volume incompressibility was also accepted. However, as it will be shown below, when an elastic/plastic or elastic//viscoplastic model is used for the rock *the ground reaction curve is not unique* since it depends on the loading history (layout sequence of excavation and support application and certainly on the mechanical properties of the support).

What concerns the mathematical model used for the response of the lining subjected to the rock pressure, all authors have assumed a linear relationship only. Below we will consider both *linear* and *nonlinear* relationship for the constitutive equation of the support. The nonlinear constitutive equations are encoutered with yieldable supports, yielding either by sliding between parts of the support or by compressing some wooden strips. For the description of such supports see for instance Woodruff (1966) Vol.2.

The consitutive equation which will be used for the rock is

either of *linear viscoelastic* or of *elastic/viscoplastic* type since as already mentioned, time is the major parameter in the rock-support interaction analysis. While the linear viscoelastic constitutive equation allows a much simpler analysis, it cannot describe the rock dilatancy (only viscoelastic volumetric compressibility) and furthermore a possible failure due to dilatancy. With the elastic/viscoplastic constitutive equation, however, one can describe failure due to dilatancy and as such this constitutive equation is instrumental for the finding of optimal excavation layout and optimal sequence procedure in the lining application, taking into account the specific properties of the rock, the depth etc.

If during the sequence operation used, at a certain moment the tunnel surface becomes stress free (due to the excavation method used: drill and blast, machine excavation etc.) then one of the main parameters involved in the rock-support interaction analysis is the *time of application of the support* after excavation. If this timing is too short, then the support may fail due to overloading; if this timing is too long, then a failure of the rock is possible due to excessive deformation. Generally the whole *loading history*, i.e. the history of the variation of the pressure exerted by the lining on the tunnel surface is of great significance on the overall evaluation of the behaviour of the rock.

In the following the rock-support interaction analysis will be presented following mainly an unpublished report by Cristescu (1985). The case of viscoelastic rock and nonlinear support was considered by Cristescu et al. (1988), while the case of elastic/viscoplastic rock and nonlinear support by Cristescu (1988b).

2. RHEOLOGICAL MODELS FOR ROCK

Two reological models will be used for rocks: an elastic/viscoplastic constitutive equation and a linear viscoelastic one.

According to the *elastic/viscoplastic model* (Cristescu (1987)) the constitutive equation is written as

$$\dot{\boldsymbol{\varepsilon}} = \left(\frac{1}{3K} - \frac{1}{2G}\right)\dot{\sigma}\mathbf{1} + \frac{1}{2G}\dot{\boldsymbol{\sigma}} + k\left\langle 1 - \frac{W^I}{H}\right\rangle \frac{\partial H}{\partial \boldsymbol{\sigma}} \qquad (2.1)$$

where the bulk modulus $K(\sigma)$ may vary gently with the mean stress σ, G is the shear modulus,

$$H(\sigma, \bar{\sigma}) = W^I(t) \qquad (2.2)$$

is the *stabilization boundary* with

$$W^I(T) = \int_0^T \boldsymbol{\sigma}(t)\cdot\dot{\boldsymbol{\varepsilon}}^I(t)dt = \int_0^T \sigma(t)\dot{\varepsilon}_v^I(t)dt+\int_0^T \boldsymbol{\sigma}'(t)\cdot\dot{\boldsymbol{\varepsilon}}'^I(t)dt=W_v^I(T)+W_D^I(T) \quad (2.3)$$

the irreversible stress work (prime stands for "deviator"),

$$\bar{\sigma}^2 = \sigma_1^2 + \sigma_2^2 + \sigma_3^2 - \sigma_1\sigma_2 - \sigma_2\sigma_3 - \sigma_3\sigma_1 \quad (2.4)$$

is the equivalent stress and $k(\sigma,\bar{\sigma},d)$ is the viscosity coefficient which may depend on the stress invariants and on a damage parameter defined by (Cristescu (1986)):

$$d(t) = W_v^I(\max) - W_v^I(t). \quad (2.5)$$

Finally $\langle A\rangle = A^+ = \frac{1}{2}(A + |A|)$.

According to (2.1) the variation of the irreversible part of the volumetric strain is

$$\dot{\varepsilon}_v = k \left\langle 1 - \frac{W^I}{H} \right\rangle \frac{\partial H}{\partial \sigma}. \quad (2.6)$$

Let us denote by $\boldsymbol{\sigma}^P = \boldsymbol{\sigma}(t_0)$ the initail "primary" stress existing in the rock and by $\boldsymbol{\sigma}(t)$ the stress field after excavation. In the points of the rock where

$$H(\boldsymbol{\sigma}(t)) < H(\boldsymbol{\sigma}^P) \quad (2.7)$$

an elastic *unloading* takes place. If $H(\boldsymbol{\sigma}(t)) = H(\boldsymbol{\sigma}^P)$ the new stress state is on the same stabilization boundary and $\dot{\boldsymbol{\varepsilon}}^I = \mathbf{0}$.

The case $H(\sigma(t)) > H(\sigma^P)$ will be called *loading* since in this case a variation of $\boldsymbol{\varepsilon}^I$ is possible. We can distinguish three subcases according to the sign of $\frac{\partial H}{\partial \sigma}$. Thus

$$\left.\begin{array}{ll} H(\boldsymbol{\sigma}(t)) > H(\boldsymbol{\sigma}^P) & \\ \frac{\partial H}{\partial \sigma} > 0 & \text{compressibility} \\ \frac{\partial H}{\partial \sigma} < 0 & \text{dilatancy} \\ \frac{\partial H}{\partial \sigma} = 0 & \text{compressibility/ dilatancy boundary.} \end{array}\right\} \text{viscoplastic deformation} \quad (2.8)$$

The second model which will be considered is a *linear viscoelastic* one of the form (Massier (1984)):

$$\dot{\varepsilon} = -k_v(\varepsilon - \frac{\sigma}{3K_o}) + \frac{1}{3K}\dot{\sigma}$$
$$\dot{\varepsilon}' = -k(\varepsilon' - \frac{\sigma'}{2G_o}) + \frac{1}{2G}\dot{\sigma}' \qquad (2.9)$$

where the material constants (all positive) are: K and G the dynamic moduli, K_o and G_o are the moduli in the relaxed state, while k_v and k are the viscosity coefficients. These coefficients must satisfy several inequalities (see Cristescu (1988a))

$$K > K_o, \quad G > G_o, \quad 2G < 3K, \quad 2G_o < 3K_o$$
$$k_v \geq k, \quad k\frac{G-G_o}{2GG_o} \geq k_v\frac{K-K_o}{3KK_o} > 0. \qquad (2.10)$$

This model can describe a compressibility of the rock but not a dilatancy.

All the constitutive constants or functions involved in (2.1) or (2.9) are determined in the laboratory (Cristescu (1988a)).

3. FORMULATION OF THE PROBLEM

Let us consider a tunnel of radius a, excavated in a rock which is squeezing (wall convergence) unless a support (lining) is properly installed in due time. The convergence rate can be reduced due to the presence of the lining. Sometimes the convergence by creep is stopped altogether by the lining, while other times the rock failure is greatly delayed by the presence of the lining. We assume that the lining has an external radius b < a and an inner radius c. Various kinds of lining will further be analysed. At time t_o elapsed from the time of excavation (here t = 0 is the time when the tunnel surface becomes stress-free due, for instance, to a blast, etc.), the rock comes in contact with the lining so that t_o is the time of initiation of the rock-support interaction. For $0 \leq t \leq t_o$ the creep of the rock is not influenced by the lining, while for $t \geq t_o$ it is.

The main aspects of the problem to be solved for $t > t_o$ are: the stress evolution in the support, the evolution of the pressure and of the displacement at the rock/lining interface, the ultimate displacement and ultimate pressure at this interface and therefore the ultimate shape of the lining (if a stabilization of the deformation occurs in a certain time interval), in what conditions a stabilization will take place and when not, which is the most economical design of the support which would avoid failure, what kind of support would be the most appropriate for the kind of rock considered, the optimal timing t_o and generally the best procedure and sequence of excavation-lining application, the amount of the rock which may deteriorate behind the lining, etc.

It is certainly assumed that the mechanical properties of the material from which the support is made, are known. For simplicity we start by assuming axial symmetry; in the first place we assume also that the far field in situ stresses are hydrostatic

$$\sigma_h \simeq \sigma_v = \gamma h \tag{3.1}$$

and that it is an equilibrium state for the constitutive equation (i.e. satisfies (2.2) for instance). In (3.1) h is the depth given in meters. We renounce at the simplifying assumption (3.1) when stability and failure of tunnels and boreholes will be considered at §6.

If an *elastic/viscoplastic* constitutive equation is used, then for the component $\dot{\varepsilon}_\theta$ this constitutive equation reads

$$\dot{\varepsilon}_\theta = \left(\frac{1}{3K} - \frac{1}{2G}\right)\dot{\sigma} + \frac{1}{2G}\dot{\sigma}_\theta + k \left\langle 1 - \frac{W^I}{H} \right\rangle \frac{\partial H}{\partial \sigma_\theta}. \tag{3.2}$$

Thus any "instantaneous" response of the rock at r = a satisfies

$$\frac{\dot{u}}{a} = \left(\frac{1}{3K} - \frac{1}{2G}\right)\dot{\sigma} + \frac{1}{2G}\dot{\sigma}_\theta \tag{3.3}$$

since $u = \varepsilon_\theta r$. For the hydrostatic far field stresses (3.1), after a fast ("instantaneous") excavation the stresses in the rock are

$$\sigma_r = -(\sigma_v - p)\frac{a^2}{r^2} + \sigma_v \qquad \sigma_z = \sigma_v \qquad \sigma_\theta = (\sigma_v - p)\frac{a^2}{r^2} + \sigma_v \tag{3.4}$$

and therefore at the interface r = a:

$$\sigma_r = p, \quad \sigma_\theta = 2\sigma_v - p, \quad \sigma_z = \sigma_v \tag{3.5}$$

with p the pressure on the surface r = a of the rock. Introducing (3.5) in (3.3) we get the relationship relating the instantaneous variation of the pressure to the displacement of the interface

$$\frac{\dot{u}}{a} = -\frac{\dot{p}}{2G}. \tag{3.6}$$

According to (3.1) any ultimate equilibrum state must satisfy

$$H(\sigma_\infty, \bar{\sigma}_\infty) = W^I(t)\big|_{t \to \infty}. \tag{3.7}$$

Thus while the instantaneous response relationship is unique, but certainly depending on initial data, the locus of ultimate states is not since it depends on the loading history as well.

Let us discuss the formulation of the *initial conditions* (see also Cristescu (1988a)), when the elastic/viscoplastic constitutive equation is used. Two cases can be distinguished depending on whether $\sigma^P = \sigma_v < \sigma_o$ or $\sigma_v > \sigma_o$ (in the model (2.1) σ_o is the pressure which closes all the microcracks existing in the rock - see fig.3.1 -). For each of these two cases, two subcases are possible: the primary stresses at the depth considered satisfy the stability equality

$$H(\sigma^P,0) = W^{IP} \tag{3.8}$$

or the stability inequality

$$H(\sigma^P,0) < W^{IP} \tag{3.9}$$

since in all cases $\bar{\sigma}^P = 0$. The four possible cases are shown in fig.3.1 a-f. The primary stress satisfying (3.8) is shown as a point P in fig.3.1 b, while the initial shape of the corresponding ultimate ground reaction curve (3.7) as a border line and that of the "instantaneous" response (3.6) at the moment of excavation is shown in fig.3.1 a as a full line.

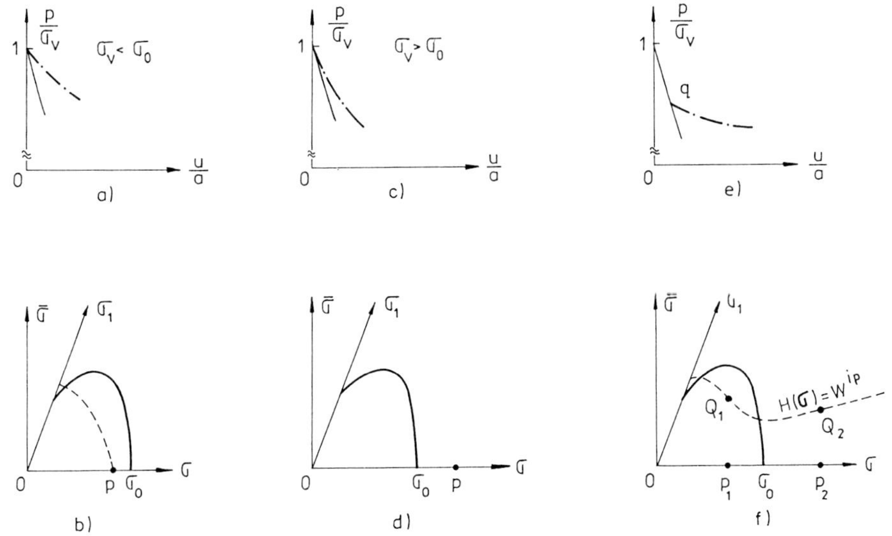

Fig.3.1. Formulation of the initial conditions for elastic/viscoplastic constitutive equations, depending on various possible primary stresses

In a similar way the case when inequality (3.9) is satisfied is shown in fig.3.1 f for both cases $\sigma_v < \sigma_o$ (point P_1) and $\sigma_v > \sigma_o$ (point P_2). A possible corresponding primary stabilization boundary is shown as dotted line. If the secondary stress states (stress after excavation) are represented by points situated above the point Q_1 which is located on the boundary (or Q_2 respectively) then the ultimate ground reaction curve (border line) and the instantaneous response curve (full line) are shown in fig.3.1. e; these two curves intersect each other at a point corresponding to the stress state in Q. The ultimate ground reaction curve is certainly not unique, but all these curves pass through point q in fig.3.1 e. If, however, the secondary stress state is represented by a point situated under the curve (3.8) which passes by point Q, then the rock will remain elastic after excavation.

An intermediate case is shown in fig.3.1 d. In this case the primary state is elastic (since $\sigma^P > \sigma_o$) and an equilibrium state, but the smallest increase of $\bar{\sigma}$ leads to an elastic/viscoplastic state. Since the passing from elastic hydrostatic state to the elastic/viscoplastic is very smooth, the ultimate ground reaction curves have an initial slope (at point $p = \sigma_v$ and $u = 0$) coinciding with the slope of the instantaneous response curve (fig.3.1. c).

For the *linear viscoelastic* constitutive equation the formulation of the initial conditions is much simpler. First, from (2.9) we have

$$\dot{\varepsilon}_\theta = -k(\varepsilon_\theta - \frac{1}{2G_o}\sigma_\theta) + \frac{1}{2G}\dot{\sigma}_\theta \tag{3.10}$$

and with $(3.4)_2$ we get the *differential equation for the rock/support interface* as

$$\dot{u}(t) + \frac{a}{2G}\dot{p}(t) = -k[u(t) + \frac{a}{2G_o}(p(t) - \sigma_v)]. \tag{3.11}$$

This equation describes the convergence of the tunnel surface when on this surface acts a pressure p(t) which may be variable (the pressure exerted by the lining).

If by excavation (a blast, say) the pressure at r = a drops instantly from σ_v to zero, then according to (3.11) the instantaneous variation of the displacement is governed by (3.6). On the other hand, if for various values of p = const. after a long time interval when $t \to \infty$ and $\dot{u} \to 0$, from (3.11) follows the relationship to be satisfied by the ultimate values of the pressure p_∞ and displacement u_∞

$$p_\infty = \sigma_v - \frac{2G_o}{a} u_\infty . \tag{3.12}$$

This is just the equation of the *ultimate ground reaction curve* which is now *unique*; it is shown in fig.3.2 as a border line. The two straight

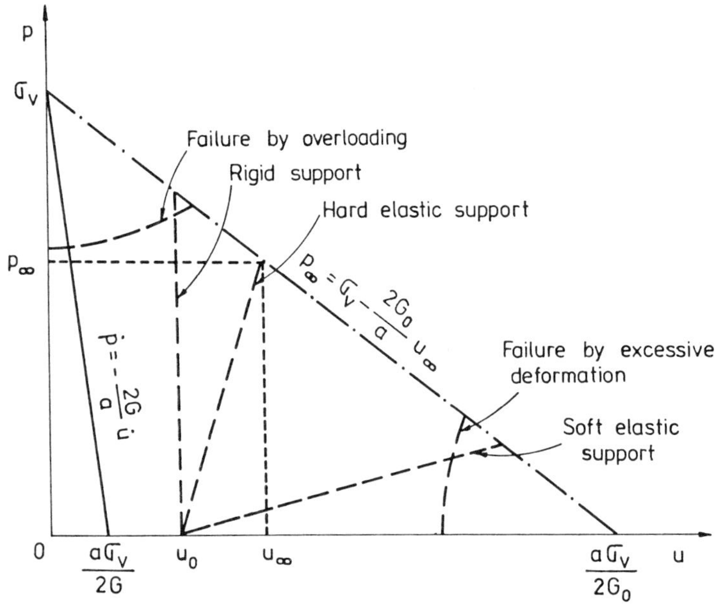

Fig.3.2. Possible pressure-displacement relationship on the tunnel surface and rock-lining interface (Cristescu et al. (1987)).

lines (3.12) and (3.6) have certainly distinct slopes; both are uniquely determined if the constitutive equation is formulated and the initial data are prescribed.

4. ROCK-SUPPORT INTERACTION ANALYSIS FOR VISCOELASTIC ROCK

For methodological reasons let us consider first the case of linear viscoelastic rock and of various kinds of linings.

4.1. Linear elastic support. Let us consider a thick tube-shaped lining of constant thickness b - c, made of linear elastic material (concrete or shotcrete, say). If p is the pressure exerted by the rock on the lining and u is the radial displacement of the lining/rock interface, then the constitutive equation for the linear elastic support is

$$p = \frac{E(b^2 - c^2)}{(1 + \nu)[(1 - 2\nu)b^2 + c^2]} \frac{u - u_o}{b} = q\frac{u - u_o}{b} \qquad (4.1)$$

where E and ν are the elastic constants of the material of the lining, q its rigidity and u_o the displacement of the rock at the moment when it comes in contact with the lining. Relatively high values of q correspond to *hard* supports while smaller values of q to *soft* supports.

If a sudden excavation is done, the pressure on the surface r = a drops suddenly to zero and the displacement of the interface r = a becomes $u = (a\sigma_v)/2G$ (see fig.3.2). Further the convergence of the walls is governed by (3.11) with p = 0, i.e.

$$u(t) = [\frac{1}{2G_o} + (\frac{1}{2G} - \frac{1}{2G_o}) \mathbf{exp}(-kt)]a\sigma_v \qquad (4.2)$$

with a possible ultimate displacement

$$u\Big|_{\substack{t\to\infty \\ p=0}} = u_\infty^o = \frac{a\sigma_v}{2G_o}. \qquad (4.3)$$

If at a conveniently chosen time t_o the support comes in contact with the rock, then from (4.1) and (3.11) we get for $t \geq t_o$ the displacement of the rock/support interface influenced by the presence of an elastic support

$$u(t) = \frac{Q}{P} + (u_o - \frac{Q}{P}) \mathbf{exp}[P(t_o - t)] \qquad (4.4)$$

with

$$P = k\frac{1 + \frac{a}{2G_o}\frac{q}{b}}{1 + \frac{a}{2G}\frac{q}{b}}, \qquad Q = \frac{ka}{2G_o}\frac{\sigma_v + \frac{q}{b}u_o}{1 + \frac{a}{2G}\frac{q}{b}} \qquad (4.5)$$

if the initial conditions

$$t = t_o: \quad u = u_o, \quad p = 0$$

are also used. When $t \to \infty$ the ultimate displacement is

$$u_\infty = a\frac{b\sigma_v + qu_o}{2G_o b + aq} \qquad (4.6)$$

and ultimate pressure yields from (3.12).

If the support is an *ideal rigid* one, i.e. for $t \geq t_o$, $u = u_o$ = const., then the increase of the pressure which acts on the support is governed by

$$p(t) = (\sigma_v - \frac{2G_o}{a} u_o)\{1 - \exp[\frac{Gk}{G_o}(t_o - t)]\}. \qquad (4.7)$$

If t_o is relatively small, then from (4.7) are obtained very high pressures, under which the support may be excepted to collapse. The same is true for elastic supports which are rigid.

The boundary in the u-p plane marking the failure by overloading (shown schematically in fig.3.2 by interrupted line) can be determined by introducing an appropriate failure condition and knowing the stress distribution in the lining

$$\sigma_r = \frac{b^2 p}{b^2 - c^2}(1 - \frac{c^2}{r^2}),$$

$$\sigma_\theta = \frac{b^2 p}{b^2 - c^2}(1 + \frac{c^2}{r^2}), \qquad (4.8)$$

$$\sigma_z = 2\nu \frac{b^2 p}{b^2 - c^2}.$$

For instance for a concrete or shotcrete support, in order to predict failure one can accept a Nadai (1950) type of failure condition

$$(\sigma_r - \sigma_\theta)^2 + (\sigma_\theta - \sigma_z)^2 + (\sigma_z - \sigma_r)^2 = 2[a_o(\sigma_r + \sigma_\theta + \sigma_z) + a_1]^2 \qquad (4.9)$$

where

$$a_o = \frac{\sigma_c - \sigma_t}{\sigma_c + \sigma_t}, \quad a_1 = \frac{2\sigma_c\sigma_t}{\sigma_c + \sigma_t} \qquad (4.10)$$

with σ_c the uniaxial compressive strength and σ_t the absolute value of the uniaxial tensile strength of the concrete. Such kind of failure condition seems appropriate for concrete since it takes into account the influence of the hydrostatic pressure as well. From (4.9) and (4.8) follows that for such a support the rock pressure must not surpass the limit value

$$p < \frac{a_1(b^2 - c^2)}{2b^2[a_o(1 + \nu) - (1 - \nu + \nu^2)^{\frac{1}{2}}]} \qquad (4.11)$$

assuming that $\nu > 0.1$ say. Conversely, if the pressure is prescribed by some other arguments, then one can determine the minimal thickness of the wall of the lining

$$c^2 < b^2[1 + \frac{2p\{(1 - \nu + \nu^2)^{\frac{1}{2}} - a_o(1 + \nu)\}}{a_1}]. \qquad (4.12)$$

The above formula can be used for instance to show how the thickness of the wall of the lining must be increased with increasing depth (increasing p). For instance in fig. 4.1 is shown the variation of the maximum pressure with the radius of the tunnel for three kinds of concrete (curves a,b,c) and two of shotcrete (curves d and e). The material constants are given in Table 4.1 and $\sigma_c = 10\sigma_t$. A first conclusion is that the Poisson coefficient of the lining has a very important influence: smaller values of ν increase the maximum pressure. An increase of σ_c increases the maximum pressure.

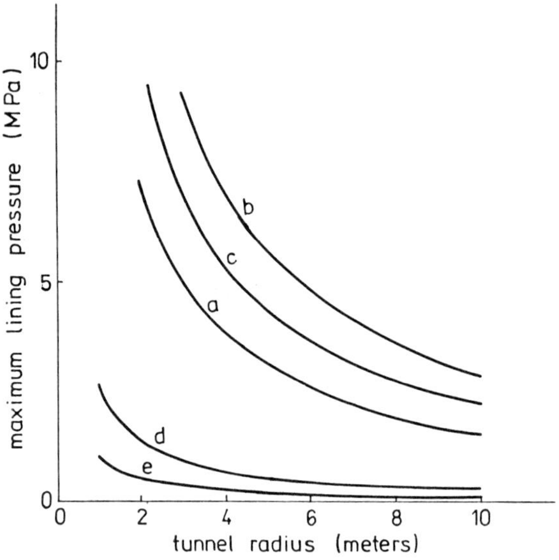

Fig.4.1. Variation of maximum pressure with tunnel radius for various linings

Table 4.1

Constants used in fig.4.1

Lining Material	ν	σ_c (MPa)	Wall thickness (cm)	Curve in fig.4.1
Concrete	0.25	35	30	a
Concrete	0.20	35	30	b
Concrete	0.25	49	30	c
Shotcrete	0.25	35	5	d
Shotcrete	0.25	14	5	e

Further in fig.4.2 is shown how the lining thickness varies with the tunnel radius for various pressures at the rock/lining interface. The example shown in this figure corresponds to the curve a shown in fig.4.1. Thus, if the rock/lining interface pressure is prescribed, one can determine for each tunnel radius the appropriate lining thickness once the material from which the lining is made, was chosen.

The boundary of the domain where rock failure may occur due to excessive deformation cannot be obtained with the present constitutive equation. This subject will be discussed in the next section.

Knowing the elastic properties of the support (i.e. knowing q) the timing t_o must be chosen in such a way that the straight line (4.1) should not cross neither of the two boundaries mentioned above.

In order to give an example of how the above formulae can be used, let us consider a circular tunnel of radius a = 200 cm excavated in a soft rock (coal) at the depth 280 m. The constitutive constants for this rock are G = = 300.8 MPa, G_o = 20 MPa, k = = $1.182 \cdot 10^{-6} s^{-q}$. The slow closure of the walls in the absence of any support is

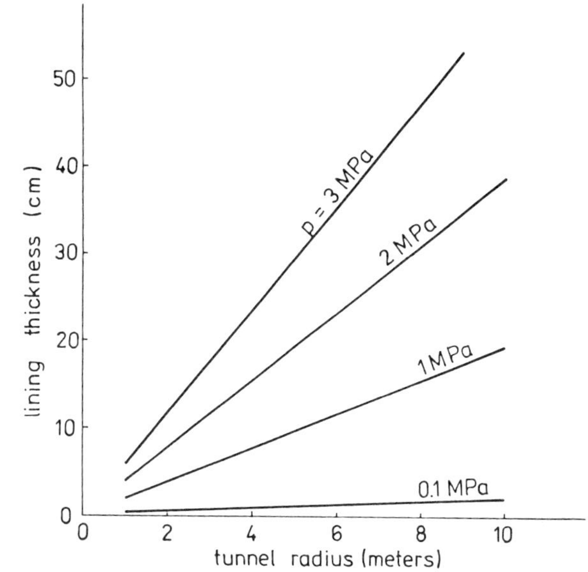

Fig.4.2. Lining thickness as function of tunnel radius for various pressures p.

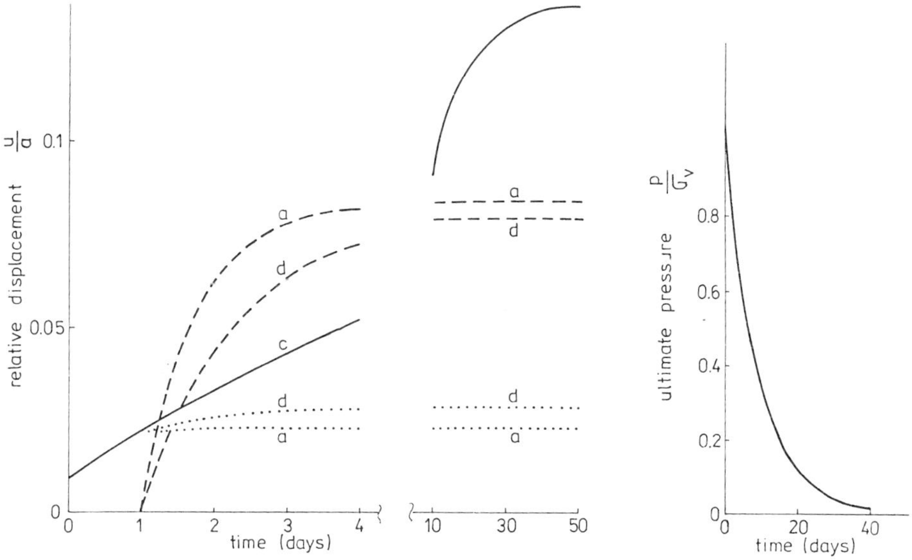

Fig.4.3. a) Variation in time of radial displacement in the absence of the support (full line), of the radial displacement in the presence of the support (dotted line) and of the pressure on the support (interrupted line) for various moments of application of the support. b) Ultimate pressure as function of the time of application of the support.

shown in fig.4.3.a by full line. This closure by creep lasts more than two months. If a support is applied after one day the closure is slowed down as shown by the two dotted lines, corresponding to curves a and d shown in fig.4.1; a shotcrete lining being softer, is certainly leading to smaller pressures but higher displacements. In fig.4.3.a by interrupted lines is shown the variation of the pressure for the two considered linings. In these examples the value of b is obtained from $b = a - u_o$ and that of u_o from (4.2) for $t_o = 1$ day. If the lining is installed after a longer time, the ultimate pressure is certainly smaller. The decrease of the ultimate pressure p_∞/σ_v with the timing t_o is shown in fig.4.3.b. This pressure decreases quite fast.

4.2. Yieldable supports. There are several kinds of yieldable supports; the relationship between p and u for such kind of supports is nonlinear. The support will be called *elastic/constant pressure support* if it satisfies the following conditions

$$\begin{aligned} t_o \le t \le t_1 &: \quad p(t) = q\frac{u(t) - u_o}{b} \\ t \ge t_1 &: \quad p = p_1 = p(t_1) = \text{const.} \end{aligned} \tag{4.13}$$

For $t < t_o$ the convergence takes place according to (4.2) while for $t_o \le t \le t_1$ the convergence is obtained from (4.4). Finally for $t > t_1$ the convergence takes place according to

$$u(t) = -\frac{a}{2G_o}(p_1 - \sigma_v) + [u_1 + \frac{a}{2G_o}(p_1 - \sigma_v)]\,\mathbf{exp}(-k(t_1 - t)) \tag{4.14}$$

if at $t = t_1$ we have $u(t_1) = u_1$ and $p(t_1) = p_1$. The ultimate displacement u_∞ is obtained either from (4.14) for $t \to \infty$ or from (3.12) for $p_1 = p_\infty$.

A *first example* of yieldable support is the one made from circular steel-yieldable shaft rings; the overlapped ends are held together by means of U-bolts clamps. If the ground pressure reaches a certain magnitude, the sliding joints are yielding so that the ground pressure is kept under control. The sliding starts when the ground pressure reaches the value (Cristescu et al. (1987))

$$p_{mi} = \frac{2n\mu_s M}{L\,R\,g\,d\,(\mathbf{sin}\,\alpha + \mu_s \mathbf{cos}\,\alpha)} \tag{4.15}$$

and stops when the pressure has dropped to

$$p_{mf} = \frac{2n\mu_d M}{L\,R\,g\,d\,(\mathbf{sin}\,\alpha + \mu_d \mathbf{cos}\,\alpha)}\ . \tag{4.16}$$

Here n is the number of U-bold clamps in a joint, M the tightening torque of a nut, L the spacing between two rings, along the tunnel, R the radius of curvature of the shaft ring, g a constant (ranging between 0.15 and 0.20), d the bolt diameter, α the angle between the friction surfaces (typical for the kind of shaft ring) and μ_s and μ_d are the static and dynamic friction coefficients respectively. These pressures are illustrated in fig. 4.4. The pressure varies between these two values until stabilization is obtained. It is easy to find the number of slip-creep cycles, the time during which a cycle takes place, the ultimate pressure etc.

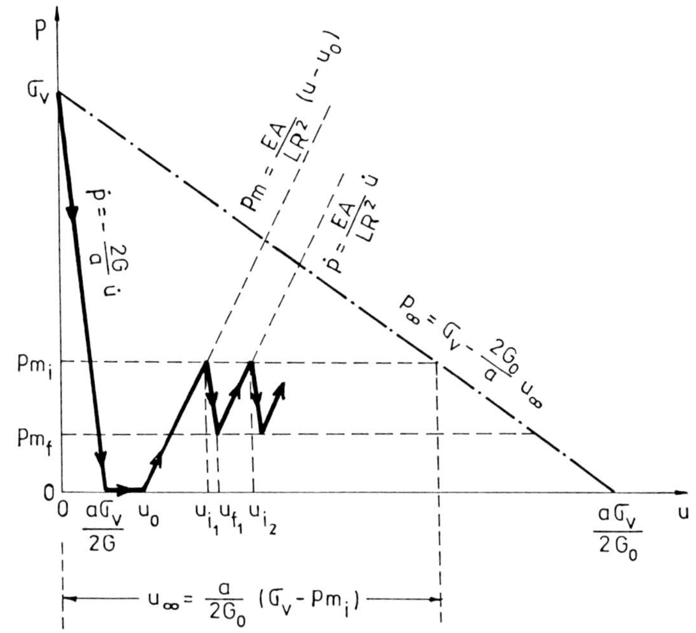

Fig.4.4. Variation of pressure and radial displacement in a yieldable steel shaft ring.

A second example considers a lining made of circular segments of reinforced concrete panels with wooden strips inserted between the longitudinal joints of the panels (see fig.4.5) (Cristescu (1988 a)). It is assumed that the reinforced concrete panels are much more rigid than the wood and for this reason the entire possible deformation of the lining is essentially due to the compressibility of the wood planks. First was studied

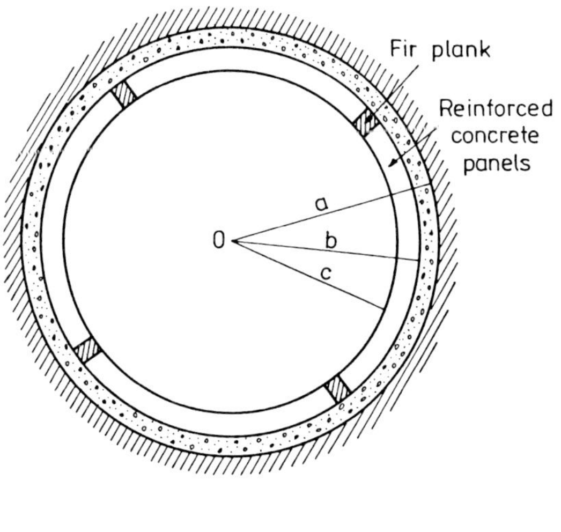

Fig.4.5. Reinforced concrete panels lining with inserted wooden strips.

the compressibility of the wood planks, when compressed perpendicular to the fibres. Three cases were considered: dry fir planks, wet planks and wet planks which were previously already once compressed. The relationship between the compressive stress and the reduction of thickness w is found to be of the form

$$\sigma = Aw^3 + Bw^2 + Cw \tag{4.17}$$

with A, B and C material constants. This relationship can be also linearized, for convenience, in the rock/support analysis. If there are n planks inserted in the lining circumference, then the relationship between the pressure at the rock/lining interface and the radial convergence of this interface become

$$p = \frac{b - c}{R}\left[A\left(\frac{2\pi(u - u_o)}{n}\right)^3 + B\left(\frac{2\pi(u - u_o)}{n}\right)^2 + C\,\frac{2\pi(u - u_o)}{n}\right] \tag{4.18}$$

and linearized versions are also possible. This is the constitutive equation of the support which can be included in the rock/support analysis. Several examples were considered. In all the ultimate pressure was prescribed and by a back analysis it was computed at what timing t_o the lining is to be installed. The results are shown in fig. 4.6; wet and dry planks were considered for two cases: 4 or 10 planks

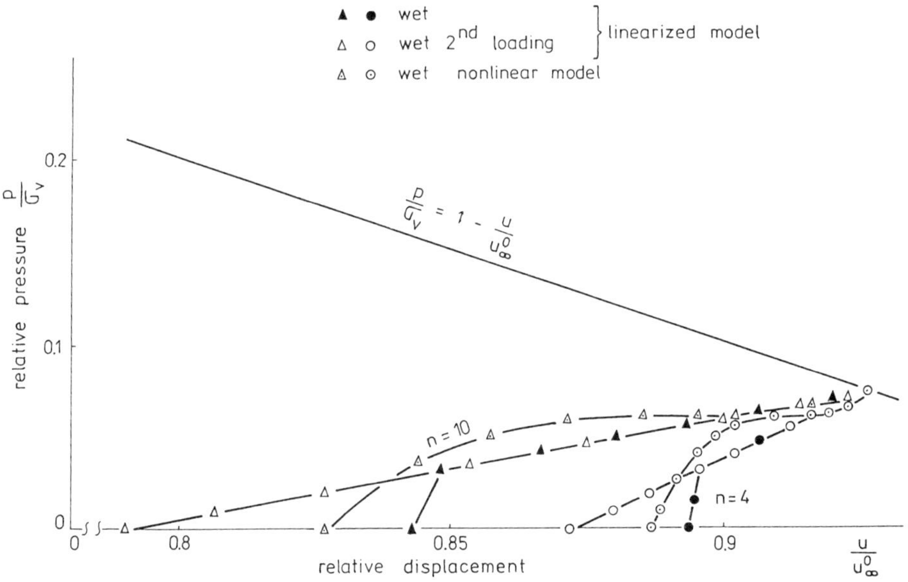

Fig.4.6. Variation of pressure and radial displacement at the rock/lining interface for 4 and 10 wood planks inserted along circumference.

inserted in the circumference. The initial thickness of the planks was 4 cm. The geometry of the tunnel excavated in coal at the depth 350 m is: a = 170 cm, b = 166 cm, c = 150 cm and R = 156 cm. For instance in the case of wet planks and a linearized version of (4.18) we obtain for the timing t_o:

$$t_o = \frac{1}{k} \ln \frac{\frac{1}{2G} - \frac{1}{2G_o}}{\frac{u_o}{a\sigma_v} - \frac{1}{2G_o}} \tag{4.19}$$

when the initial radial displacement due to creep without support was

$$u_o = \frac{R}{q_1}[(\frac{q_1}{R} + \frac{2G_o}{a})u_\infty - \sigma_v + p_1 - \frac{b - c}{R} S_1 w_1]. \tag{4.20}$$

Thus a complete rock/support analysis can be done. A major conclusion is that the most convenient lining, from the point of view of yielding under the rock pressure is that obtained with a greater number of planks (or thicker) which were previously already compressed (loading followed by unloading) and which are wet. In this case the rigidity of the support (q_1 in formula (4.20)) is the smallest from all the cases considered. All the material constants involved in the above formulae are determined in laboratory tests.

5. ROCK-SUPPORT INTERACTION ANALYSIS FOR ELASTIC/VISCOPLASTIC ROCK

For elastic/viscoplastic consitutive equation (2.1), from (3.5) and (3.2) we get the *differential equation describing the motion of the lining/rock interface* as

$$\frac{\dot{u}}{a} = - \frac{\dot{p}}{2G} + k \left\langle 1 - \frac{W^I}{H} \right\rangle \frac{\partial H}{\partial \sigma_\theta} . \tag{5.1}$$

Thus the instantaneous response (sudden blast) is governed by (3.6) while the ultimate ground reaction satisfies (3.7).

Two solutions of the problem were given (Cristescu (1988b)). A solution called *simplified* was obtained with the assumption that the stresses in the rock surrounding the tunnel vary at excavation but further remain constant during creep. With this assumption the wall convergence in the absence of a lining (i.e. p = 0) or when a support of constant pressure is in place (p = const.) is governed by

$$\frac{u(t)}{a} = -\frac{p-\sigma_v}{2G} + \frac{H\left(1-\frac{W^{IP}}{H}\right)\frac{\partial H}{\partial \sigma_\theta}}{\frac{\partial H}{\partial \boldsymbol{\sigma}}\cdot\boldsymbol{\sigma}}\left\{1 - \mathbf{exp}\left[\frac{k}{H}\frac{\partial H}{\partial \boldsymbol{\sigma}}\cdot\boldsymbol{\sigma}(t_c - t)\right]\right\} \quad (5.2)$$

assuming that p remains constant during creep: here W^{IP} is the value of W^I for the primary stress state and t_c is the moment when the deformation by creep begins. If in (5.2) for $p = p_\infty$ = const. we make $t \to \infty$ we obtain the *equation of the ultimate ground reaction curve*

$$\frac{u_\infty}{a} = -\frac{p_\infty - \sigma_v}{2G} + \left.\frac{H\left(1-\frac{W^{IP}}{H}\right)\frac{\partial H}{\partial \sigma_\theta}}{\frac{\partial H}{\partial \boldsymbol{\sigma}}\cdot\boldsymbol{\sigma}}\right|_{p = p_\infty} \quad (5.3)$$

or the rock/support interface stabilization curve corresponding to the particular loading history considered, i.e. due to a fast excavation a sudden decrease of p from σ_v to a certain value p_∞ which afterwards remains constant. It is obvious that the $u_\infty - p_\infty$ relationship (5.3) is not a straight line and that this line is passing by the point u = = 0, $p = \sigma_v$. As an example in fig. 5.1 is shown by border line the ultimate ground reaction obtained for coal for which

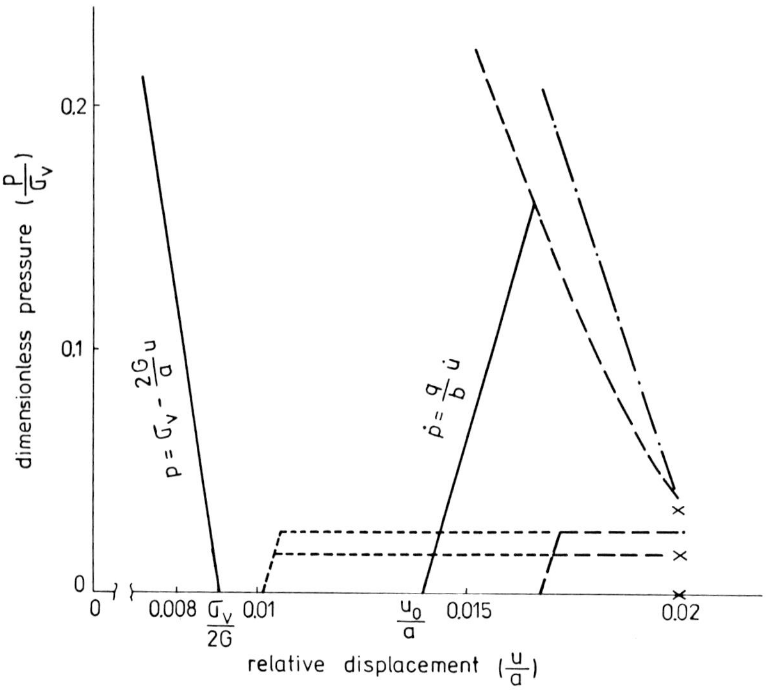

Fig.5.1. Variation of pressure and radial displacement of the rock/lining interface for yieldable steel shaft ring and various loading histories. Crosses mark incipient failure, while the border line and the dotted line are two possible ultimate ground reaction curves (Cristescu (1988b)).

$$H(\sigma,\bar{\sigma}) \equiv \frac{a_o}{a_2\dfrac{\sigma}{\sigma_*}+a_1}\left(\frac{\bar{\sigma}}{\sigma_*}\right)^2 + b_o\frac{\bar{\sigma}}{\sigma_*} + \begin{cases} c_o\sin\left(\omega\dfrac{\sigma}{\sigma_*}+\phi\right)+c_1 & \text{if } 0\le\sigma\le\sigma_o \\ c_o+c_1 & \text{if } \sigma_o\le\sigma \end{cases} \tag{5.4}$$

with $a_o = 7.65\cdot10^{-4}$ MPa, $a_1 = 0.55$, $a_2 = 8.1599\cdot10^{-3}$, $b_o = 0.001$ MPa, $c_o = 4.957\cdot10^{-4}$ MPa, $c_1 = 4.8955\cdot10^{-4}$ MPa, $\omega = 171.927^o$, $\phi = -99.068^0$, $\sigma_o = 1.0996$ MPa, $\sigma_* = 1$ MPa, $K = 769.6$ MPa, $G = 300.8$ MPa, $k = 6\cdot10^{-6}\ s^{-1}$ and $d_{cr} = 4.48\cdot10^3\ J\ m^{-3}$.

Other history of variation of p will certainly lead to some other ultimate ground reaction (since the loading history is involved in $W^I(t)$). If we reduce the pressure p in several successive stages we obtain ultimate values u_∞ which are bigger than that obtained in the previous history of variation of p.

The constitutive equation (2.1) can be used to estimate when the threshold of a *failure process due to excessive deformation by dilatancy* may be reached. For this purpose we must calculate the value of $W_v^I(t)$ (see (2.3)) for the considered history of variation of p. When this quantity reaches a certain critical value $W_v^I(cr) = W_v^I(t_{cr}) - W_v^{IP} = -d_{cr}$ then failure starts taking place from this time t_{cr} on. This timing is obtained from

$$t_{cr} = t_c - \frac{H}{k\dfrac{\partial H}{\partial \boldsymbol{\sigma}}\cdot\boldsymbol{\sigma}}\ \mathbf{ln}\left\{1 - \frac{\dfrac{1}{H}\dfrac{\partial H}{\partial \boldsymbol{\sigma}}\cdot\boldsymbol{\sigma}}{\left(1-\dfrac{W^{IP}}{H}\right)\dfrac{\partial H}{\partial \sigma}\sigma}W_v^I(cr)\right\} \tag{5.5}$$

where again t_c is the moment when creep of the rock has started. The failure by dilatacy is a progressive failure, some kind of erosion. If such a failure threshold is obtained in the case when the pressure p is decreased in a single step from σ_v to its ultimate value (during which the creep of the rock has taken place) then it can be shown that if this decrease is progressive (in several steps, say), then failure may be avoided on the expense of a slightly greater radial displacement u_∞. Also histories of variation of p in which p is never reduced to zero, are more favourable from the point of view of avoiding failure. By stars is shown in fig.5.1 the line of incipient failure according to the criterion given above and for the decrease of p in a single step.

The case when by sudden excavation (blast) the pressure is reduced to zero for a certain time interval during which the rock creeps and afterwards at time t_o a yieldable support is installed, was also considered. First, by computing W_v^I and W^I from

$$dW_v^I = \left(\frac{b}{qa} + \frac{1}{2G}\right) \frac{\frac{\partial H}{\partial \sigma}\sigma}{\frac{\partial H}{\partial \sigma_\theta}} dp$$

$$dW^I = \left(\frac{b}{qa} + \frac{1}{2G}\right) \frac{\frac{\partial H}{\partial \sigma} \cdot \sigma}{\frac{\partial H}{\partial \sigma_\theta}} dp \tag{5.6}$$

along the straight line $(4.13)_1$ we can determine if in the time interval $t_o \leq t \leq t_1$ a stabilization of the creep or a failure threshold will take place. In the first case the ultimate ground reaction is certainly different from the previous one. For instance in fig.5.1 is shown by interrupted line the ultimate ground reaction obtained with the present loading program. The threshold of failure is quite close to the line marked by stars.

If a support made of circular steel-yieldable shaft rings is installed after two hours and $p_{mi}/\sigma_v = 0.0161$, $p_{mf}/\sigma_v = 0.0143$, the pressure-displacement variation at the rock/lining interface is shown as lower dotted line in fig.5.1. The upper dotted line corresponds to a higher tightening torque of the nuts for which $p_{mi}/\sigma_v = 0.025$ and $p_{mf}/\sigma_v = 0.0232$.

The horizontal interrupted line corresponds to the installation of the support after one day.

The solution given above is but an approximate one since it was obtained with the assumption of stress constancy during creep and also since the compatibility equation to be satisfied by the strain components was disregarded. A rigorous solution can be obtained by a numerical procedure. One must integrate the system of partial differential equations

$$\frac{\partial \sigma_r}{\partial r} + \frac{\sigma_r - \sigma_\theta}{r} = 0$$

$$\frac{\partial \varepsilon_\theta}{\partial r} + \frac{\varepsilon_\theta - \varepsilon_r}{r} = 0$$

$$\dot{\varepsilon}_r = \left(\frac{1}{3K} - \frac{1}{2G}\right)\dot{\sigma} + \frac{1}{2G}\dot{\sigma}_r + k\left\langle 1 - \frac{W^I}{H}\right\rangle \frac{\partial H}{\partial \sigma_r} \tag{5.7}$$

$$\dot{\varepsilon}_\theta = \left(\frac{1}{3K} - \frac{1}{2G}\right)\dot{\sigma} + \frac{1}{2G}\dot{\sigma}_\theta + k\left\langle 1 - \frac{W^I}{H}\right\rangle \frac{\partial H}{\partial \sigma_\theta}$$

$$0 = \left(\frac{1}{3K} - \frac{1}{2G}\right)\dot{\sigma} + \frac{1}{2G}\dot{\sigma}_z + k\left\langle 1 - \frac{W^I}{H}\right\rangle \frac{\partial H}{\partial \sigma_z}.$$

This system was integrated with appropriate initial and boundary conditions using finite difference method (Cristescu (1988a)). The initial conditions are obtained from the elastic solution since it is assumed that due to a fast excavation, in a very short time interval the rock response is essentially elastic. The boundary conditions are formulated at infinity (at far distances stresses and strains are those from the primary state) and at the tunnel/support interface (the pressure varies according to the relationship characteristic for the kind of lining considered).

If the pressure at the tunnel surface decreases suddenly from σ_v to a certain value which is afterwards kept constant, then the ultimate ground reaction curve which is obtained for a tunnel excavated in coal at the depth 350 m, is shown in fig.5.2.a by curve b (dash-circle). The ultimate ground reaction obtained with the simplified solution is shown as line a (dash-dot). Thus the ultimate displacement obtained with the numerical method is more significant than the one obtained with the simplified solution. What concerns the lines marking the threshold of failure, the two solutions lead practically to the same result.

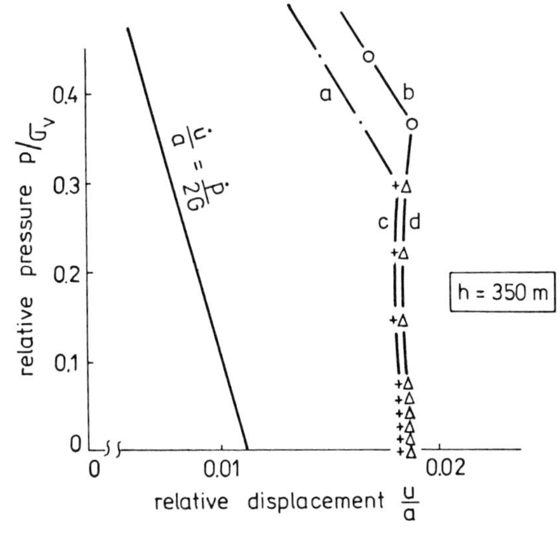

Fig.5.2. a) Ultimate ground reaction curves obtained with simplified method (curve a) and with numerical method (curve b), onset of failure curves obtained with simplified method (curve c) and with numerical method (curve d), at the depth of 350 m.

Though for $p/\sigma_v < 0.3$ a failure behind the lining of constant pressure is taking place, one can ask the question how important is the volume of the rock which will be ultimately damaged by microcracking and how big is the additional loading on the lining. The amount of this failed rock is shown in fig. 5.2.b. For relatively small values of p the amount of the damaged rock may be significant mainly if there are also

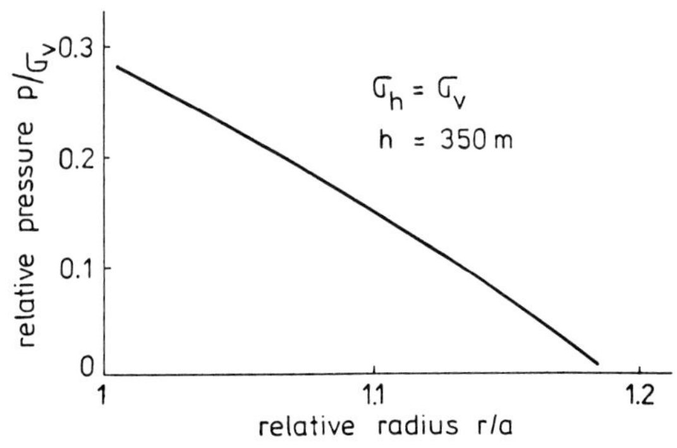

Fig.5.2. b) The amount of damaged rock behind the lining.

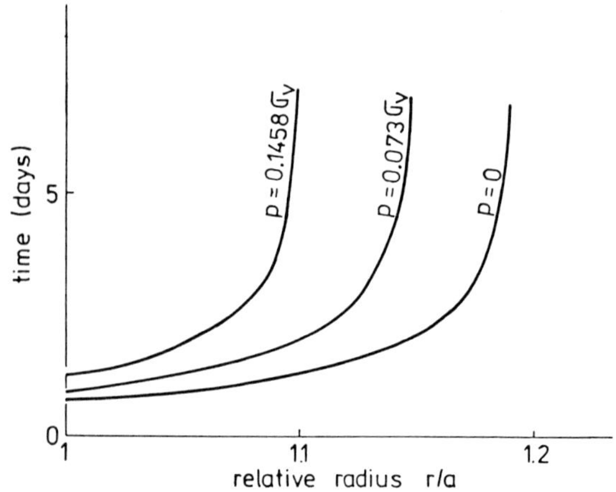

Fig.5.3. Propagation of the damage in the rock surrounding the tunnel surface.

some weak planes, other structural weaknesses etc. How fast propagates the damage in the rock is shown in fig.5.3 for three levels of pressure. In the absence of the lining the amount of the damaged rock is quite significant: it starts at t = = 0.78 days and lasts for more than 5 days, though the fast deterioration takes place mainly in the first two days. The presence of a constant pressure is certainly reducing the amount of the damaged rock behind the lining.

6. TUNNEL OR BOREHOLE STABILITY

In order to study possible failure and loosing of stability of the rocks surrounding a tunnel (in the presence of a lining) or a borehole, let us consider a case which is very unfavourable. The most important factor influencing possible failure and stability is the ratio between the horizontal σ_h and vertical σ_v far field stresses. Let us consider a case when this ratio is quite distinct from unity, for instance $\sigma_h = 0.2\sigma_v$. The depth of the tunnel is certainly another factor influencing significantly failure and stability, besides the mechanical properties of the rock.

In order to make an analysis of the failure, it is first necessary to study the stress distribution around the circumference in the first moments following excavation. Assuming that after a fast excavation the rock response is elastic, the stress distribution is obtained from the well known formulae

$$\sigma_{rr} = p\frac{a^2}{r^2} + \frac{\sigma_h + \sigma_v}{2}\left(1 - \frac{a^2}{r^2}\right) + \frac{\sigma_h - \sigma_v}{2}\left(1 - \frac{4a^2}{r^2} + \frac{3a^4}{r^4}\right)\cos 2\theta$$

$$\sigma_{\theta\theta} = -p\frac{a^2}{r^2} + \frac{\sigma_h + \sigma_v}{2}\left(1 + \frac{a^2}{r^2}\right) - \frac{\sigma_h - \sigma_v}{2}\left(1 + \frac{3a^4}{r^4}\right)\cos 2\theta \qquad (6.1)$$

$$\sigma_{r\theta} = \frac{\sigma_h - \sigma_v}{2}\left(-1 - \frac{2a^2}{r^2} + \frac{3a^4}{r^4}\right) \sin 2\theta$$
$$\sigma_{zz} = \sigma_h - \nu(\sigma_h - \sigma_v)\frac{2a^2}{r^2} \cos 2\theta \tag{6.1}$$

and the displacements from (see Cristescu (1988a))

$$u_r = \frac{1+\nu}{E}\left\{-p\frac{a^2}{r} + \frac{\sigma_h + \sigma_v}{2}\frac{a^2}{r} + \frac{\sigma_h - \sigma_v}{2}\left[4(1-\nu)\frac{a^2}{r} - \frac{a^4}{r^3}\right] \cos 2\theta\right\}$$
$$u_\theta = -\frac{1+\nu}{E}\frac{\sigma_h - \sigma_v}{2}\left[2(1-2\nu)\frac{a^2}{r} + \frac{a^4}{r^3}\right] \sin 2\theta . \tag{6.2}$$

It is useful to represent this stress distribution in the plane of the invariants $\sigma O\bar{\sigma}$. For instance for coal (see (5.4)) at the depth h = 350 m, $\sigma_h = 0.2\sigma_v$ and $\sigma_v = \gamma h$ ($\gamma = 0.0196$) we obtain the stress distribution shown in fig.6.1. The primary stress is marked by point P and the initial stabilization boundary $H(\sigma) = H(\boldsymbol{\sigma}^P)$ by interrupted line. In the neighbourhood of the crown we get tensile stresses, while along the sidewalls the stresses are far in the dilatant domain. Thus two kinds of failure are possible (Cristescu (1986)). At the crown the stress state is elastic while the maximum principal stress is negative and surpasses the tensile strength of the rock (here σ_t = 0.5 MPa).Along the walls the stress state is visco-plastic and a significant dilatancy takes place; thus an evolutive failure is possible. This is shown in fig.6.2.a for p = 0. The domain at the crown where failure by

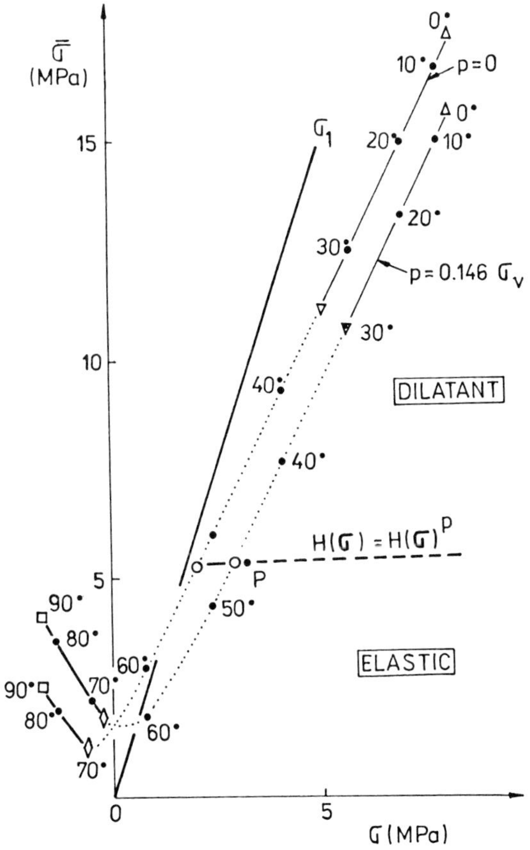

Fig.6.1. Stress distribution along the circumference of a tunnel

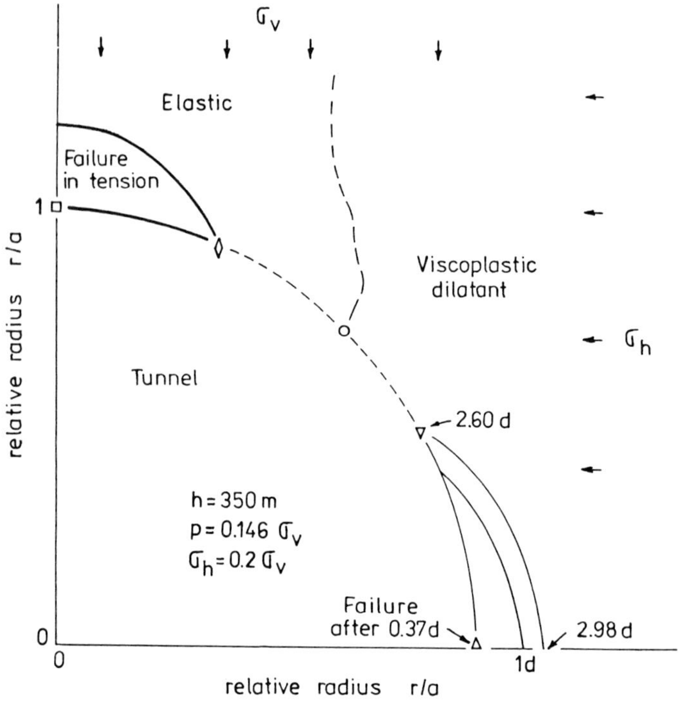

Fig.6.2 a) Failure in tension and by dilatancy along the circumference of a borehole.

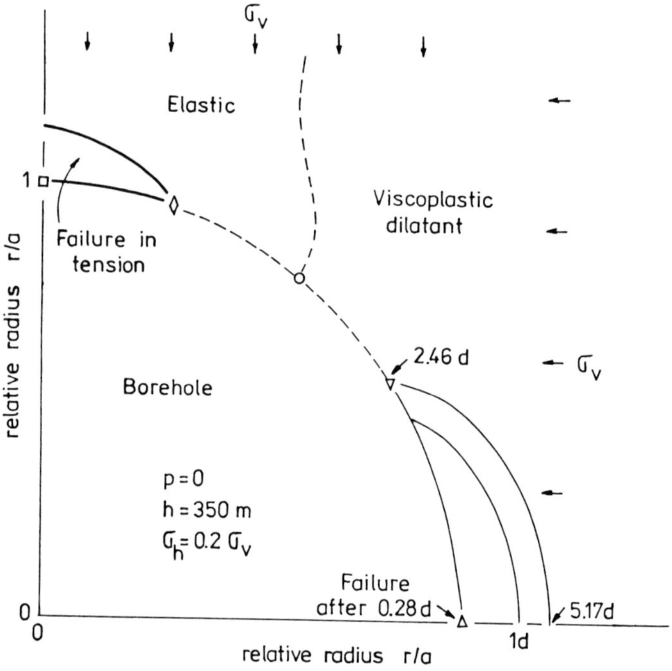

Fig.6.2. b) Failure in tension and by dilatancy along the circumference of a lined tunnel

tensile stresses is to be expected "instantaneously" is shown surrounded by thick lines. The evolutive damage of the rock starts at $r = a$ and $\theta = 0^o$ after 0.28 days if the beginning of damage is characterised by the value $d_{cr} = 4.48 \cdot 10^3 \, J \, m^{-3}$. This evolutive damage progresses slowly in time and reaches the ultimate position after 5.17 days at $\theta = 0^o$ and $r = 1.22\,a$, but after 2.46 days at $r = a$ and $\theta = 33^o$. These boundaries, i.e. lines of equal damage, are shown in fig.6.2 as thin full lines. If the rock is nonhomogeneous or has some structure weaknesses, then the damage domain can be much larger. The portion of the contour shown by dotted lines is not affected by any failure. The same convention for the lines is used also in fig.6.1.

In the case of a lined tunnel, with a constant pressure support of pressure $p = 0.146\,\sigma_v$ say, the domain at the crown where failure by tensile stresses is possible increases in size (fig.6.2.b). The domains where a progressive damage is possible are however shrinking in size.

It is interesting to mention that while the vertical diameter of the opening decreases suddenly at excavation and continues to decrease further slowly in time, the horizontal diameter first slightly increases at excavation but afterwards decreases slowly in time. This is shown in fig. 6.3. If $p = 0$ the initial decrease of u/a is only 5% from the increase which follows, while for $p = 0.146\,\sigma_v$, u/a decreases initially with 18% from the maximum increase which follows.

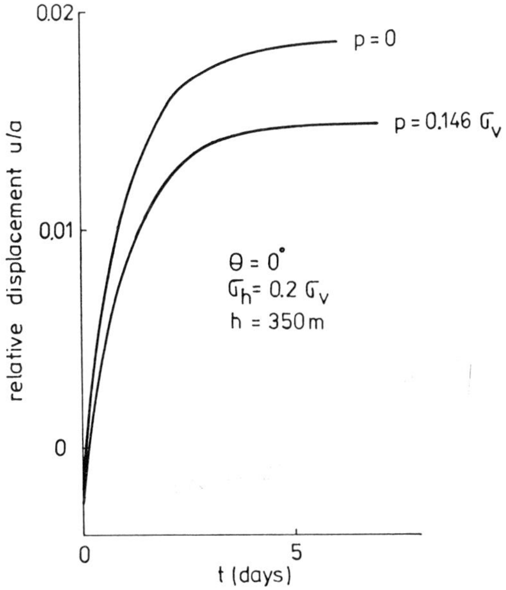

Fig.6.3. Variation in time of the displacement of the wall.

We arrived at the conclusion that an increase in depth increases the danger of failure and damage by dilatancy. If however, the ratio σ_h/σ_v is close to unity, this danger decreases. The damage by dilatancy depends on the accurate determination of d_{cr}, which is not an easy task. Also the exact determination of the reference configuration for the exact estimation of the damage parameter at the primary state is still an open problem. Therefore the above examples are but illustrative.

7. CONCLUSION

The main conclusion which yields from the above is that an analysis of the rock-support interaction can be done assuming a quite general rheological model for the rock and a non linear relationship for the support (possibly a yieldable one). Any other kind of support can be considered in a similar way and possibly also a more involved sequence of several kinds of supports succesively applied to the tunnel.

After the formulation of a mathematical model for the rock-support interaction, an optimization of the support installation procedure can be obtained by mathematical methods. If the rock is quite soft and squeezing very much, a general conclusion would be that in order to avoid failure by excessive deformation the appropriate design of tunnel support set up is the one in which the excavation lay out and sequence would never reduce simultaneously to zero the pressure all along the entire circumference of the future surface of the tunnel. In other words, the final pressure p_∞ is to be reached by steadily decreasing step by step the far field stresses to the ultimate pressure p_∞, possibly by a decrease done succesively along the contour.

REFERENCES

Adachi,T., Tamura,T. Shinkawa,M. 1982, Analytical and Experimental Study on Tunnel Support System. *Fourth Internat. Conf. on Numerical Methods in Geomechanics*, Edmonton, Balkema, Rotterdam, 513-522.

Ardashev,K.A., Amusin,B.Z., Koshelev,V.F. 1985, Numerical methods of analysis of underground constructions. *Fifth Internat.Conf. on Numerical Methods in Geomechanics*, Nagoya, Balkema, Rotterdam--Boston, 1077-1084.

Baoshen,L. 1979, A study of the mechanism of support. *Fourth Internat. Congr. on Rock Mechanics*, Montreux, 1979, Balkema Books, I, 157-160.

Berest,P., Nguyen,M.D. 1983, Time-dependent behaviour of lined tunnels in soft rocks. *Eurotunnel'83 Conference*, Basel, Switzerland, 57-62.

Cividini,A., Gioda,G., Barla,G. 1985, Calibration of a rheological model on the basis of field measurements. *Fifth Internat. Conf. on Numerical Methods in Geomechanics*, Nagoya, Balkema, Rotterdam--Boston, 1621-1628.

Cristescu,N. 1985, Mathematical modelling of the mining support and of the rock-support interaction. Report, University of Bucharest.

Cristescu,N. 1986, Damage and Failure of Viscoplastic Rock-Like Materials. *Internat. J. Plasticity*, **2**, 2, 189-204.

Cristescu,N. 1987, Elastic-Viscoplastic Constitutive Equations for Rock. *Internat. J. Rock Mech. Min. Sci. & Geomech. Abstr.* **24**, 5, 271-282.

Cristescu,N. 1988a, *Rock Rheology*, Martinus Nijhoff, The Hague (in press).

Cristescu,N. 1988b, Viscoplastic creep of rocks around a lined tunnel. *Internat. J. Plasticity* (submitted for publication).

Cristescu,N. Fotă,D., Medveş,E. 1987 Tunnel Suport Analysis incorporating Rock Creep. *Internat. J. Rock Mech. Min. Sci. & Geomech. Abstr.*, **24**, 6.

Detournay,E., Fairhurst,C. 1982, Generalization of the ground reaction curve concept. *23rd U.S. Symp. on Rock Mechanics*, Berkeley, 926-936.

Detournay,E., Vardoulakis,I. 1985, Determination of the Ground Reaction Curve Using the Hodograph Method. *Internat. J. Rock Mech. Min. Sci & Geomech. Abstr.*, **22**, 3, 173-176.

Hoek,E., Brown,E.T. 1980, *Underground Excavation in Rock.* Institution of Mining and Metallurgy, London.

Kaiser,P.K. 1981, A New Concept to Evaluate Tunnel Performance--Influence of Excavation Procedure. *22nd U.S. Rock Mech. Symp.*, Boston, 264-271.

Kimura,F., Okabayashi,N., Ono,K., Kawamoto,T. 1985, Rock-mechanical discussion for the mechanism of supporting system in severe swelling rock tunnel. *Fifth Internat. Conf. on Numerical Methods in Geomechanics*, Nagoya, Balkema, Rotterdam-Boston, 1265-1271.

Massier,D. 1984, private communcation.

Moore,I.D., Booker,J.R. 1982, A Circular Boundary Element for the Analysis of Deep Underground Openings. *Fourth Internat. Conf. on Numerical Methods in Geomechanics*, Edmonton, Balkema, Rotterdam, 53-60.

Mühlhaus,H.-B. 1985, Lower Bound Solutions for Circular Tunnels in Two and Three Dimensions. *Rock Mechanics and Rock Engineering*, **18**, 37-52.

Nadai,A. 1950, *Theory of Flow and Fracture of Solids*, McGraw Hill, New York, Toronto, London, Vol.1.

Nguyen,M.D., Berest,P., Bergues,J. 1983, Analyse du comportement differe des ouvrages souterrains. *5th Internat. Congr. on Rock Mechanics*, Melbourne, Preprints Section D, D233-D239.

Nguyen,M.D., Habib,P., Guerpillon,Y. 1984, Time dependent behaviour of a pilot tunnel drive in hard marls. *Design and performance of underground excavation.* ISRM/BGS, Cambridge, 453-459.

Panet,M. 1979, Time-Dependent Deformations in Underground Works. *Fourth Internat. Congr. on Rock Mechanics*, Montreux, Balkema Books, I, 279-289.

Popović,B., Marković,O., Manojlović,M. 1979, Stresses and Strains at the Contact of Rigid Tunnel Lining and Soft Rock. *Fourth Internat. Congr. on Rock Mechanics*, Montreux, Balkema Books, I, 525-531.

Rodriguez-Roa,F. 1985, Lining ground interaction in circular tunnels. *Fifth Internat. Conf. on Numerical Methods in Geomechanics*, Nagoya, Balkema, Rotterdam-Boston, 1257-1264.

Rowe,R.K., Lo,K.Y., Tham,L.G. 1982, The Analysis of Tunnels and Shafts in Dense (Oil) Sand. *Fourth Internat.Conf. on Numerical Methods in Geomechanics*, Edmonton, Balkema, Rotterdam, 587-596.

Sharma,K.G., Varadarajan,A., Srivastava,R.K., 1985,Elasto-viscoplastic finite element analysis of tunnels. *Fifth Internat. Conf. on Numerical Methods in Geomechanics*, Nagoya, Balkema, Rotterdam-Boston, 1141-1148.

Sun,J., Lee,Y.S. 1985, A viscous elasto-plastic numerical analysis of the underground structure interacted with families of multilaminate rock mass using FFM. *Fifth Internat. Conf. on Numerical Methods in Geomechanics*, Nagoya, Rotterdam-Boston, Balkema, 1127-1134.

Swoboda,G. 1985, Interpretation of field measurements under consideration of the three-dimensional state of stress, the visco-elasticity of shotcrete, and the viscpoplastic behaviour of rock. *Fifth Internat. Conf. on Numerical Methods in Geomechanics*, Nagoya, Balkema, Rotterdam-Boston, 1651-1658.

Vardoulaskis,I., Detournay,E. 1982, Determination of the Ground Reaction Curve in Deep Tunnels Using Biot's Hodograph Method. *Fourth Internat. Conf. on Numerical Methods in Geomechanics*, Edmonton, Balkema, Rotterdam, 619-624.

Woodruff,S.D. 1966, *Methods of Working Coal and Metal Mines*, Pergamon Press, Vol.2.

Yufin,S.A., Postolskaya,O.K., Shavarchko,I,R., Titkov,V.I. 1985, Some aspects of underground structure mechanism in the finite element method analysis. *Fifth Internat. Conf. on Numerical Methods in Geomechanics*, Nagoya, Balkema, Rotterdam-Boston, 1093-1100.

DEFORMATION OF LAMINATED LACUSTRINE SEDIMENTS OF THE DEAD SEA

Y. Arkin
Geological Survey of Israel, Jerusalem

The Dead Sea (Figure 1) is the lowest point on the earth's surface, at more than 400m below the level of the oceans, representing

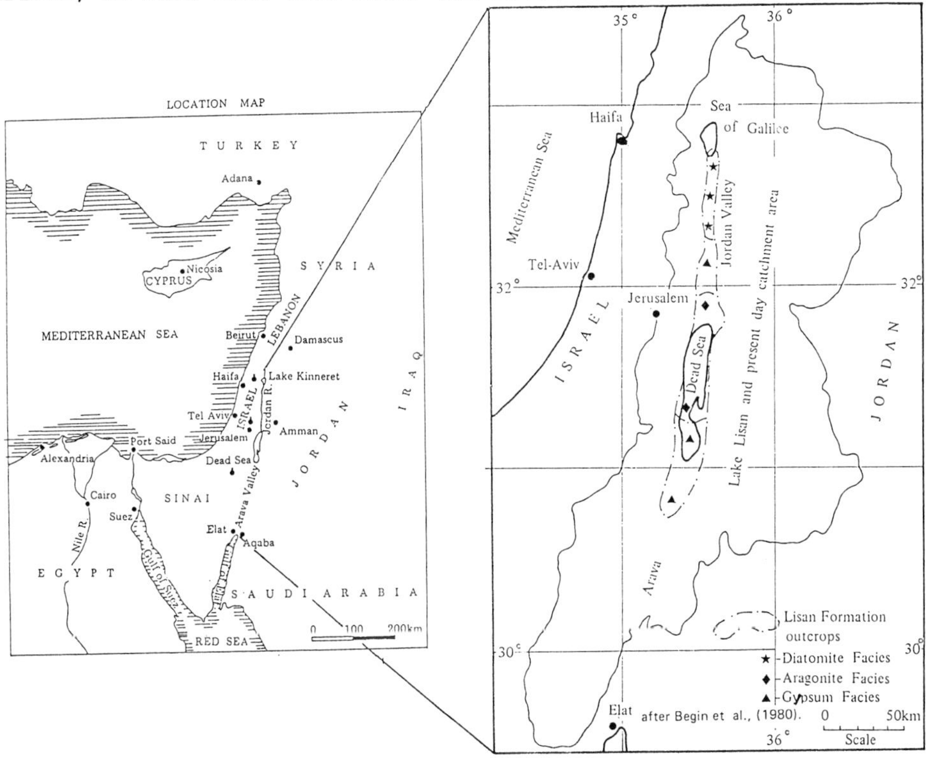

FIGURE 1

the modern day phase of flood and evaporite activity which began in the Miocene tectonic, Arava - Jordan Rift Valley (Picard, 1942).

Upper Pleistocene lacustrine, laminated sediments, known as the Lisan Formation (Neev and Emery, 1967), are found surrounding the Dead Sea and extending northwards within the rift valley to Lake Kinneret (Sea of Galilee). These sediments are part of a fluvial clastic - evaporitic sequence which was inpart formed by seasonal winter floods discharging into the rift valley by way of the main wadis, and periodic high evaporation. Three main facies, in a north to south distribution, are recognised as the Diatomite, Aragonite and Gypsum Facies (Begin et al., 1980) with the latter two found around the Dead Sea (Figure 1).

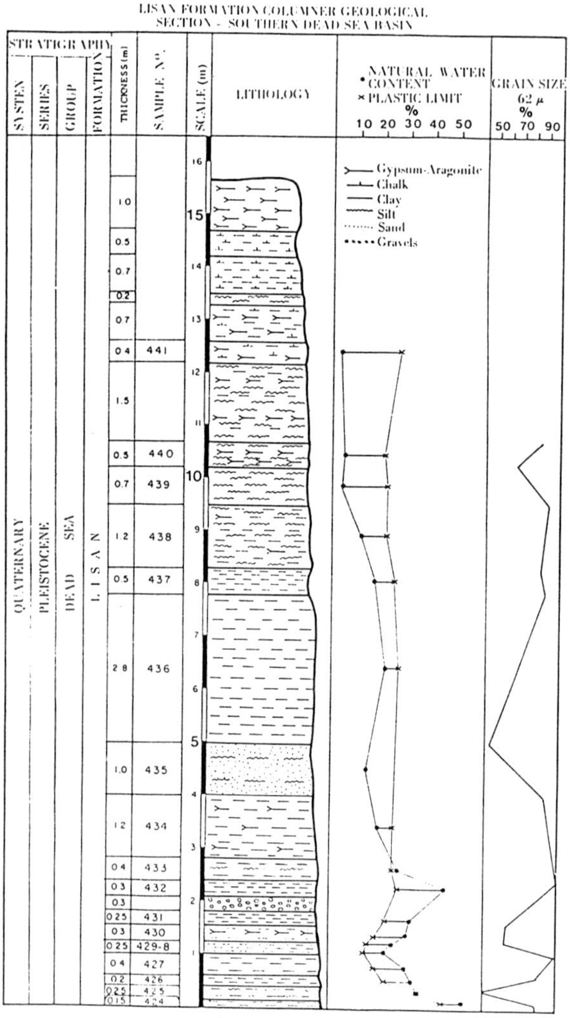

FIGURE 2

The Lisan Formation studied (Figure 2) was deposited in a brackish environment and consists of, essentially, a laminated sequence of detrital carbonate (65-80%) and clay minerals (20-35%) (Kaolinite, illite, montmorillonite and some palygorskite), quartz and dolomite, generally in this order of abundance (Arkin, 1980). This is overlain by a varved sequence of gypsum and aragonite with minor silt and clay.

Deformation of the sequence in mainly governed by the composition, porosity and geochemistry of pore water as affected by the amount of post-depositional consolidation (Arkin and Starinsky, 1982). This consolidation is considered to be minimal since the only overburden pressure that has been applied is the load of several meters of varved gypsum and aragonite.

The Lisan sediment may have a natural water content exceeding 42%, with a porosity ranging from 33-47%. The total dissolved salts in the pore water is 40 g/l as compared to 320 g/l in the present day Dead Sea (Arkin and Starinsky, 1982). The shear strength imparted by the above conditions changes according to the range in relative content of clay - silt - sand and the salinity of the pore water. The overall range is 90 to 190 kN/m2.

In general, the shear strength is less in the plane of

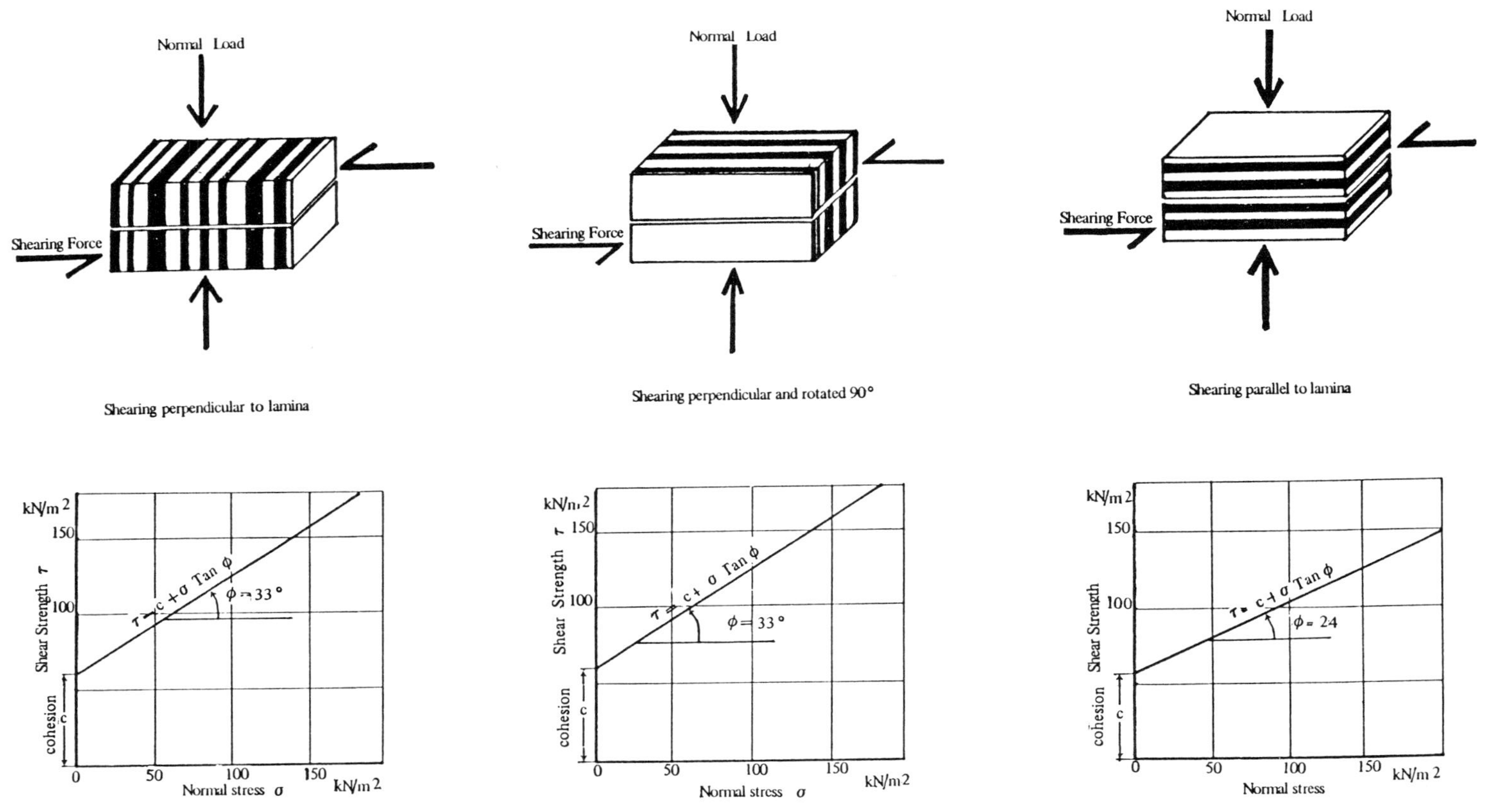
Normal Load
Shearing Force
Shearing perpendicular to lamina
Normal Load
Shearing Force
Shearing perpendicular and rotated 90°
Normal Load
Shearing Force
Shearing parallel to lamina
kN/m 2
150
100
Shear Strength τ
cohesion
c
τ = c + σ Tan φ
φ = 33°
0
50
100
150
Normal stress σ
kN/m 2
kN/m 2
150
100
Shear Strength τ
cohesion
c
τ = c + σ Tan φ
φ = 33°
0
50
100
150
Normal stress σ
kN/m 2
kN/m 2
150
100
Shear Strength
cohesion
c
τ = c + σ Tan φ
φ = 24
0
50
100
150
Normal stress
kN/m 2

FIGURE 3

FIGURE 4

Outcrop of Lisan Fm., showing convolute lamina in a specific bundle of lamina. Folds are tilted downstream.

FIGURE 5

Enlarged view of a convolute fold.

Note thrusting at the forward end of fold.

FIGURE 6

Low angle fault in a specific bundle of lamina. Note undeformed overlying and underlying lamina.

deposition than at an angle to it, and the form of deformation depends on the direction of the stress (Figure 3). Shear stress in the plane of deposition leads to convolute folding and thrust features in specific bundles of lamina (Figures 4,5,6). Shear stress at an angle to the plane of deposition gives rise to normal type faults which dissipate into the laminae overlying and underlying the specific bundle (Arkin and Michaeli, in press).

The restriction of deformation to specific bundles of lamina and the accompanying sedimentary structures suggest that deformation was the result of flash, freshwater floods (external load) causing remoulding and dilution of saline pore water and consequent loss of shear strength. This later characteristic is often related to sensitive clays (Skempton and Northey, 1952). However, in the present case only an apparent sensitivity can be determined for the Lisan Formation, since complete dilution of saline pore water is difficult to achieve. Thus flow conditions leading to deformation seem to be dependent on a two-fold process of dilution of saline pore water and extrusion by an external load, in the present case the weight of the body of water of the flash flood.

REFERENCES

Arkin, Y. 1980. Underconsolidated sensitive clay in the Lisan Formation. Sedom, Southern Dead Sea Basin. *5th Conf. Min. Eng. Qiryat Anavim*, Israel.

Arkin, Y., and **Michaeli, L.** (in press). The significance of shear strength in the deformation of laminated sediments. *Dead Sea area. Spec. Issue, Eng. Geol. Isr. J. of Earth Sci.*

Arkin, Y., and **Starinsky, A.** 1982. Lisan sediment porosity and pore water as indicators of original Lake Lisan composition. Current Research 1981 Geological Survey of Israel.

Begin, Z.B., **Nathan, Y.**, **Ehrlich, A.** 1980. Stratigraphy and facies distribution in the Lisan Formation, new evidence from the south of the Dead Sea, Israel. *Isr. J. of Earth Sci.* Vol. **29**, pp. 182-189.

Neev, D., and **Emery, O.K.** 1967. *The Dead Sea, depositional processes and environments of evaporites.* Bull. No. **41**. Isr. Geol. Surv.

Picard, L. 1942. *Structure and evolution of Palestine, with comparative notes on neighbouring countries.* Geol. Dept. Hebrew Univ., Jerusalem.

Skempton, A.W., and **Northey, R.D.** 1952. The sensitivity of clays. *Geotechnique* Vol. **3**. No.1.

ON THE CONSTRUCTION OF A CONSTITUTIVE EQUATION OF SOIL BY MAKING USE OF THE DLS MODEL

Roman Traczyk
Institute of Geotechnique, Technical University of Wrocław,
pl. Grunwaldzki 9, 50-377 Wroclaw, Poland

1. INTRODUCTION

In the past three decades many different rheological models have been suggested to describe the properties of soil. Despite a great number of investigations, these models have not received wide acceptance in engineering practice. There is a major limitation to their uses - specifically for linear models. Apart from approximate approaches, soil cannot regarded as a body of a linear behaviour. On the other hand, many of the linear models that had been proposed so far, yielded solutions with a great number of constants which cannot be determined in a standard test apparatus. A few years ago, S. Dmitruk, B. Lysik and H. Suchnicka have developed a nonlinear rheological model of soil (DLS) which eliminates these disadvantages.

The concept of representing a soil model in the form of a physical model was developed by Dmitruk, Suchnicka and Lysik in 1969 [1, 2]. This resulted in the formulation of a mathematical model [1, 2, 3]. Such a method of constructing soil models raised great interest on one hand and distrust on the other hand, as the approach suggested by the authors differed from the conventional one involving notions and definitions from classical mechanics and rheology of soils. And that is why intensive experimental studies have been initiated before 1977 [4-8, to name some of these].

The objective of the study reported here was to answer the question of whether or not a consitutive equation of soil might be established by making use of the DLS model. It was considered an elementary area of soil [1] equivalent to a soil sample which had been exposed to loading in a triaxial compression apparatus. The considerations involved the experimental results obtained by the author himself, as well as literature data, specifically those reported by A. Kwaśnik-Piaścik [4-6].

2. CONCEPT OF THE IDENTIFYING PROCEDURE

While many reports are available on the assumptions to the physical and mathematical model, these are mostly written in Polish. It seems, therefore, advisable to present the major formulae involved so that no confusion occurs as to what is included.

The equation of state of the DLS model may be written as

$$\gamma = A(g^{R_o} + g^{R\xi}) \tag{1}$$

where g^{R_o} is density of external stress, which takes the form

$$g^{R_o} = \sum_{i=1}^{n} \frac{\Delta\sigma_i(\xi)H(t - \xi_i)}{t - \xi_i} \quad \text{or} \quad g^{R_o} = \int_{0+\varepsilon}^{t} \frac{\sigma'(\xi)H(t - \xi)}{t - \xi}d\xi \tag{2}$$

and $g^{R\xi}$ denotes density of internal changes, which is defined as

$$g^{R\xi} = \sum_{i=1}^{n} \frac{\Delta\theta_i(\xi)H(t - \xi_i)}{t - \xi_i} \quad \text{or} \quad g^{R\xi} = \int_{0+\varepsilon}^{t} \frac{\theta'(\xi)H(t - \xi)}{t - \xi}d\xi \tag{3}$$

where t indicates time of observation of loading effects, u shows time of variation in the state of stress, H is Heaviside function, and $\theta(\xi)$ is used for the function of unbalanced internal changes, A and $\theta(\xi)$ being reserved for the parameters of the model.

When the loads acting on the soil are smaller than the value of k_{min} defined in the physical model, the density of internal changes will be zero. Thus, for $\sigma = \text{const} \leq k_{min}$ equation (1) becomes

$$\dot{\gamma} = A\frac{\sigma}{t} \tag{4}$$

The investigations reported in the literature [4] have shown that k_{min} may be, in principal, identified with long-term strength.

The concept of the identification procedure was developed by A. Kwaśnik-Piaścik [4]. The procedure involved the determination of the model parameters, A and $\theta(\xi)$, by making use of experimental data. To evaluate A, the authoress availed herself of the results of creeping tests wherein the loads applied were small and stain tended to a constant value after a sufficiently long time of loading i.e. strain rate tended to zero. The determination of A enabled assessment of the $\theta(\xi)$ function on the basis of triaxial compression test results, which had been obtained at various kinds of stresses. The expressions for A and $\theta(\xi)$, the formulae involved and their derivation are given in the literature [4-6].

Laboratory investigations reported by Kwaśnik-Piaścik were carried out on two types of samples. Some of these consisted of

Jaroszów clay of an undisturbed structure, others were prepared from bentonite. While the values of stress intensity were taken as a measure of the state of stress, the intensity of strain was used for describing the state of strain. The results of plastometric tests have shown a scattering of the A values. Kwasnik-Piascik [4] proved by making use of regression analysis that A was independent of time and only slightly depending on stress intensity and water content, if at all. Thus, Kwaśnik-Piaścik believes that coefficient A may be regarded as a characteristic constant for a given type of soil. Taking into account the small variability interval for the moisture content of Jaroszów clay samples ($\Delta w = 2.5\%$) on one hand, and the greater scattering of the results obtained for prepared samples, on the other hand, it becomes obvious that the anticipation of A being a constant ought to be reconsidered.

According to the assumptions of the model, the form and the values of the $\theta(\xi)$ function were found to vary and depend on the variation of the state of stresses. Kwaśnik-Piaścik has suggested that $\theta(\xi)$ be described by the integral of g^{R_o} which takes the form of G^τ and may be written as

$$G^\tau = \sum_{i=0}^{n-1} g_i^{R_o} \Delta\xi \quad \text{or} \quad G^\tau = \int_{0+\varepsilon}^{t} g^{R_o}\, d\xi. \tag{5}$$

Analysis of regressions $\theta(\xi) = f(G^\tau)$ revealed that they had a similar plot, irrespective of what kind of loading had been applied.

Although the investigations discussed here seem to corroborate the high probability of success when constructing the constitutive equation of soil in terms of the DLS model, they do not enable an accurate formulation. What accounts for this shortcoming is 1 – the considerable scattering of results for prepared soil samples, and 2 – the unusually small difference between the physical parameters of the undisturbed Jaroszów clay samples which made them resistant to volumetric changes during exposure to isotropic stress. Thus, it becomes obvious that similar investigations were needed to examine a wider variability interval for the physical properties of soil.

The fact, that volume variations occurred, accounted for the necessity of distinguishing the effects of the spherical tensor and those of the deviator. Hence, the parameter of the model were defined as

$$A = f(w, e, \sigma_o(\xi), t) \tag{6}$$

$$\theta = f(A, w, e, \tau(\xi), t) \tag{7}$$

where w indicates water content, e denotes porosity factor, $\sigma_o(\xi)$

describes variation of isotropic stress, $\tau(\xi)$ is used for variation of deviator stress, ξ stands for time of stress variation, and t is time during which stress effects are observed.

The choice of the terms in parentheses incorporated in equations (6) and (7) results from the assumption that they exert a significant effect on the values of A and θ. To determine this influence and, consequently, to examine the possibilities of predicting the A and θ values, a series of triaxial compression tests (with outflow) were carried out.

3. LABORATORY TESTS

The available data sets show that it is convenient to describe the state of stress in terms of stress intensity. Hence, we have

$$\hat{\tau} = \frac{1}{\sqrt{2}}\sqrt{(\sigma_1 - \sigma_2)^2 + (\sigma_2 - \sigma_3)^2 + (\sigma_3 - \sigma_1)^2} \qquad (8)$$

which becomes

$$\hat{\tau} = (\sigma_1 - \sigma_3) \qquad (8a)$$

for axially symmetrical stress.

To describe the state of strain the author availed himself of the term of strain intensity. Thus,

$$\hat{\gamma} = \frac{\sqrt{2}}{2}\sqrt{(\varepsilon_1 - \varepsilon_2)^2 + (\varepsilon_2 - \varepsilon_3)^2 + (\varepsilon_3 - \varepsilon_1)^2} \qquad (9)$$

and for axially symmetrical strain

$$\hat{\gamma} = \frac{3}{4}(\varepsilon_1 - \varepsilon_3). \qquad (9a)$$

Equations (7) and (8) are valid at the assumption that no variations occur in the volume of the sample. Hence, it became necessary to perform measurements of volumetric changes in the course of the triaxial compression tests.

The laboratory investigations involved samples which had been prepared from clay collected in the Edmund openpit of Jaroszów, Poland. The method of preparation applied in the study enabled high homogeneity, thus contributing to the abatement of deformation during consolidation. The creeping tests were carried out by making use for two loading schemes.

Scheme 1

The samples were consolidated at different pressures. Following completion of the consolidation procedure water outflow was closed and axial loading was applied. The results were used for determining the relations A = f(e) and A = f(t).

Scheme 2

The tests aimed at defining the nature of the relationship between A and $\sigma_o(\xi)$. Samples were cut from the same block and consolidated at 0.20 MN/m^2 for 12 days in order to obtain specimens of identical physical properties. After the consolidation procedure had been completed, the pressure in the apparatus was increased to 0.35 MN/m^2. Axial loading was applied after 0.5 h, 24 h, 72 h and 144 h form the moment at which pressure began to rise. Water outflow from the sample was cut off prior to loading.

The determination of A = f(e) makes it possible to define $\theta(\xi)$ for soil samples of different porosity by conducting triaxial compression tests.

Failure tests were carried out by the following two methods: shearing at a constant rate of table motion (method I), and shearing at a controlled rate of loading (method II).

4. RESULTS

4.1. COEFFICIENT A

Creeping tests involving scheme 1 revealed the linear behaviour of A = f(e). The results are plotted in Fig.1. To investigate the

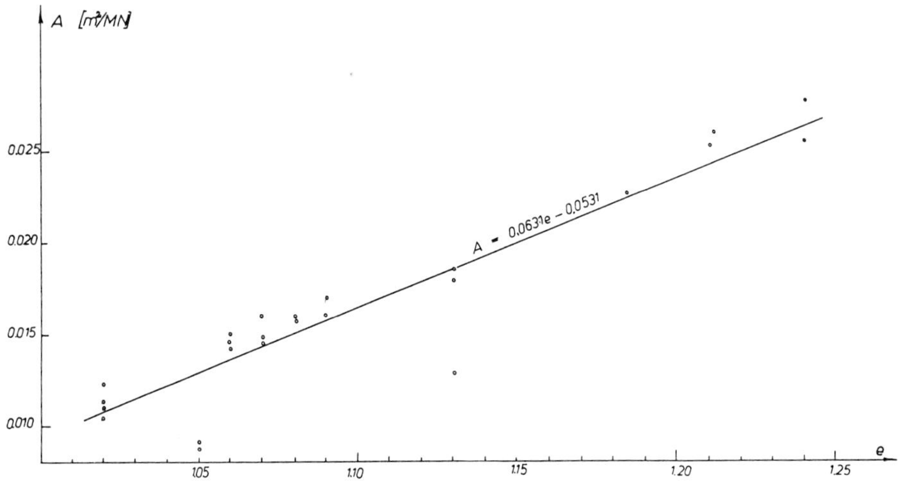

Figure 1. Coefficient A as a function of porosity factor for Jaroszów clay samples.

behaviour of $A = f(t)$ at a constant porosity factor and a constant pressure in the apparatus, the values of this function were subject to regression analysis. Analyzed were creeping processes with time of observation, t, longer than 300 h. There were no reasons for ignoring the hypothesis of lack of correlation.

The investigation of the $A = f[\sigma_o(\xi)]$ relationship requires a large number of experiments on samples of identical initial properties. Obtaining the number of the results desired when using the method described in [7] takes a long time and calls for further study. Despite these, it is possible to describe the dependence of A on $\sigma_o(\xi)$. Oedometric investigations performed earlier show that it is convenient to use the integral of isotropic stress density to describe the effect of this kind of stress. Thus, the relation between coefficient A and variation of the state of stress may be written as follows

$$A[\sigma_o(\xi)] = A_o - f[G^{\sigma_o}]. \qquad (10)$$

For the test performed according to scheme 2, the following relation was established

$$A[\sigma_o,t] = A_o - CG^{\sigma_o}, \qquad (11)$$

where C is constant. Hence, for clays normally consolidated and for determined variation of stresses, the A - G^{σ_o} relation is linear.

4.2. FUNCTION OF UNBALANCED INTERVAL CHANGES

The relationship between $\theta(\xi)$ and G^{τ} was analyzed by using plots in logarithmic scale. The plot of function θ for method I is given in Fig.2. The considerable error should be attributed primarily to the necessarily inaccurate determination of the moment at which the process beings. This is best illustrated by the plot of the θ-function for sample no.14 which shows values of internal changes markedly lower than do samples of a lower porosity factor. It was for this reason that method II replaced method I.

The application of method II made it possible to define the exact moment at which the process began, but the $\hat{\gamma} - \hat{\tau}$ relation included an error at the initial stage of the investigation. The error is easy to eliminate only if the force is measured in side the apparatus. Unfortunately, the apparatus available to the author did not yield sufficiently accurate indications until higher stresses were

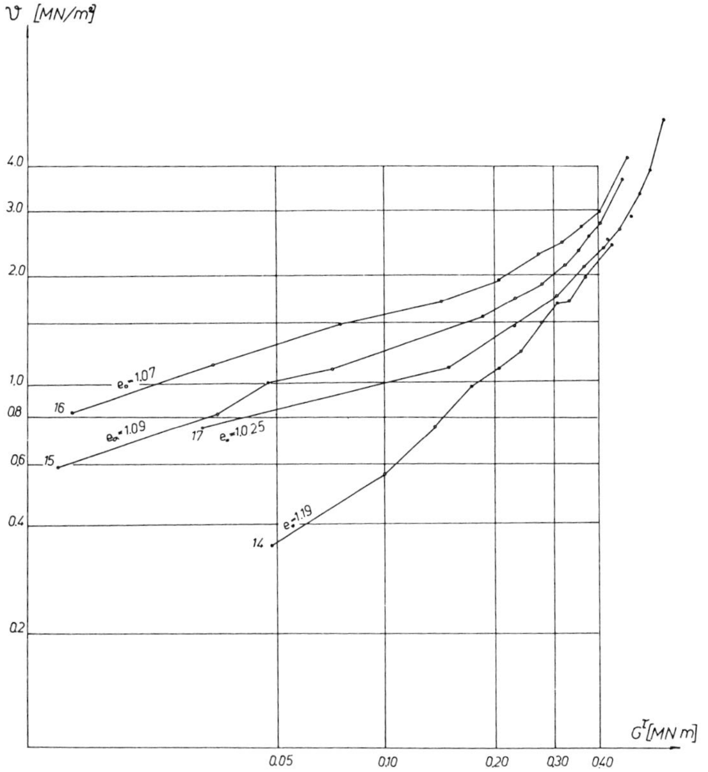

Figure 2. Function of unbalanced internal changes versus G^τ determined by triaxial compression tests involving method I (at constant rate of table motion).

applied. That is why two methods of calculating $\Delta\theta(\xi)$ seemed to be appropriate. One of these made use of the assumption that the $\hat{\gamma} - \hat{\tau}$ relationship was linear in the initial stage of the testing procedure. In the other method, the first two degrees of loading were neglected as they showed values lower than k_{min} ([1]). Hence, it has been assumed that no internal changes occur for there two degrees of loading ($k_{min} < 0.030$ MN/m^2).

When the $\hat{\gamma} - \hat{\tau}$ relation is assumed to be linear, the plot behaves as shown by Fig.3. The dependence of θ on G^τ may be described with great accuracy as

$$\theta = C_1(G^\tau)C_2 + C_3. \tag{12}$$

While C_1 increased with the increasing porosity factor, C_2 varied only slightly. The values of C_3 showed no relation to porosity, so they should be regarded as incidental, taking into account the measuring problems dealt with at the initial stage of shear.

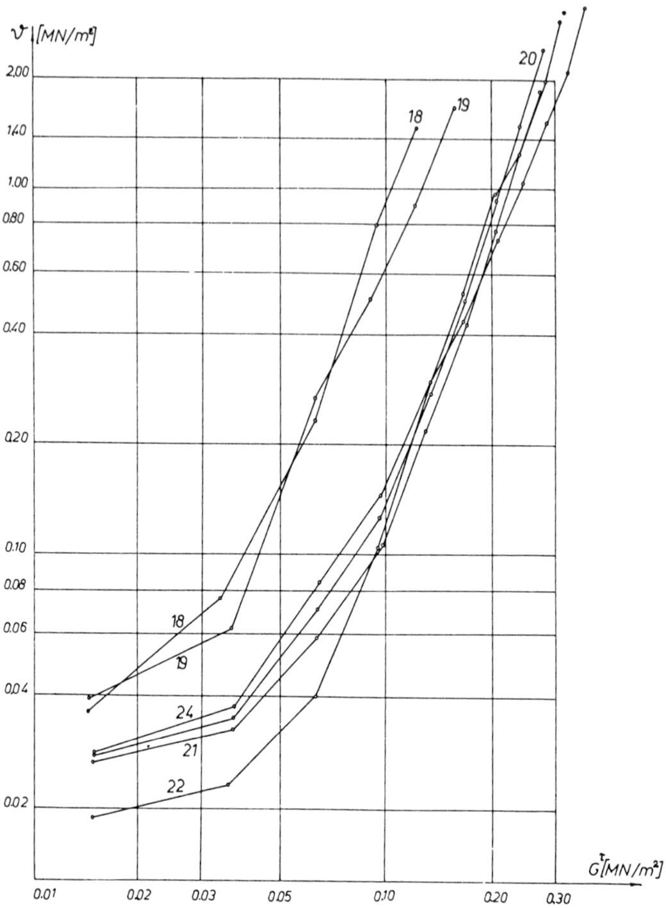

Figure 3. Function $\theta(\xi)$ versus G^{τ} determined by triaxial compression tests involving method II (at constant increase of intensity) and making use of the assumption that the $\hat{\tau}$ - $\hat{\gamma}$ relation is linear at the beginning of the process.

Neglecting the first two degrees of loading gave the plots of Fig.4. These may be interpreted as a pencil of straight lines either parallel or convergent, which leads to expression (13) and expression (14), respectively,

$$\theta = C_1 (G^{\tau})^{C_2} \tag{13}$$

$$\theta = C_1 (G^{\tau} + C_2)^{C_0}. \tag{14}$$

Using relation (13), the C_1 values were found to increase with the increasing porosity factor, whereas the values of C_2 underwent slight variations only. The determination of the C_1, C_2, C_3 constants by virtue of (14) for the plots of Fig.4 failed, as it was impossible to separate the quantities occurring in parentheses. It has only been

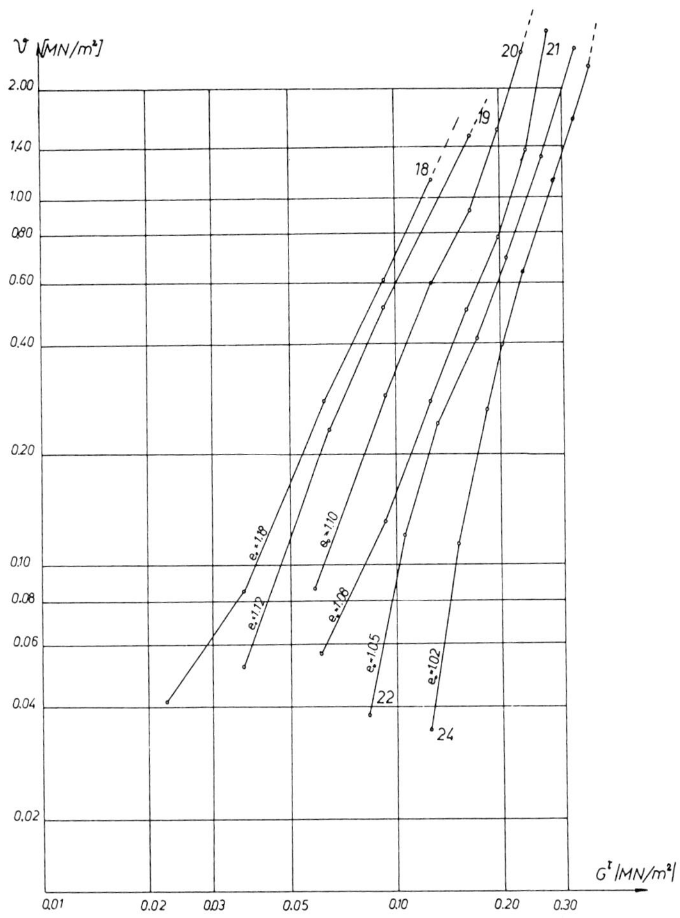

Figure 4. Function $\theta(\xi)$ versus G^{τ} determined by method II at the assumption that no internal changes occur at the beginning of the process.

shown that, if C_1 and C_3 are assumed constant, the values of C_2 increase with a rise of the porosity factor.

5. CONCLUDING COMMENTS

Analysis of results enables the $\theta(\xi)$ function to be described in terms of $(G^{\tau} - G^{k_{min}})$, $G^{k_{min}}$ being given by

$$G^{k_{min}} = \sum_{i=1}^{n} \frac{k_{min}}{i\Delta\xi}\Delta\xi \quad \text{or} \quad G^{k_{min}} = \int_{0+\varepsilon}^{t} \frac{k_{min}}{t}dt. \tag{15}$$

Relation $\theta = f(G^{\tau})$ is then likely to take the form

$$\theta = C_1(G^\tau - G^{k_{min}})^{C_2} H(G^\tau - G^{k_{min}}), \tag{16}$$

where $H(G^\tau - G^{k_{min}})$ is Heaviside function.

Taking into account the plot of relation (16), we may conclude on the invariability of C_1, C_2 for a wide range of porosity. Thus, if our interpretation makes use of the statement that the variability of θ can be represented by the quantity $G^{k_{min}}$, we may also avail ourselves of the statement that the change of porosity is included.

Fig.5 gives the plots of θ for assumed k_{min}, C_1, C_2 at

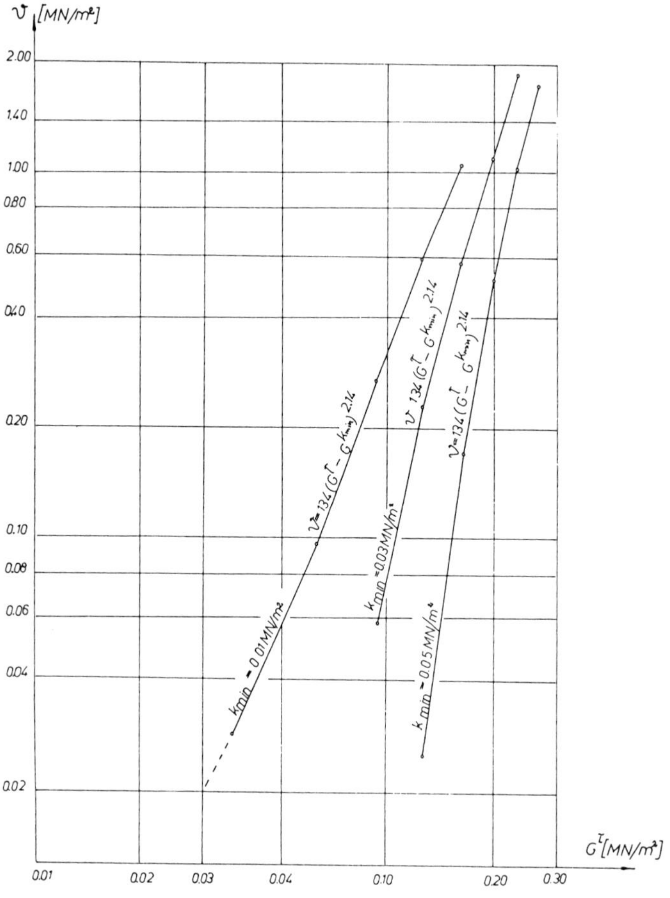

Figure 5. Function $\theta(\xi)$ versus G^τ calculated by virtue of equation (16).

$\Delta\tau = 0.015\,MN/m^2$ and $\Delta\xi = 0.166\,h$.

6. SUMMARY

The considerations presented in this paper indicate that the results of the investigations performed so far deserve attention despite the fact that the problem of establishing a constitutive equation of soil in terms of the DLS model has not yet been solved adequately. But it seems to be of particular importance that the parameters of the equation may be determined, using standard laboratory apparatus.

REFERENCES

1. **S. Dmitruk, B. Lysik, H. Suchnicka,** Fundamental problems of soil strength, *Archiwum Hydrotechniki*, 1973, Vol. **20**, fasc. 4 (in Polish).
2. **S. Dmitruk, B. Lysik, H. Suchnicka,** Problems of physical relations in soil mechanics, *Studia Geotechnica*, 1973, Vol. **4**, fasc. 1.
3. **S. Dmitruk,** Problems of Representing Geological and Engineering Processes in Openpit Mining. Warszawa, Wydawnictwa Geologiczne, 1984 (in Polish).
4. **A. Kwaśnik-Piaścik,** Some problems dealt with in the identification of the DLS model. Ph.D. thesis. Technical University of Wroclaw, Institute of Geotechnique, 1978. PWr I-10/K-242/78 (in Polish).
5. **A. Kwaśnik-Piaścik,** On the identification of the DLS model for a granular medium, *Archiwum Hydrotechniki*, 1981, Vol. **28**, fasc. 1 (in Polish).
6. **A. Kwaśnik-Piaścik,** On the identification of the DLS model for a granular medium, *Archiwum Hydrotechniki*, 1981, Vol. **28**, fasc. 2 (in Polish).
7. **R. Traczyk,** Analysis of the indentification procedure for the DLS model. Ph.D. Thesis, Technical University of Wrocław, Institute of Geotechnique (in Polish).
8. **R. Traczyk,** On the identification procedure for the DLS model, *Archiwum Hydrotechniki*, 1982, Vol. **29**, fasc. 1-2 (in Polish).